Synthesis Lectures on Electromagnetics

Series Editor

Akhlesh Lakhtakia, Department of Engineering Science and Mechanics, Pennsylvania State University, University Park, PA, USA

This series of short books focuses on a wide array of applications on electromagnetics, particularly in relation to design and interactions with advanced materials and devices. Topics include cutting-edge applications in bioengineering and biomaterials, optics, nanotechnology, and metamaterials.

Hamad M. Alkhoori

Concise Introduction to Electromagnetic Fields

Springer

Hamad M. Alkhoori
Department of Electrical and Communication
Engineering
United Arab Emirates University
Al Ain, Abu Dhabi, UAE

ISSN 2691-5448 ISSN 2691-5456 (electronic)
Synthesis Lectures on Electromagnetics
ISBN 978-3-031-60330-3 ISBN 978-3-031-60331-0 (eBook)
https://doi.org/10.1007/978-3-031-60331-0

This Springer imprint is published by the registered company Springer Nature Switzerland AG
The registered company address is: Gewerbestrasse 11, 6330 Cham, Switzerland

Preface

As the title suggests, this book serves as an introductory book to the subject of electromagnetics. Moreover, the word 'concise' from the book's title stresses that the book covers the fundamental topics a reader needs to acquire before consulting advanced-level references on electromagnetics. The book is intended to be used in an undergraduate-level course, whether in physics curriculum, or in electrical engineering curriculum. It can, as well, be used as a reference for researchers who wish to solidify their understanding of the subject.

Like any phenomenon being described by governing equations, electromagnetic phenomenon is described by Maxwell equations. In many undergraduate-level textbooks of electromagnetics, Maxwell equations are derived starting from Coulomb law, Biot-Savart law, Faraday law, and, lastly, Maxwell's correction to Ampere law. Consequently, the reader has to be exposed to the subjects of electrostatics and magnetostatics before discussing full-version electromagnetics in which time character emerges. While this approach is perfectly fine, I adopt a different approach in this book, whereby the full-version electromagnetics is introduced first. Then, electrostatics and magnetostatics can be regarded as special cases. It is to be mentioned that this approach is somewhat similar, in terms of the sequence, at least, to the approach adopted by Landau and Lifshitz, in which Maxwell equations are introduced from relativistic principles. However, due to the complexity of such an approach to be taught at an undergraduate-level course, Maxwell equations in this book are postulated in the beginning.

Electromagnetics can be divided into two regimes. These are (i) time-dependent regime and (ii) time-independent regime. This book is divided into four parts. Part I is about some required mathematical background, and an introduction to electromagnetics. Part II is about time-independent electromagnetics, namely, electrostatics and magnetostatics. Then, Part III discusses time-dependent electromagnetics in source-free regions. Finally, Part IV is about time-dependent electromagnetics in source regions. It should be emphasized that the division into source-free and source regions is made for the sake of facilitating the presentation. Once the concepts from source-free-region problems are grasped, transition to source-region problems becomes smoother.

In terms of chapters, this book comprises 14 chapters distributed in the four aforementioned parts. Chapter 1 presents a brief revision on vector algebra and vector calculus. Chapter 2 introduces Maxwell equations and divides the theory into two regimes. Chapter 3 is about electrostatics in which a static charge distribution gives rise to an electric field. Currents (i.e., moving charges) is discussed in Chap. 4, followed by magnetostatics, in which a current distribution gives rise to a magnetic field, in Chap. 5. Then, Chap. 6 is about the transition from time-independent regime to time-dependent regime. Chapter 7 discusses the propagation of electromagnetic fields in an unbounded, source-free region. This is followed by the propagation in the presence of an infinite-extent obstacle in Chap. 8. Chapters 9 and 10 treat the problem of propagation of electromagnetic fields in guided structures. Transition to source regions, namely, radiation problem, is discussed in Chap. 11. Chapter 12 discusses radiators, well known as antennas, and their properties. Chapter 13 discusses simple antenna structures. Finally, Chap. 14 briefly discusses the analysis of group of antennas, well known as antenna arrays.

Since the usage of computer programs to validate analytical procedures, to tackle problems not amendable to analytical solutions, or at least to gain a better visualization has significantly increased, some chapters are supplied with an appendix containing useful *Mathematica* computer programs. These can be used for the purpose of validating the solutions of end-of-chapter problems, validating the solutions of problems from other textbooks, or even validating the solutions of problems a reader can propose and solve. Furthermore, these computer programs can be used by researchers to produce various forms of plots (e.g., two-dimensional and three-dimensional plots for scalars, two-dimensional and three-dimensional streamline plots for vectors, etc.).

The reader is assumed to have some background in standard topics taught in junior undergraduate-level courses, or even in high school, such as differentiation and integration. Also, an exposure to elementary physics courses might be beneficial, though not necessary.

Al Ain, UAE Hamad M. Alkhoori

Contents

Part I
Introduction to Essential Mathematics and Electromagnetics

The first part of the book gives a brief review on mathematical topics needed in this book. Then, it gives an overview on electromagnetics from a system perspective. This part consists of two chapters. Chapter 1 discusses vector algebra and vector calculus in three coordinate systems (Cartesian, cylindrical, and spherical), as well as in a general curvilinear system. Chapter 2 presents Maxwell equations as governing equations of electromagnetics. Specialization to electrostatics and magnetostatics is discussed then as special cases from the general setting.

Vector Algebra and Vector Calculus

1

This chapter is devoted to vector algebra and vector calculus. In Sect. 1.1, we give an overview on vector algebra, including definition, Cartesian bases and vector expansion, vector arithmetic operators, and position and distance vectors. Then, Sect. 1.2 discusses the various coordinates systems encountered in this book (e.g., Cartesian, cylindrical, and spherical), as well as transformation among them. We then discuss vector calculus in Sect. 1.3, including vector integral calculus, and vector differential calculus. These are discussed first in Cartesian, cylindrical, and spherical coordinate systems, and then are extended to a general curvilinear coordinate system in Sect. 1.4. Finally, time-harmonic vectors is discussed in Sect. 1.5. Useful computer programs are given in the appendix at the end of the chapter.

1.1 Vector Algebra

1.1.1 Definition and Expansion

A scalar is a quantity that has a magnitude only (e.g., mass, charge, temperature, etc.), whereas a vector is a quantity that has a magnitude and a direction (e.g., velocity, acceleration, force, momentum, etc.). The magnitude of a vector $\mathbf{A}$ is written as $|\mathbf{A}|$, or A, and its direction is written as $\hat{\mathbf{A}}$ given by

$$\hat{\mathbf{A}} = \frac{\mathbf{A}}{|\mathbf{A}|}. \tag{1.1}$$

The vector $\hat{\mathbf{A}}$ is called a unit vector because its magnitude is unity. A vector $\mathbf{A}$ can be expanded into Cartesian unit vectors as

$$\mathbf{A} = A_x\hat{\mathbf{x}} + A_y\hat{\mathbf{y}} + A_z\hat{\mathbf{z}}, \tag{1.2}$$

H. M. Alkhoori, *Concise Introduction to Electromagnetic Fields*, Synthesis Lectures on Electromagnetics, https://doi.org/10.1007/978-3-031-60331-0_1

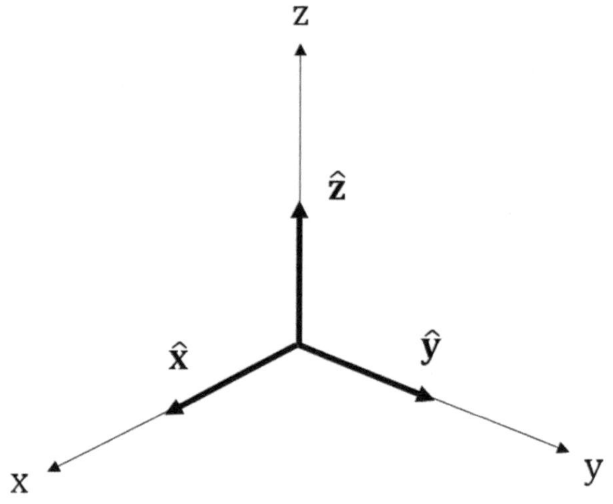

Fig. 1.1 Cartesian bases vectors

where $\hat{\mathbf{x}}$, $\hat{\mathbf{y}}$, and $\hat{\mathbf{z}}$ are unit vectors in the direction of the x axis, the y axis, and the z axis, respectively; see Fig. 1.1. These unit vectors can be called Cartesian bases. The scalars A_x, A_y, and A_z are components of the vector $\mathbf{A}$ in the direction of the x axis, the y axis, and the z axis, respectively.

1.1.2 Vector Addition and Subtraction

In addition to $\mathbf{A}$, let us define the vectors $\mathbf{B} = B_x\hat{\mathbf{x}} + B_y\hat{\mathbf{y}} + B_z\hat{\mathbf{z}}$ and $\mathbf{C} = C_x\hat{\mathbf{x}} + C_y\hat{\mathbf{y}} + C_z\hat{\mathbf{z}}$. Addition between two vectors $\mathbf{A}$ and $\mathbf{B}$ can be done using

$$\mathbf{A} + \mathbf{B} = (A_x + B_x)\hat{\mathbf{x}} + (A_y + B_y)\hat{\mathbf{y}} + (A_z + B_z)\hat{\mathbf{z}}. \tag{1.3}$$

Subtraction between two vectors $\mathbf{A}$ and $\mathbf{B}$ can be done as $\mathbf{A} + (-\mathbf{B})$. Addition is commutative (i.e., $\mathbf{A} + \mathbf{B} = \mathbf{B} + \mathbf{A}$), associative [i.e., $(\mathbf{A} + \mathbf{B}) + \mathbf{C} = \mathbf{A} + (\mathbf{B} + \mathbf{C})$], and distributive (i.e., $\alpha(\mathbf{A} + \mathbf{B}) = \alpha\mathbf{A} + \alpha\mathbf{B}$), where α is a scalar..

1.1.3 The Dot Product

The dot product between two vectors $\mathbf{A}$ and $\mathbf{B}$ can be performed as

$$\mathbf{A} \cdot \mathbf{B} \equiv |\mathbf{A}||\mathbf{B}| \cos\theta, \tag{1.4}$$

where θ (in rad) is the angle between $\mathbf{A}$ and $\mathbf{B}$; see Fig. 1.2.

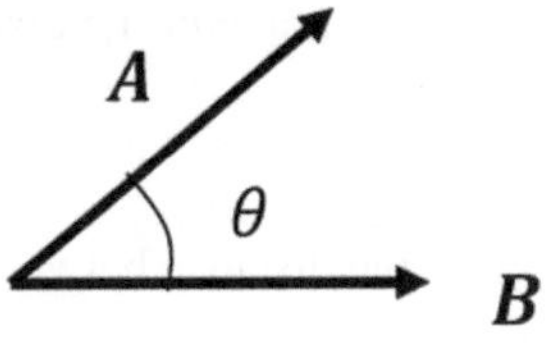

Fig. 1.2 Dot product

When (i) $\theta = 0°$, **A** and **B** are coparallel to each other, (ii) when $\theta = 90°$, **A** and **B** are perpendicular (normal or orthogonal) to each other, and (iii) when $\theta = 180°$, **A** and **B** are antiparallel to each other. Note that $\hat{\mathbf{x}} \bullet \hat{\mathbf{x}} = \hat{\mathbf{y}} \bullet \hat{\mathbf{y}} = \hat{\mathbf{z}} \bullet \hat{\mathbf{z}} = 1$, whereas $\hat{\mathbf{x}} \bullet \hat{\mathbf{y}} = \hat{\mathbf{x}} \bullet \hat{\mathbf{z}} = \hat{\mathbf{y}} \bullet \hat{\mathbf{z}} = 0$. In component form, the dot product can be written as

$$\mathbf{A} \bullet \mathbf{B} = A_x B_x + A_y B_y + A_z B_z. \tag{1.5}$$

From Eqs. (1.4) and (1.5), we see that

$$\mathbf{A} \bullet \mathbf{A} = |\mathbf{A}|^2 = A_x^2 + A_y^2 + A_z^2. \tag{1.6}$$

Hence, the magnitude of the vector **A** is

$$|\mathbf{A}| = \sqrt{\mathbf{A} \bullet \mathbf{A}} = \sqrt{A_x^2 + A_y^2 + A_z^2}. \tag{1.7}$$

The dot product is commutative (i.e., $\mathbf{A} \bullet \mathbf{B} = \mathbf{B} \bullet \mathbf{A}$), and associative (i.e., $\mathbf{A} \bullet (\mathbf{B} + \mathbf{C}) = \mathbf{A} \bullet \mathbf{B} + \mathbf{A} \bullet \mathbf{C}$).

Example 1.1 Let $\mathbf{A} = \hat{\mathbf{x}} + 2\,\hat{\mathbf{y}} + 5\,\hat{\mathbf{z}}$ and $\mathbf{B} = -\,\hat{\mathbf{x}} + 2\,\hat{\mathbf{y}} + 3\,\hat{\mathbf{z}}$. Find the angle θ between **A** and **B**.

Solution. We have $\mathbf{A} \bullet \mathbf{B} = 18$, $|\mathbf{A}| = 5.47$, and $|\mathbf{B}| = 3.74$. Therefore, $\theta = \cos^{-1}\left(\dfrac{18}{5.47 \times 3.74}\right) = 0.498\ (28.56°)$. ◁

Position and Distance Vectors

The position vector of a point represented by coordinates (x, y, z) is given in Cartesian coordinates by

$$\mathbf{r} = x\hat{\mathbf{x}} + y\hat{\mathbf{y}} + z\hat{\mathbf{z}}. \tag{1.8}$$

Suppose that another point is represented by (x', y', z'), with a corresponding position vector

$$\mathbf{r}' = x'\hat{\mathbf{x}} + y'\hat{\mathbf{y}} + z'\hat{\mathbf{z}}. \tag{1.9}$$

Then, the distance vector between the two points is given by

$$\mathbf{R} = \mathbf{r} - \mathbf{r}'. \tag{1.10}$$

The magnitude of the distance vector

$$R = |\mathbf{r} - \mathbf{r}'| = \sqrt{|\mathbf{r}|^2 + |\mathbf{r}'|^2 - 2\mathbf{r} \bullet \mathbf{r}'} \tag{1.11}$$

gives the distance between two points.

Example 1.2 Find the distance between the two points $(1, -1, 3)$ and $(5, 0, 3)$.

Solution. Let $\mathbf{r} = \hat{\mathbf{x}} - \hat{\mathbf{y}} + 3\,\hat{\mathbf{z}}$, and $\mathbf{r}' = 5\,\hat{\mathbf{x}} + 3\,\hat{\mathbf{z}}$. Then, $\mathbf{R} = \mathbf{r} - \mathbf{r}' = -4\,\hat{\mathbf{x}} - \hat{\mathbf{y}}$. So, $R = |\mathbf{R}| = 4.12$. ◁

Component of a Vector

The scalar component (projection) of a vector **A** in the direction of a vector **B** is written as A_B, which is given by

$$A_B = \mathbf{A} \bullet \hat{\mathbf{B}}. \tag{1.12}$$

The vector component of a vector **A** in the direction of a vector **B** is written as $\mathbf{A}_B$, which is given by

$$\mathbf{A}_B = (\mathbf{A} \bullet \hat{\mathbf{B}})\hat{\mathbf{B}}. \tag{1.13}$$

This can be used in defining the normal and tangential components of a vector with respect to a surface characterized by a unit normal $\hat{\mathbf{n}}$. Given a surface with a unit normal $\hat{\mathbf{n}}$, the vector component normal to the surface, denoted by $\mathbf{A}_\perp$, is

$$\mathbf{A}_\perp = (\mathbf{A} \bullet \hat{\mathbf{n}})\hat{\mathbf{n}}, \tag{1.14}$$

whereas the the vector component tangential to the surface, denoted by $\mathbf{A}_\parallel$, is

$$\mathbf{A}_\parallel = \mathbf{A} - (\mathbf{A} \bullet \hat{\mathbf{n}})\hat{\mathbf{n}}. \tag{1.15}$$

1.1.4 The Cross Product

The cross product between two vectors **A** and **B** is performed as

$$\mathbf{A} \times \mathbf{B} = |\mathbf{A}||\mathbf{B}| \sin\theta\, \hat{\mathbf{n}}, \tag{1.16}$$

where $\hat{\mathbf{n}}$ is a unit vector perpendicular to both **A** and **B**. In component form, the cross product can be written as

$$\mathbf{A} \times \mathbf{B} = \begin{vmatrix} \hat{\mathbf{x}} & \hat{\mathbf{y}} & \hat{\mathbf{z}} \\ A_x & A_y & A_z \\ B_x & B_y & B_z \end{vmatrix} = \hat{\mathbf{x}}(A_y B_z - A_z B_y) + \hat{\mathbf{y}}(A_z B_x - A_x B_z) + \hat{\mathbf{z}}(A_x B_y - A_y B_x). \tag{1.17}$$

If we set $\mathbf{A} = \hat{\mathbf{x}}$ and $\mathbf{B} = \hat{\mathbf{y}}$ in Eq. (1.17), we find that $\hat{\mathbf{x}} \times \hat{\mathbf{y}} = \hat{\mathbf{z}}$. By a similar approach, it can be seen that $\hat{\mathbf{y}} \times \hat{\mathbf{z}} = \hat{\mathbf{x}}$ and $\hat{\mathbf{z}} \times \hat{\mathbf{x}} = \hat{\mathbf{y}}$. The cross product is distributive (i.e., $\mathbf{A} \times (\mathbf{B} + \mathbf{C}) = \mathbf{A} \times \mathbf{B} + \mathbf{A} \times \mathbf{C}$), but it is not commutative (i.e., $\mathbf{B} \times \mathbf{A} = -\mathbf{A} \times \mathbf{B}$).

A combination of dot and cross products is also encountered in electromagnetics. These are scalar triple product

$$\mathbf{A} \cdot (\mathbf{B} \times \mathbf{C}) = \mathbf{B} \cdot (\mathbf{C} \times \mathbf{A}) = \mathbf{C} \cdot (\mathbf{A} \times \mathbf{B}), \tag{1.18}$$

and vector triple product

$$\mathbf{A} \times (\mathbf{B} \times \mathbf{C}) = \mathbf{B}(\mathbf{C} \cdot \mathbf{A}) - \mathbf{C}(\mathbf{A} \cdot \mathbf{B}). \tag{1.19}$$

1.2 Coordinate Systems and Transformations

A coordinate system in three dimensions is comprised of spatial variables (i.e., coordinates) $\{v_1, v_2, v_3\}$, as well as bases $\{\hat{\mathbf{v}}_1, \hat{\mathbf{v}}_2, \hat{\mathbf{v}}_3\}$. We discuss in this section the most common three coordinate systems. These are Cartesian coordinates, circular cylindrical (or simply cylindrical) coordinates, and spherical coordinates. A general curvilinear coordinate system is discussed in Sect. 1.4 after learning vector calculus.

1.2.1 Cartesian Coordinates

Cartesian coordinates $\{x, y, z\}$, where $x \in (-\infty, \infty)$, $y \in (-\infty, \infty)$, and $z \in (-\infty, \infty)$, constitute the simplest coordinate system. The following Cartesian surfaces arise when one coordinate is fixed.

- The equation $x = x_0$ defines an infinite plane on the yz plane with a coordinate $x = x_0$.
- The equation $y = y_0$ defines an infinite plane on the xz plane with a coordinate $y = y_0$.
- The equation $z = z_0$ defines an infinite plane on the xy plane with a coordinate $z = z_0$.

Notice that all of the aforementioned surfaces are infinite because only one coordinate is specified, while the range of the other two are not. If the range of each one of the other two coordinates is specified and is finite, then, the resulting plane will no more be infinite. Figure 1.3 shows Cartesian surfaces.

Cartesian bases $\{\hat{\mathbf{x}}, \hat{\mathbf{y}}, \hat{\mathbf{z}}\}$ can be used for expanding a vector $\mathbf{A}$ as

$$\mathbf{A} = A_x\hat{\mathbf{x}} + A_y\hat{\mathbf{y}} + A_z\hat{\mathbf{z}}. \tag{1.20}$$

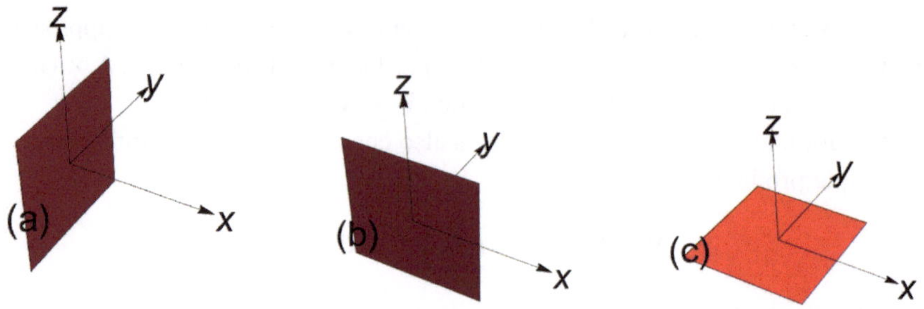

Fig. 1.3 **a** $x = 0$, **b** $y = 0$, and **c** $z = 0$

Here, we can regard A_x as the projection of $\mathbf{A}$ into $\hat{\mathbf{x}}$, A_y is the projection of $\mathbf{A}$ into $\hat{\mathbf{y}}$, and A_z is the projection of $\mathbf{A}$ into $\hat{\mathbf{z}}$. It is to be noted that, each of A_x, A_y, and A_z is a scalar that can be a function of the variables x, y, and z. That is,

$$\mathbf{A}(x, y, z) = A_x(x, y, z)\,\hat{\mathbf{x}} + A_y(x, y, z)\,\hat{\mathbf{y}} + A_z(x, y, z)\,\hat{\mathbf{z}}. \tag{1.21}$$

For shorthand notation, we can let $(x, y, z) \to \mathbf{r}$. Then, Eq. (1.21) can be written as

$$\mathbf{A}(\mathbf{r}) = A_x(\mathbf{r})\,\hat{\mathbf{x}} + A_y(\mathbf{r})\,\hat{\mathbf{y}} + A_z(\mathbf{r})\,\hat{\mathbf{z}}.^1 \tag{1.22}$$

We already saw that Cartesian bases satisfy

$$\left.\hat{\mathbf{x}} \bullet \hat{\mathbf{x}} = \hat{\mathbf{y}} \bullet \hat{\mathbf{y}} = \hat{\mathbf{z}} \bullet \hat{\mathbf{z}} = 1 \qquad \hat{\mathbf{x}} \times \hat{\mathbf{y}} = \hat{\mathbf{z}} \qquad \hat{\mathbf{y}} \times \hat{\mathbf{z}} = \hat{\mathbf{x}} \qquad \hat{\mathbf{z}} \times \hat{\mathbf{x}} = \hat{\mathbf{y}}\right\}. \tag{1.23}$$

Figure 1.4 shows streamline plots of Cartesian bases. These bases can be represented more simply as in Fig. 1.1. Notice that Cartesian bases are constant vectors (i.e., do not depend on either x, y, or z). Also, we see that $\hat{\mathbf{x}}$ points in the direction of increase of x, $\hat{\mathbf{y}}$ points in the direction of increase of y, and $\hat{\mathbf{z}}$ points in the direction of increase of z. As to been seen in Sect. 1.3.3, this is not a mere coincidence.

1.2.2 Cylindrical Coordinates

Cylindrical coordinates $\{\rho, \phi, z\}$, as well as cylindrical bases $\{\hat{\boldsymbol{\rho}}, \hat{\boldsymbol{\phi}}, \hat{\mathbf{z}}\}$ are shown in Fig. 1.5. Cylindrical coordinates are related to the Cartesian coordinates through

$$\left.x = \rho\cos\phi \qquad y = \rho\sin\phi \qquad z = z\right\}. \tag{1.24}$$

Given a point P in space, we see that $\rho \in [0, \infty)$ is the distance between the z axis and the point P, and $\phi \in [0, 2\pi)$ (called the azimuthal angle) is measured from the x axis to the

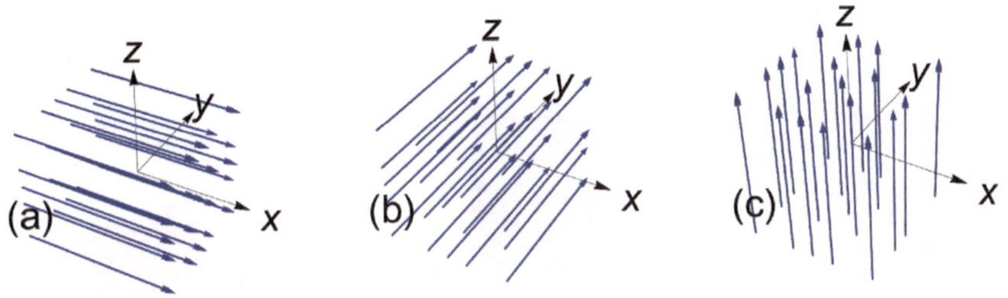

Fig. 1.4 **a** $\hat{\mathbf{x}}$, **b** $\hat{\mathbf{y}}$, and **c** $\hat{\mathbf{z}}$

Fig. 1.5 Cylindrical coordinates and bases

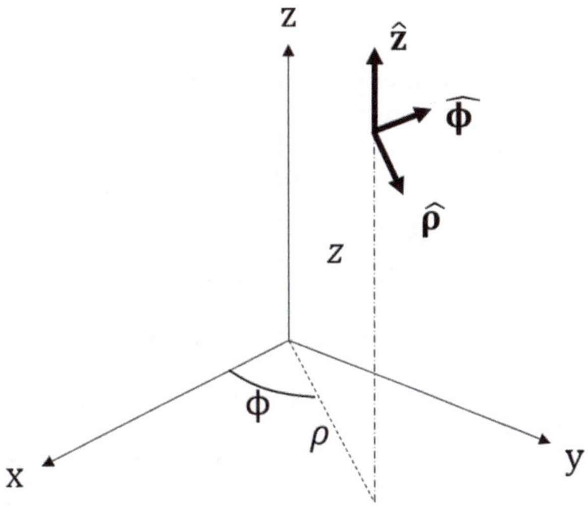

projection of the point P on the xy plane. Equations (1.24) can be used when converting from Cartesian coordinates to cylindrical coordinates. Conversion from cylindrical coordinates to Cartesian coordinates can be done upon inverting Eqs. (1.24), which gives

$$\left.\rho = \sqrt{x^2 + y^2} \qquad \phi = \tan^{-1}\frac{y}{x} \qquad z = z\right\} . \tag{1.25}$$

The following cylindrical surfaces arise when one coordinate is fixed.

- The equation $\rho = \rho_0$ is the equation of an infinite cylinder with a radius ρ_0. Notice that if the range of z is specified and is finite, then the cylinder will no more be infinite.
- The equation $\phi = \phi_0$ is the equation of a semi-infinite plane making an angle ϕ_0 with respect to the positive x axis. Notice that if the ranges of ρ and z are specified and are finite, then the plane will no more be semi infinite.
- The equation $z = z_0$ is the equation of an infinite plane on the xy plane with a coordinate $z = z_0$.

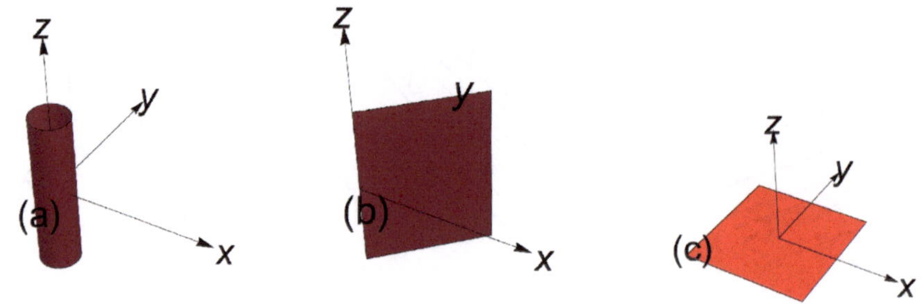

Fig. 1.6 **a** $\rho = 1$, **b** $\phi = \pi/4$, and **c** $z = 0$

Figure 1.6 shows examples of cylindrical surfaces.

Cylindrical bases $\{\hat{\boldsymbol{\rho}}, \hat{\boldsymbol{\phi}}, \hat{\mathbf{z}}\}$ can be used for expanding a vector $\mathbf{A}$ as

$$\mathbf{A}(\mathbf{r}) = A_\rho(\mathbf{r})\,\hat{\boldsymbol{\rho}} + A_\phi(\mathbf{r})\,\hat{\boldsymbol{\phi}} + A_z(\mathbf{r})\,\hat{\mathbf{z}}, \tag{1.26}$$

where A_ρ is the projection of $\mathbf{A}$ into $\hat{\boldsymbol{\rho}}$, A_ϕ is the projection of $\mathbf{A}$ into $\hat{\boldsymbol{\phi}}$, and A_z is the projection of $\mathbf{A}$ into $\hat{\mathbf{z}}$. Those cylindrical bases satisfy

$$\left.\hat{\boldsymbol{\rho}}\bullet\hat{\boldsymbol{\rho}} = \hat{\boldsymbol{\phi}}\bullet\hat{\boldsymbol{\phi}} = \hat{\mathbf{z}}\bullet\hat{\mathbf{z}} = 1 \qquad \hat{\boldsymbol{\rho}}\times\hat{\boldsymbol{\phi}} = \hat{\mathbf{z}} \qquad \hat{\boldsymbol{\phi}}\times\hat{\mathbf{z}} = \hat{\boldsymbol{\rho}} \qquad \hat{\mathbf{z}}\times\hat{\boldsymbol{\rho}} = \hat{\boldsymbol{\phi}}\right\}. \tag{1.27}$$

Consequently, like Cartesian coordinates, the dot and the cross products in cylindrical coordinates can be done, respectively, as

$$\mathbf{A}\bullet\mathbf{B} = A_\rho B_\rho + A_\phi B_\phi + A_z B_z, \tag{1.28}$$

and

$$\mathbf{A}\times\mathbf{B} = \begin{vmatrix} \hat{\boldsymbol{\rho}} & \hat{\boldsymbol{\phi}} & \hat{\mathbf{z}} \\ A_\rho & A_\phi & A_z \\ B_\rho & B_\phi & B_z \end{vmatrix}. \tag{1.29}$$

Figure 1.7 shows streamline plots of cylindrical bases. These bases can be represented more simply as in Fig. 1.5. Notice that the cylindrical bases $\hat{\boldsymbol{\rho}}$ and $\hat{\boldsymbol{\phi}}$ are not constant vectors. Also, we see that $\hat{\boldsymbol{\rho}}$ points in the direction of increase of ρ, and $\hat{\boldsymbol{\phi}}$ points in the direction of increase of ϕ.

Cylindrical bases can be transformed into Cartesian bases as follows. The basis $\hat{\boldsymbol{\rho}}$, like any vector, can be expanded into Cartesian bases as

$$\hat{\boldsymbol{\rho}} = \alpha\,\hat{\mathbf{x}} + \beta\,\hat{\mathbf{y}} + \gamma\,\hat{\mathbf{z}}, \tag{1.30}$$

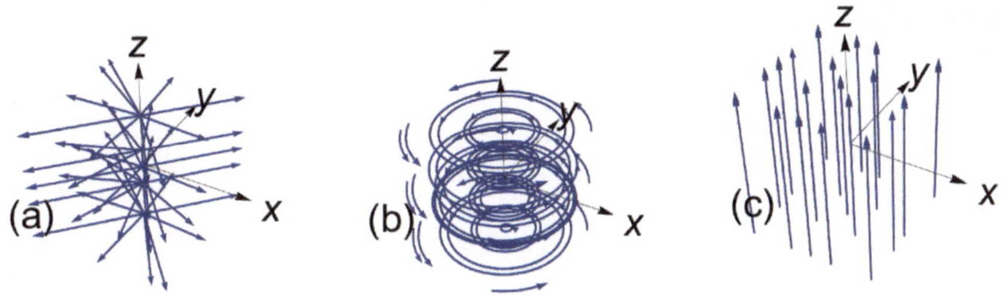

Fig. 1.7 **a** $\hat{\rho}$, **b** $\hat{\boldsymbol{\phi}}$, and **c** $\hat{\mathbf{z}}$

where α, β, and γ are unknown that have to be determined. Using the fact that Cartesian bases are orthogonal, one can find that $\alpha = \hat{\rho} \cdot \hat{\mathbf{x}} = \cos\phi$, $\beta = \hat{\rho} \cdot \hat{\mathbf{y}} = \sin\phi$, and $\gamma = \hat{\rho} \cdot \hat{\mathbf{z}} = 0$. Therefore,

$$\hat{\rho} = \cos\phi\,\hat{\mathbf{x}} + \sin\phi\,\hat{\mathbf{y}}. \tag{1.31}$$

Equation (1.31) transforms the cylindrical basis $\hat{\rho}$ into the Cartesian bases $\hat{\mathbf{x}}$ and $\hat{\mathbf{y}}$. Transforming $\hat{\boldsymbol{\phi}}$ can be done similarly. Hence, transformation between cylindrical bases to Cartesian bases can be written in matrix form as

$$\begin{pmatrix} \hat{\rho} \\ \hat{\boldsymbol{\phi}} \\ \hat{\mathbf{z}} \end{pmatrix} = \begin{pmatrix} \cos\phi & \sin\phi & 0 \\ -\sin\phi & \cos\phi & 0 \\ 0 & 0 & 1 \end{pmatrix} \begin{pmatrix} \hat{\mathbf{x}} \\ \hat{\mathbf{y}} \\ \hat{\mathbf{z}} \end{pmatrix}. \tag{1.32}$$

Notice that, unlike Cartesian bases, cylindrical bases $\hat{\rho}$ and $\hat{\boldsymbol{\phi}}$ depend on the coordinate ϕ. Conversion relation from Cartesian bases to cylindrical bases can be established upon inverting the square matrix appearing in Eq. (1.32). Since this matrix is orthogonal, its inverse is simply its transpose. Therefore,

$$\begin{pmatrix} \hat{\mathbf{x}} \\ \hat{\mathbf{y}} \\ \hat{\mathbf{z}} \end{pmatrix} = \begin{pmatrix} \cos\phi & -\sin\phi & 0 \\ \sin\phi & \cos\phi & 0 \\ 0 & 0 & 1 \end{pmatrix} \begin{pmatrix} \hat{\rho} \\ \hat{\boldsymbol{\phi}} \\ \hat{\mathbf{z}} \end{pmatrix}, \tag{1.33}$$

1.2.3 Spherical coordinates

Spherical coordinates $\{r, \theta, \phi\}$, as well as spherical bases $\{\hat{\mathbf{r}}, \hat{\boldsymbol{\theta}}, \hat{\boldsymbol{\phi}}\}$ are shown in Fig. 1.8. Spherical coordinates are related to the Cartesian coordinates through

$$x = r\sin\theta\cos\phi \qquad y - r\sin\theta\sin\phi \qquad z = r\cos\theta\}. \tag{1.34}$$

Given a point P in space, we see that $r \in [0, \infty)$ is the distance between the origin and the point P, $\theta \in [0, \pi]$, called the colatitude (or polar) angle, is an angle drawn from the z axis

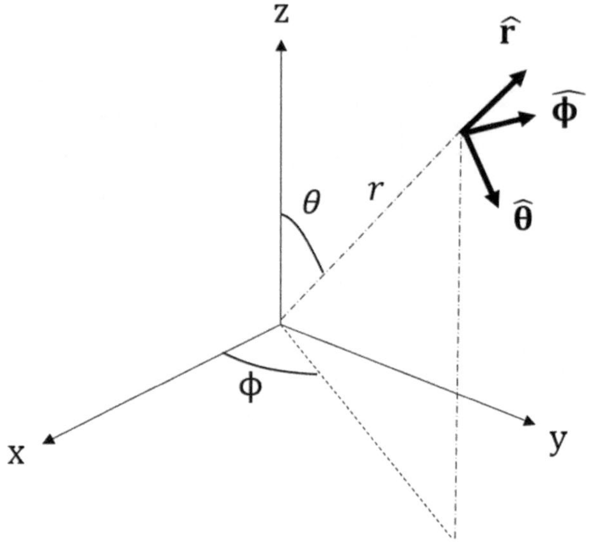

Fig. 1.8 Spherical coordinates and bases

to the line formed by the origin and the point P, and ϕ is defined same as before. Equations (1.34) can be used when converting from Cartesian coordinates to spherical coordinates. Conversion from spherical coordinates to Cartesian coordinates can be done upon inverting Eqs. (1.34), which gives

$$\left. r = \sqrt{x^2 + y^2 + z^2} \qquad \theta = \tan^{-1} \frac{\sqrt{x^2 + y^2}}{z} \qquad \phi = \tan^{-1} \frac{y}{x} \right\}. \tag{1.35}$$

The following spherical surfaces arise when one coordinate is fixed.

- The equation $r = r_0$ is the equation of a sphere with a radius r_0.
- The equation $\theta = \theta_0$ is the equation of an infinite cone of an angle θ_0. Notice that if the range of r is specified and is finite, then the cone will no more be infinite.
- The equation $\phi = \phi_0$ is the equation of a semi-infinite plane making an angle ϕ_0 with respect to the positive x axis. Notice that if the ranges of r and θ are specified, and the range of r is finite, then the plane will no more be semi infinite.

Spherical bases $\{\hat{\mathbf{r}}, \hat{\boldsymbol{\theta}}, \hat{\boldsymbol{\phi}}\}$ can be used for expanding a vector $\mathbf{A}$ as

$$\mathbf{A}(\mathbf{r}) = A_r(\mathbf{r})\, \hat{\mathbf{r}} + A_\theta(\mathbf{r})\, \hat{\boldsymbol{\theta}} + A_\phi(\mathbf{r})\, \hat{\boldsymbol{\phi}}, \tag{1.36}$$

where A_r is the projection of $\mathbf{A}$ into $\hat{\mathbf{r}}$, A_θ is the projection of $\mathbf{A}$ into θ, and A_ϕ is the projection of $\mathbf{A}$ into $\hat{\boldsymbol{\phi}}$. Those spherical bases satisfy

$$\left. \hat{\mathbf{r}} \cdot \hat{\mathbf{r}} = \hat{\boldsymbol{\theta}} \cdot \hat{\boldsymbol{\theta}} = \hat{\boldsymbol{\phi}} \cdot \hat{\boldsymbol{\phi}} = 1 \qquad \hat{\mathbf{r}} \times \hat{\boldsymbol{\theta}} = \hat{\boldsymbol{\phi}} \qquad \hat{\boldsymbol{\theta}} \times \hat{\boldsymbol{\phi}} = \hat{\mathbf{r}} \qquad \hat{\boldsymbol{\phi}} \times \hat{\mathbf{r}} = \hat{\boldsymbol{\theta}} \right\}. \tag{1.37}$$

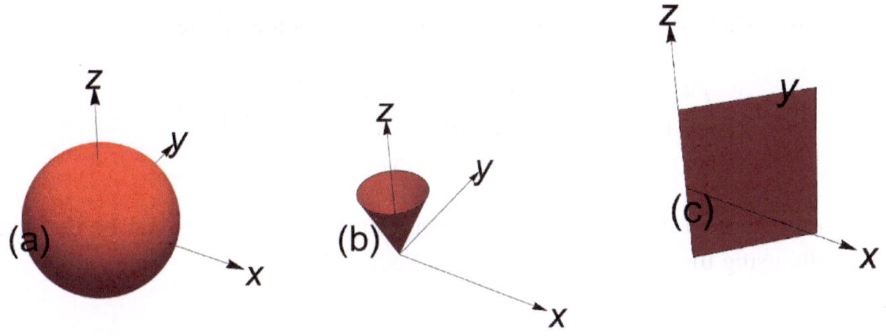

Fig. 1.9 **a** $r = 4$, **b** $\theta = \pi/8$, and **c** $\phi = \pi/4$

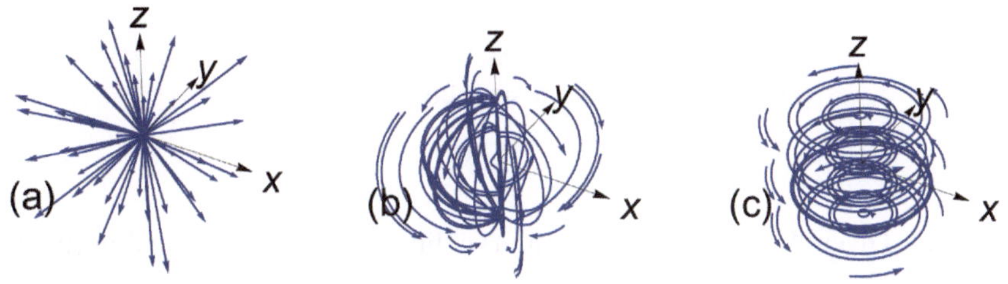

Fig. 1.10 **a** $\hat{\mathbf{r}}$, **b** $\hat{\boldsymbol{\theta}}$, and **c** $\hat{\boldsymbol{\phi}}$

Consequently, like Cartesian coordinates, the dot and the cross products in spherical coordinates can be done, respectively, as

$$\mathbf{A} \cdot \mathbf{B} = A_r B_r + A_\theta B_\theta + A_\phi B_\phi, \tag{1.38}$$

and

$$\mathbf{A} \times \mathbf{B} = \begin{vmatrix} \hat{\mathbf{r}} & \hat{\boldsymbol{\theta}} & \hat{\boldsymbol{\phi}} \\ A_r & A_\theta & A_\phi \\ B_r & B_\theta & B_\phi \end{vmatrix}. \tag{1.39}$$

Figure 1.10 shows streamline plots of spherical bases. These bases can be represented more simply as in Fig. 1.8. Notice that spherical bases are not constant vectors. Also, we see that $\hat{\mathbf{r}}$ points in the direction of increase of r, and $\hat{\boldsymbol{\theta}}$ points in the direction of increase of θ.

Using the procedure used in cylindrical coordinates, spherical bases can related to the Cartesian bases using Fig. 1.8 as

$$\begin{pmatrix} \hat{\mathbf{r}} \\ \hat{\boldsymbol{\theta}} \\ \hat{\boldsymbol{\phi}} \end{pmatrix} = \begin{pmatrix} \sin\theta\cos\phi & \sin\theta\sin\phi & \cos\theta \\ \cos\theta\cos\phi & \cos\theta\sin\phi & -\sin\theta \\ -\sin\phi & \cos\phi & 0 \end{pmatrix} \begin{pmatrix} \hat{\mathbf{x}} \\ \hat{\mathbf{y}} \\ \hat{\mathbf{z}} \end{pmatrix}. \tag{1.40}$$

Conversion from Cartesian bases to spherical bases can then be done by

$$\begin{pmatrix} \hat{\mathbf{x}} \\ \hat{\mathbf{y}} \\ \hat{\mathbf{z}} \end{pmatrix} = \begin{pmatrix} \sin\theta\cos\phi & \cos\theta\cos\phi & -\sin\phi \\ \sin\theta\sin\phi & \cos\theta\sin\phi & \cos\phi \\ \cos\theta & -\sin\theta & 0 \end{pmatrix} \begin{pmatrix} \hat{\mathbf{r}} \\ \hat{\boldsymbol{\theta}} \\ \hat{\boldsymbol{\phi}} \end{pmatrix}. \tag{1.41}$$

It should be mentioned that transformation between bases formulas can be derived more systematically using the gradient operator (see Sect. 1.3.3).

Example 1.3 Given $f(x, y, z) = x^2 y + \dfrac{z}{x}$, express $f(x, y, z)$ in spherical coordinates [i.e., obtain $f(r, \theta, \phi)$].

Solution.

$$f(r, \theta, \phi) = r^3 \sin^3\theta \sin\phi \cos^2\phi + \cot\theta \sec\phi.$$

◁

Example 1.4 Given $\mathbf{A} = y\hat{\mathbf{x}}$, express $\mathbf{A}$ in (a) cylindrical coordinates and in (b) spherical coordinates.

Solution.

(a)

$$\mathbf{A} = y\hat{\mathbf{x}} = \rho\sin\phi(\cos\phi\,\hat{\boldsymbol{\rho}} - \sin\phi\,\hat{\boldsymbol{\phi}}) = \rho\sin\phi\cos\phi\,\hat{\boldsymbol{\rho}} - \rho\sin^2\phi\,\hat{\boldsymbol{\phi}}.$$

(b)

$$\begin{aligned} \mathbf{A} = y\hat{\mathbf{x}} &= r\sin\theta\sin\phi(\sin\theta\cos\phi\,\hat{\mathbf{r}} + \cos\theta\cos\phi\,\hat{\boldsymbol{\theta}} - \sin\phi\,\hat{\boldsymbol{\phi}}) \\ &= r\sin^2\theta\sin\phi\cos\phi\,\hat{\mathbf{r}} + r\sin\theta\cos\theta\sin\phi\cos\phi\,\hat{\boldsymbol{\theta}} - r\sin\theta\sin^2\phi\,\hat{\boldsymbol{\phi}}. \end{aligned}$$

◁

Example 1.5 Given $\mathbf{A} = A_0\hat{\mathbf{r}}$, where A_0 is a constant, determine $\mathbf{A} \cdot \hat{\mathbf{x}}$.

Solution. We have $\mathbf{A} \cdot \hat{\mathbf{x}} = A_0\hat{\mathbf{r}} \cdot \hat{\mathbf{x}}$. From

$$\begin{pmatrix} \hat{\mathbf{x}} \\ \hat{\mathbf{y}} \\ \hat{\mathbf{z}} \end{pmatrix} = \begin{pmatrix} \sin\theta\cos\phi & \cos\theta\cos\phi & -\sin\phi \\ \sin\theta\sin\phi & \cos\theta\sin\phi & \cos\phi \\ \cos\theta & -\sin\theta & 0 \end{pmatrix} \begin{pmatrix} \hat{\mathbf{r}} \\ \hat{\boldsymbol{\theta}} \\ \hat{\boldsymbol{\phi}} \end{pmatrix},$$

we see that $\hat{\mathbf{x}} = \sin\theta\cos\phi\,\hat{\mathbf{r}} + \cos\theta\cos\phi\,\hat{\boldsymbol{\theta}} - \sin\phi\,\hat{\boldsymbol{\phi}}$. Thus,

$$\hat{\mathbf{r}} \cdot \hat{\mathbf{x}} = (\sin\theta\cos\phi\,\hat{\mathbf{r}} + \cos\theta\cos\phi\,\hat{\boldsymbol{\theta}} - \sin\phi\,\hat{\boldsymbol{\phi}}) \cdot \hat{\mathbf{r}} = \sin\theta\cos\phi.$$

Therefore, $\mathbf{A} \cdot \hat{\mathbf{x}} = A_0 \sin\theta\cos\phi$. ◁

Example 1.6 Given $\mathbf{A} = \hat{\rho} + 3\hat{\boldsymbol{\phi}} + 6\hat{\mathbf{z}}$, determine the scalar component of **A** parallel to the y axis.

Solution. We have

$$\mathbf{A} \cdot \hat{\mathbf{y}} = (\hat{\rho} + 3\hat{\boldsymbol{\phi}} + 6\hat{\mathbf{z}}) \cdot \hat{\mathbf{y}}.$$

Using

$$\begin{pmatrix} \hat{\mathbf{x}} \\ \hat{\mathbf{y}} \\ \hat{\mathbf{z}} \end{pmatrix} = \begin{pmatrix} \cos\phi & -\sin\phi & 0 \\ \sin\phi & \cos\phi & 0 \\ 0 & 0 & 1 \end{pmatrix} \begin{pmatrix} \hat{\rho} \\ \hat{\boldsymbol{\phi}} \\ \hat{\mathbf{z}} \end{pmatrix},$$

we write $\hat{\mathbf{y}} = \sin\phi\,\hat{\rho} + \cos\phi\hat{\boldsymbol{\phi}}$. Then,

$$\mathbf{A} \cdot \hat{\mathbf{y}} = (\hat{\rho} + 3\hat{\boldsymbol{\phi}} + 6\hat{\mathbf{z}}) \cdot (\sin\phi\,\hat{\rho} + \cos\phi\hat{\boldsymbol{\phi}}) = \sin\phi + 3\cos\phi.$$

Notice that $\phi = \pi/2$ along the y axis. Hence, $\mathbf{A} \cdot \hat{\mathbf{y}}\Big|_{\phi=\pi/2} = 1.$ ◁

Example 1.7 Given $\mathbf{A} = \rho z \cos^2\phi\,\hat{\rho} + \sin\phi\,\hat{\boldsymbol{\phi}} + \rho\,\hat{\mathbf{z}}$, determine (a) the scalar component of **A** parallel to the x axis, (b) the vector component of **A** normal to the surface $\rho = 1$ [i.e., $(\mathbf{A} \cdot \hat{\rho})\hat{\rho}$ when $\rho = 1$], and (c) the vector component of **A** tangential to the plane $z = 0$.

Solution.

(a) We have

$$\mathbf{A} \cdot \hat{\mathbf{x}} = (\rho z \cos^2\phi\,\hat{\rho} + \sin\phi\,\hat{\boldsymbol{\phi}} + \rho\,\hat{\mathbf{z}}) \cdot \hat{\mathbf{x}}.$$

Using

$$\begin{pmatrix} \hat{\mathbf{x}} \\ \hat{\mathbf{y}} \\ \hat{\mathbf{z}} \end{pmatrix} = \begin{pmatrix} \cos\phi & -\sin\phi & 0 \\ \sin\phi & \cos\phi & 0 \\ 0 & 0 & 1 \end{pmatrix} \begin{pmatrix} \hat{\rho} \\ \hat{\boldsymbol{\phi}} \\ \hat{\mathbf{z}} \end{pmatrix},$$

we write $\hat{\mathbf{x}} = \cos\phi\,\hat{\rho} - \sin\phi\hat{\boldsymbol{\phi}}$. Thus,

$$\mathbf{A} \cdot \hat{\mathbf{x}} = \rho z \cos^3\phi - \sin^2\phi.$$

But since $\phi = 0$ along the x axis, we get $\mathbf{A} \cdot \hat{\mathbf{x}}\Big|_{\phi=0} = \rho z.$

(b) We have

$$\mathbf{A}_\perp = (\mathbf{A} \cdot \hat{\rho})\hat{\rho} = \rho z \cos^2\phi\,\hat{\rho}.$$

But this has to be evaluated at $\rho = 1$. Hence, $\mathbf{A}_\perp\Big|_{\rho=1} = z\cos^2\phi\,\hat{\rho}.$

(c) We have

$$\mathbf{A}_\parallel = \mathbf{A} - \mathbf{A}_\perp = \mathbf{A} - (\mathbf{A} \cdot \hat{\mathbf{z}})\hat{\mathbf{z}} = \rho z \cos^2\phi\,\hat{\rho} + \sin\phi\,\hat{\boldsymbol{\phi}}.$$

But this has to be evaluated at $z = 0$. Hence, $\mathbf{A}_\parallel\Big|_{z=0} = \sin\phi\,\hat{\boldsymbol{\phi}}.$ ◁

Example 1.8 Given $\mathbf{A} = r\hat{\mathbf{r}} - \sin\theta\cos\phi\hat{\boldsymbol{\theta}} + \cos\phi\hat{\boldsymbol{\phi}}$, determine (a) the vector component of $\mathbf{A}$ tangential to the surface $r = 1$, and (b) the vector component normal to the surface $\theta = \pi/2$.

Solution. (a) We determine the vector component of $\mathbf{A}$ tangential to the surface $r = 1$ using

$$\mathbf{A}_{\parallel} = \mathbf{A} - \mathbf{A}_{\perp} = \mathbf{A} - (\mathbf{A} \bullet \hat{\mathbf{r}})\hat{\mathbf{r}} = -\sin\theta\cos\phi\hat{\boldsymbol{\theta}} + \cos\phi\hat{\boldsymbol{\phi}}.$$

Since this is independent of r, we get $\mathbf{A}_{\parallel}\Big|_{r=1} = -\sin\theta\cos\phi\hat{\boldsymbol{\theta}} + \cos\phi\hat{\boldsymbol{\phi}}$.

(b) We determine the vector component of $\mathbf{A}$ normal to the surface $\phi = \pi/2$ using

$$\mathbf{A}_{\perp} = (\mathbf{A} \bullet \hat{\boldsymbol{\theta}})\hat{\boldsymbol{\theta}} = -\sin\theta\cos\phi\hat{\boldsymbol{\theta}}.$$

But this has to be evaluated at $\theta = \pi/2$. Hence, $\mathbf{A}_{\perp}\Big|_{\theta=\pi/2} = -\cos\phi\hat{\boldsymbol{\theta}}$. ◁

Example 1.9 Express the position vector $\mathbf{r} = x\,\hat{\mathbf{x}} + y\,\hat{\mathbf{y}} + z\,\hat{\mathbf{z}}$ in (a) cylindrical coordinates, and in (b) spherical coordinates.

Solution. (a) Upon conversion of Cartesian variables to cylindrical variables, the position vector in cylindrical coordinates becomes

$$\mathbf{r} = \rho\cos\phi\,\hat{\mathbf{x}} + \rho\sin\phi\,\hat{\mathbf{y}} + z\,\hat{\mathbf{z}} = \rho(\cos\phi\,\hat{\mathbf{x}} + \sin\phi\,\hat{\mathbf{y}}) + z\,\hat{\mathbf{z}} = \rho\,\hat{\boldsymbol{\rho}} + z\,\hat{\mathbf{z}}.$$

(b) Upon conversion of Cartesian variables to spherical variables, the position vector in spherical coordinates becomes

$$\mathbf{r} = r\sin\theta\cos\phi\,\hat{\mathbf{x}} + r\sin\theta\sin\phi\,\hat{\mathbf{y}} + r\cos\theta\,\hat{\mathbf{z}} = r(\sin\theta\cos\phi\,\hat{\mathbf{x}} + \sin\theta\sin\phi\,\hat{\mathbf{y}} + \cos\theta\,\hat{\mathbf{z}}) = r\,\hat{\mathbf{r}}.$$

◁

Example 1.10 Obtain (a) $\hat{\boldsymbol{\rho}} \bullet \hat{\boldsymbol{\rho}}'$, and (b) $\hat{\mathbf{r}} \bullet \hat{\mathbf{r}}'$.

Solution. (a) After noting that $\hat{\boldsymbol{\rho}} = \cos\phi\,\hat{\mathbf{x}} + \sin\phi\,\hat{\mathbf{y}}$ and $\hat{\boldsymbol{\rho}}' = \cos\phi'\,\hat{\mathbf{x}} + \sin\phi'\,\hat{\mathbf{y}}$, we get

$$\hat{\boldsymbol{\rho}} \bullet \hat{\boldsymbol{\rho}}' = \cos\phi\cos\phi' + \sin\phi\sin\phi' = \cos(\phi - \phi').$$

(b) After noting that $\hat{\mathbf{r}} = \sin\theta\cos\phi\,\hat{\mathbf{x}} + \sin\theta\sin\phi\,\hat{\mathbf{y}} + \cos\theta\,\hat{\mathbf{z}}$ and $\hat{\mathbf{r}}' = \sin\theta'\cos\phi'\,\hat{\mathbf{x}} + \sin\theta'\sin\phi'\,\hat{\mathbf{y}} + \cos\theta'\,\hat{\mathbf{z}}$, we get

$$\hat{\mathbf{r}} \bullet \hat{\mathbf{r}}' = \sin\theta\sin\theta'\cos(\phi - \phi') + \cos\theta\cos\theta'.$$

◁

Example 1.11 Express the distance vector $R = |\mathbf{r} - \mathbf{r}'|$ in (a) cylindrical coordinates, and in (b) spherical coordinates.

Solution. (a) In cylindrical coordinates, we have $\mathbf{r} = \rho\,\hat{\boldsymbol{\rho}} + z\,\hat{\mathbf{z}}$ and $\mathbf{r}' = \rho'\,\hat{\boldsymbol{\rho}}' + z'\,\hat{\mathbf{z}}$. Then, the distance vector becomes

$$R = |\mathbf{r} - \mathbf{r}'| = \sqrt{|\mathbf{r}|^2 + |\mathbf{r}'|^2 - 2\mathbf{r} \bullet \mathbf{r}'} = \sqrt{(\rho^2 - 2\rho\rho' \cos(\phi - \phi') + \rho'^2) + (z - z')^2}.$$

(b) In spherical coordinates, we have $\mathbf{r} = r\,\hat{\mathbf{r}}$ and $\mathbf{r}' = r'\,\hat{\mathbf{r}}'$. Then, the distance vector becomes

$$R = |\mathbf{r} - \mathbf{r}'| = \sqrt{|\mathbf{r}|^2 + |\mathbf{r}'|^2 - 2\mathbf{r} \bullet \mathbf{r}'} = \sqrt{r^2 - 2rr'\left(\sin\theta \sin\theta' \cos(\phi - \phi') + \cos\theta \cos\theta'\right) + r'^2}.$$

◁

1.3 Vector Calculus

In this section, we first discuss elementary quantities needed in integral and differential calculus in the coordinate systems defined in Sect. 1.2. These are differential length, surface, and volume elements. Then, we discuss integral calculus, including line integrals, surface integrals, and volume integrals. We then discuss differential calculus, including gradient, divergence, curl, and Laplacian operators. Finally, we discuss classifications of vectors.

1.3.1 Differential Elements

Cartesian coordinates

The position vector in Cartesian coordinates is

$$\mathbf{r} = x\hat{\mathbf{x}} + y\hat{\mathbf{y}} + z\hat{\mathbf{z}}. \tag{1.42}$$

Since $\hat{\mathbf{x}}$, $\hat{\mathbf{y}}$, and $\hat{\mathbf{z}}$ are constant vectors, the differential position vector $d\mathbf{r}$ becomes

$$d\mathbf{r} = dx\hat{\mathbf{x}} + dy\hat{\mathbf{y}} + dz\hat{\mathbf{z}}. \tag{1.43}$$

Here, we call $d\mathbf{r}$ the differential length element instead of the differential position vector, and is represented by $d\mathbf{l}$ instead of $d\mathbf{r}$. Therefore, the differential length element in Cartesian coordinates is given by

$$d\mathbf{l} = dx\hat{\mathbf{x}} + dy\hat{\mathbf{y}} + dz\hat{\mathbf{z}}. \tag{1.44}$$

The differential surface element for a surface is defined as

$$d\mathbf{S} = dS\hat{\mathbf{n}}, \tag{1.45}$$

where dS is the area of the surface, and $\hat{\mathbf{n}}$ is a unit vector normal to the surface. The differential surface element can be found from the differential length element as follows. The element $x = x_o$, where x_o is a constant, has a unit normal $\hat{\mathbf{x}}$. Then, the area of it can be found from $d\mathbf{l}$ upon eliminating the $\hat{\mathbf{x}}$ component, and multiplying the $\hat{\mathbf{y}}$ and $\hat{\mathbf{z}}$ components (i.e., $dS = dydz$). Therefore, $d\mathbf{S}\Big|_{x=x_o} = dydz\hat{\mathbf{x}}$. This can be used for other elementary surfaces, as well. Therefore,

$$\left. d\mathbf{S}\Big|_{x=x_o} = dydz\hat{\mathbf{x}} \quad d\mathbf{S}\Big|_{y=y_o} = dxdz\hat{\mathbf{y}} \quad d\mathbf{S}\Big|_{z=z_o} = dxdy\hat{\mathbf{z}} \right\}, \tag{1.46}$$

where y_o and z_o are constants.

Like the differential surface element, the differential volume can be found from the differential length element upon multiplying the $\hat{\mathbf{x}}$, $\hat{\mathbf{y}}$, and $\hat{\mathbf{z}}$ components. This gives

$$d\mathcal{V} = dxdydz. \tag{1.47}$$

Cylindrical coordinates

In cylindrical coordinates, the position vector is

$$\mathbf{r} = \rho\,\hat{\boldsymbol{\rho}} + z\,\hat{\mathbf{z}}. \tag{1.48}$$

Then, the associated differential length element becomes

$$d\mathbf{l} = d(\rho\,\hat{\boldsymbol{\rho}}) + d(z\,\hat{\mathbf{z}}) = d\rho\,\hat{\boldsymbol{\rho}} + \rho d\hat{\boldsymbol{\rho}} + z\,\hat{\mathbf{z}} = d\rho\,\hat{\boldsymbol{\rho}} + \rho d\phi\,\hat{\boldsymbol{\phi}} + z\,\hat{\mathbf{z}}, \tag{1.49}$$

where use has been made of

$$d\hat{\boldsymbol{\rho}} = \frac{d\hat{\boldsymbol{\rho}}}{d\phi}d\phi = d\phi\,\hat{\boldsymbol{\phi}}. \tag{1.50}$$

Using the procedure used in Cartesian coordinates, the differential surface elements of the various cylindrical surfaces are

$$\left. d\mathbf{S}\Big|_{\rho=\rho_o} = \rho d\phi dz\hat{\boldsymbol{\rho}} \quad d\mathbf{S}\Big|_{\phi=\phi_o} = d\rho dz\hat{\boldsymbol{\phi}} \quad d\mathbf{S}\Big|_{z=z_o} = \rho d\rho d\phi\hat{\mathbf{z}} \right\}, \tag{1.51}$$

where ρ_o, ϕ_o, and z_o are constants.

Using the procedure used in Cartesian coordinates, the differential volume element in cylindrical coordinates is

$$d\mathcal{V} = \rho d\rho d\phi dz. \tag{1.52}$$

Spherical coordinates
In spherical coordinates, the position vector is

$$\mathbf{r} = r\,\hat{\mathbf{r}}. \tag{1.53}$$

Then, the associated differential length element becomes

$$d\mathbf{l} = d(r\,\hat{\mathbf{r}}) = dr\,\hat{\mathbf{r}} + r\,d\hat{\mathbf{r}} = dr\hat{\mathbf{r}} + rd\theta\hat{\boldsymbol{\theta}} + r\sin\theta d\phi\hat{\boldsymbol{\phi}}, \tag{1.54}$$

where use has been made of

$$d\hat{\mathbf{r}} = \frac{\partial\hat{\mathbf{r}}}{\partial\theta}d\theta + \frac{\partial\hat{\mathbf{r}}}{\partial\phi}d\phi = d\theta\,\hat{\boldsymbol{\theta}} + \sin\theta\,\hat{\boldsymbol{\phi}}. \tag{1.55}$$

Using the procedure used in Cartesian coordinates, the differential surface elements of the various spherical surfaces are

$$\left.d\mathbf{S}\right|_{r=r_o} = r^2\sin\theta d\theta d\phi\hat{\mathbf{r}}\quad \left.d\mathbf{S}\right|_{\theta=\theta_o} = r\sin\theta dr d\phi\hat{\boldsymbol{\theta}}\quad \left.d\mathbf{S}\right|_{\phi=\phi_o} = rdrd\theta\hat{\boldsymbol{\phi}}\Big\}, \tag{1.56}$$

where r_o, ϕ_o, and z_o are constants.

Using the procedure used in Cartesian coordinates, the differential volume element in spherical coordinates becomes

$$d\mathcal{V} = r^2\sin\theta dr d\theta d\phi. \tag{1.57}$$

1.3.2 Integral Calculus

Here, we discuss line integrals, surface integrals, and volume integrals.

Line integrals
A line integral is an integral whose domain is a curve in space. There are two types of line integrals: scalar line integrals and vector line integrals.

• **Scalar line integrals**: The scalar line integral of a vector **A** is given by

$$\int_{\mathcal{L}} \mathbf{A}\cdot d\mathbf{l}, \tag{1.58}$$

where $\mathcal{L}$ is the domain of integration, and $d\mathbf{l}$ is tangential to the curve $\mathcal{L}$. This implies that only the tangential component of **A** contributes to this integral. If $\mathcal{L}$ represents a closed path, the line integral becomes

$$\oint_{\mathcal{L}} \mathbf{A}\cdot d\mathbf{l}. \tag{1.59}$$

Such an integral is interpreted as the circulation of **A** around $\mathcal{L}$.

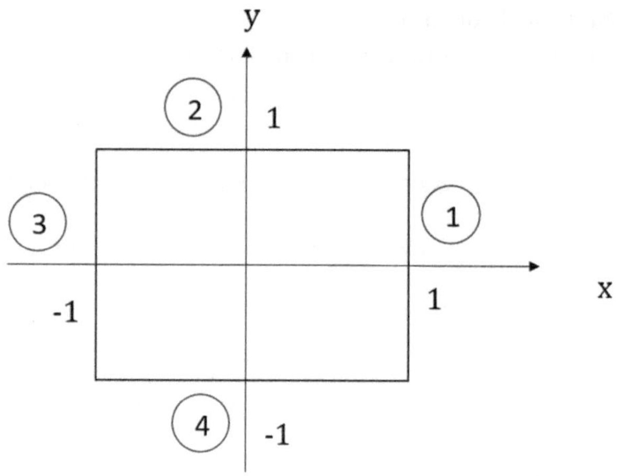

Fig. 1.11 For Example 1.12

Example 1.12 Calculate $\oint_{\mathcal{L}} \mathbf{A} \cdot d\mathbf{l}$, where $\mathbf{A} = x^2(1-z)\hat{\mathbf{x}} - yz\hat{\mathbf{y}} - e^x\hat{\mathbf{z}}$ and $\mathcal{L}$ is the square $\{z = 0, x \in [-1, 1], y \in [-1, 1]\}$ in the counter-clockwise direction (Fig. 1.11).

Solution. Since $\mathcal{L}$ is piecewise smooth, it cannot be described by a single differential length element. Instead, we break $\mathcal{L}$ into 4 segments, each of which a $d\mathbf{l}$ can be assigned. Therefore,

$$\oint_{\mathcal{L}} \mathbf{A} \cdot d\mathbf{l} = \int_{①} \mathbf{A} \cdot d\mathbf{l} + \int_{②} \mathbf{A} \cdot d\mathbf{l} + \int_{③} \mathbf{A} \cdot d\mathbf{l} + \int_{④} \mathbf{A} \cdot d\mathbf{l}.$$

Then,

$$\int_{①} \mathbf{A} \cdot d\mathbf{l} = \int_{-1}^{1} \mathbf{A} \cdot dy\hat{\mathbf{y}} \Big| \{x = 1, z = 0\} = 0,$$

$$\int_{②} \mathbf{A} \cdot d\mathbf{l} = \int_{1}^{-1} \mathbf{A} \cdot dx\hat{\mathbf{x}} \Big| \{y = 1, z = 0\} = -2/3,$$

$$\int_{③} \mathbf{A} \cdot d\mathbf{l} = \int_{1}^{-1} \mathbf{A} \cdot dy\hat{\mathbf{y}} \Big| \{x = -1, z = 0\} = 0,$$

and

$$\int_{④} \mathbf{A} \cdot d\mathbf{l} = \int_{-1}^{1} \mathbf{A} \cdot dx\hat{\mathbf{x}} \Big| \{y = -1, z = 0\} = 2/3.$$

So, $\oint_{\mathcal{L}} \mathbf{A} \cdot d\mathbf{l} = 0$. ◁

Fig. 1.12 For Example 1.13

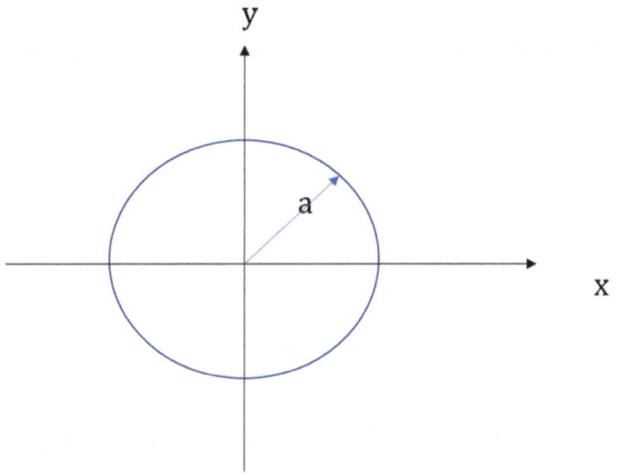

Example 1.13 Calculate $\oint_{\mathcal{L}} \mathbf{A} \cdot d\mathbf{l}$, where $\mathbf{A} = z\cos\phi\hat{\boldsymbol{\rho}} + \rho\sin^2\phi\hat{\boldsymbol{\phi}}$ and $\mathcal{L}$ is the circle $\{\rho = a, z = 0\}$ (in the counter-clockwise direction) (Fig. 1.12).

Solution.

$$\oint_{\mathcal{L}} \mathbf{A} \cdot d\mathbf{l} = \int_0^{2\pi} \mathbf{A} \cdot \rho d\phi\, \hat{\boldsymbol{\phi}} \Big| \{\rho = a, z = 0\} = a^2 \int_0^{2\pi} \sin^2\phi d\phi = \pi a^2.$$

◁

- **Vector line integrals**: The vector line integral of a vector **A** is given by

$$\int_{\mathcal{L}} \mathbf{A}\, d\mathcal{L}. \tag{1.60}$$

Another form that is frequently encountered is

$$\int_{\mathcal{L}} \mathbf{A} \times d\mathbf{l}. \tag{1.61}$$

For these integrals, we always need to make sure that the vector **A** is expressed in Cartesian bases. ◁

Example 1.14 Evaluate (a) $\int_0^1 \hat{\mathbf{x}}\,dx$,(b) $\int_0^{2\pi} \hat{\boldsymbol{\rho}}\,d\phi$, and (c) $\int_0^{2\pi} \cos\phi\,\hat{\boldsymbol{\rho}}\,d\phi$.

Solution.

(a)

$$\int_0^1 \hat{\mathbf{x}}\,dx = \hat{\mathbf{x}}\int_0^1 dx = \hat{\mathbf{x}}.$$

(b)

$$\int_0^{2\pi} \hat{\boldsymbol{\rho}}\,d\phi = \int_0^{2\pi} (\cos\phi\,\hat{\mathbf{x}} + \sin\phi\,\hat{\mathbf{y}})\,d\phi = \hat{\mathbf{x}}\int_0^{2\pi} \cos\phi\,d\phi + \hat{\mathbf{y}}\int_0^{2\pi} \sin\phi\,d\phi = \mathbf{0}.$$

(c)

$$\int_0^{2\pi} \cos\phi\,\hat{\boldsymbol{\rho}}\,d\phi = \int_0^{2\pi} (\cos^2\phi\,\hat{\mathbf{x}} + \sin\phi\cos\phi\,\hat{\mathbf{y}})\,d\phi = \hat{\mathbf{x}}\int_0^{2\pi} \cos^2\phi\,d\phi + \hat{\mathbf{y}}\int_0^{2\pi} \sin\phi\cos\phi\,d\phi = \pi\hat{\mathbf{x}}.$$

◁

Surface Integrals

A surface integral is an integral whose domain is a surface in space. There are two types of surface integrals: scalar surface integrals and vector surface integrals.

• **Scalar surface integrals**: The scalar surface integral of a vector **A** is given by

$$\int_S \mathbf{A} \cdot d\mathbf{S}, \tag{1.62}$$

where S is the domain of integration, and $d\mathbf{S}$ is normal to the surface S. Note that only the normal component of **A** contributes to this integral. If S represents a closed surface, the surface integral becomes

$$\oint_S \mathbf{A} \cdot d\mathbf{S}. \tag{1.63}$$

Such an integral is interpreted as the flux of **A** from S. ◁

Example 1.15 Determine $\oint_S \mathbf{A} \cdot d\mathbf{S}$, where $\mathbf{A} = \hat{\boldsymbol{\rho}} + \hat{\mathbf{z}}$, and S is the cylinder $\{\rho = 1, 0 \leq z \leq 1\}$.

Solution. Since the cylinder is finite, we decompose it into three parts as

$$\oint_S \mathbf{A} \cdot d\mathbf{S} = \int_{①} \mathbf{A} \cdot d\mathbf{S} + \int_{②} \mathbf{A} \cdot d\mathbf{S} + \int_{③} \mathbf{A} \cdot d\mathbf{S},$$

where ① is $z = 1$ with $\hat{\mathbf{n}} = \hat{\mathbf{z}}$, ② is $\rho = 1$ with $\hat{\mathbf{n}} = \hat{\boldsymbol{\rho}}$, and ③ is $z = 0$ with $\hat{\mathbf{n}} = -\hat{\mathbf{z}}$. So,

$$\int_{①} \mathbf{A} \cdot d\mathbf{S} = \int_0^{2\pi}\int_0^1 \mathbf{A} \cdot \rho d\rho d\phi\, \hat{\mathbf{z}}\Big|_{z=1} = \pi,$$

$$\int_{②} \mathbf{A} \cdot d\mathbf{S} = \int_0^{1}\int_0^{2\pi} \mathbf{A} \cdot \rho d\phi dz\, \hat{\boldsymbol{\rho}}\Big|_{\rho=1} = 2\pi,$$

and

$$\int_{③} \mathbf{A} \cdot d\mathbf{S} = \int_0^{2\pi}\int_0^1 \mathbf{A} \cdot \rho d\rho d\phi\, (-\hat{\mathbf{z}})\Big|_{z=0} = -\pi.$$

Therefore, $\oint_S \mathbf{A} \cdot d\mathbf{S} = 2\pi$. ◁

- **Vector surface integrals**: The vector surface integral of a vector **A** is given by

$$\int_S \mathbf{A}\, dS. \tag{1.64}$$

Another form that is frequently encountered is

$$\int_S \mathbf{A} \times d\mathbf{S}. \tag{1.65}$$

Again, for these integrals, we always need to make sure that the vector **A** is expressed in Cartesian bases $\hat{\mathbf{x}}$, $\hat{\mathbf{y}}$, and $\hat{\mathbf{z}}$.

Volume Integrals

The volume integral is an integral whose domain is a volume in space. The volume integral of a vector **A** is given by

$$\int_{\mathcal{V}} \mathbf{A}\, d\mathcal{V}, \tag{1.66}$$

where $\mathcal{V}$ is the domain of integration.

Example 1.16 Determine $\int_{\mathcal{V}} \mathbf{A}\, d\mathcal{V}$, where $\mathbf{A} = z(\rho\hat{\boldsymbol{\rho}} + \hat{\mathbf{z}})$, and $\mathcal{V}$ is the region enclosed by the cylinder $\{0 \le \rho \le 1, 0 \le z \le 1\}$.

Solution.

$$\int_{\mathcal{V}} \mathbf{A}\, d\mathcal{V} = \int_0^1\int_0^{2\pi}\int_0^1 z(\rho\hat{\boldsymbol{\rho}} + \hat{\mathbf{z}})\, \rho d\rho d\phi dz = \mathbf{0} + \hat{\mathbf{z}}\int_0^1\int_0^{2\pi}\int_0^1 z\rho d\rho d\phi dz = \frac{\pi}{2}\hat{\mathbf{z}}.$$

◁

1.3.3 Differential Calculus

The del operator (∇) is a vector differential operator that can act on a scalar or on a vector. In Cartesian, cylindrical, and spherical coordinates, the del operator is written, respectively, as

$$\nabla = \frac{\partial}{\partial x}\hat{\mathbf{x}} + \frac{\partial}{\partial y}\hat{\mathbf{y}} + \frac{\partial}{\partial z}\hat{\mathbf{z}}, \tag{1.67}$$

$$\nabla = \frac{\partial}{\partial \rho}\hat{\boldsymbol{\rho}} + \frac{1}{\rho}\frac{\partial}{\partial \phi}\hat{\boldsymbol{\phi}} + \frac{\partial}{\partial z}\hat{\mathbf{z}}, \tag{1.68}$$

and

$$\nabla = \frac{\partial}{\partial r}\hat{\mathbf{r}} + \frac{1}{r}\frac{\partial}{\partial \theta}\hat{\boldsymbol{\theta}} + \frac{1}{r\sin\theta}\frac{\partial}{\partial \phi}\hat{\boldsymbol{\phi}}. \tag{1.69}$$

In what follows, we discuss four differential operators involving the del operator. These are the gradient of a scalar, divergence of a vector, curl of a vector, and Laplacian of a scalar and of a vector.

Gradient

In single-variable calculus, the total derivative for a scalar $V(x)$ is given by

$$dV(x) = \frac{dV(x)}{dx}dx. \tag{1.70}$$

In three-variable calculus, the total derivative for a scalar $V(\mathbf{r})$ is given by

$$dV(\mathbf{r}) = \frac{\partial V(\mathbf{r})}{\partial x}dx + \frac{\partial V(\mathbf{r})}{\partial y}dy + \frac{\partial V(\mathbf{r})}{\partial z}dz, \tag{1.71}$$

which can be written as

$$dV(\mathbf{r}) = \nabla V(\mathbf{r}) \bullet d\mathbf{l}, \tag{1.72}$$

where

$$\nabla V(\mathbf{r}) = \frac{\partial V(\mathbf{r})}{\partial x}\hat{\mathbf{x}} + \frac{\partial V(\mathbf{r})}{\partial y}\hat{\mathbf{y}} + \frac{\partial V(\mathbf{r})}{\partial z}\hat{\mathbf{z}} \tag{1.73}$$

is the gradient of $V(\mathbf{r})$. The gradient operates on a scalar and gives a vector whose magnitude is the highest rate of change of the scalar, and whose direction points to the highest rate of increase (or decrease) of the scalar. In cylindrical and spherical coordinates, the gradient is, respectively, given by

$$\nabla V = \frac{\partial V}{\partial \rho}\hat{\boldsymbol{\rho}} + \frac{1}{\rho}\frac{\partial V}{\partial \phi}\hat{\boldsymbol{\phi}} + \frac{\partial V}{\partial z}\hat{\mathbf{z}}, \tag{1.74}$$

and

$$\nabla V = \frac{\partial V}{\partial r}\hat{\mathbf{r}} + \frac{1}{r}\frac{\partial V}{\partial \theta}\hat{\boldsymbol{\theta}} + \frac{1}{r\sin\theta}\frac{\partial V}{\partial \phi}\hat{\boldsymbol{\phi}}. \tag{1.75}$$

As an application of the gradient, given a surface described by $f(x, y, z) = 0$, the unit vector normal to the surface is given by

$$\hat{\mathbf{n}} = \frac{\nabla f}{|\nabla f|}. \tag{1.76}$$

Example 1.17 Determine the unit normal to the surfaces (a) $x = a$, (b) $\rho = a$, and (c) $r = a$.

Solution. (a) $\hat{\mathbf{n}} = \hat{\mathbf{x}}$, (b) $\hat{\mathbf{n}} = \hat{\rho}$, and (c) $\hat{\mathbf{n}} = \hat{\mathbf{r}}$.

From this example, we see that in general, a surface $v = v_o$ has a unit normal vector $\hat{\mathbf{v}}$ associated with it. ◁

Example 1.18 Starting from Eqs. (1.24), derive Eqs. (1.33).

Solution. Taking the gradient of both sides of Eqs. (1.33) and dividing by the magnitude, one gets

$$\left.\begin{aligned} \frac{\nabla x}{|\nabla x|} &= \frac{\nabla(\rho\cos\phi)}{|\nabla(\rho\cos\phi)|} \\ \frac{\nabla y}{|\nabla y|} &= \frac{\nabla(\rho\sin\phi)}{|\nabla(\rho\sin\phi)|} \\ \frac{\nabla z}{|\nabla z|} &= \frac{\nabla z}{|\nabla z|} \end{aligned}\right\}.$$

After performing the gradient of the left side in Cartesian coordinates, and the one on the right side in cylindrical coordinates, one gets Eqs. (1.33). Notice that this approach can be used to derive bases transformation formulas between Cartesian bases and any coordinate system bases, provided that the transformation formulas from Cartesian variables to the other coordinate system variables are provided. Also, although division by the magnitude of the gradient was not necessary in this particular coordinate system, it might become necessary in other coordinate systems. ◁

Divergence

The divergence of a vector $\mathbf{A}$ is represented by $\nabla \bullet \mathbf{A}$. Divergence indicates how much a vector $\mathbf{A}$ spreads out (diverges) from a point P. Mathematically, the divergence can be described by

$$\nabla \bullet \mathbf{A} = \lim_{\Delta v \to 0} \frac{\oint_S \mathbf{A} \bullet d\mathbf{S}}{\Delta v}. \tag{1.77}$$

Equation (1.77) indicates that the divergence of a vector $\mathbf{A}$ at a point P is the outward flux per unit volume as the volume shrinks to the point P. Divergence can be positive (P is a source), negative (P is a sink), or zero. In Cartesian, cylindrical, and spherical coordinates, respectively, the divergence is written as

$$\nabla \bullet \mathbf{A} = \frac{\partial A_x}{\partial x} + \frac{\partial A_y}{\partial y} + \frac{\partial A_z}{\partial z}, \tag{1.78}$$

$$\nabla \bullet \mathbf{A} = \frac{1}{\rho}\frac{\partial}{\partial \rho}(\rho A_\rho) + \frac{1}{\rho}\frac{\partial A_\phi}{\partial \phi} + \frac{\partial A_z}{\partial z}, \tag{1.79}$$

and

$$\nabla \bullet \mathbf{A} = \frac{1}{r^2}\frac{\partial}{\partial r}(r^2 A_r) + \frac{1}{r \sin\theta}\frac{\partial}{\partial \theta}(A_\theta \sin\theta) + \frac{1}{r\sin\theta}\frac{\partial A_\phi}{\partial \phi}. \tag{1.80}$$

The divergence theorem: The divergence theorem states that the total outward flux of a vector **A** through a closed surface $\mathcal{S}$ is equal to the volume integral of the divergence of **A** over the volume $\mathcal{V}$ formed by $\mathcal{S}$. Mathematically,

$$\oint_{\mathcal{S}} \mathbf{A} \bullet d\mathbf{S} = \int_{\mathcal{V}} \nabla \bullet \mathbf{A}\, d\mathcal{V}. \tag{1.81}$$

Example 1.19 Verify the divergence theorem using Example 1.15.

Solution. We found that $\oint_{\mathcal{S}} \mathbf{A} \bullet d\mathbf{S} = 2\pi$. Now,

$$\nabla \bullet \mathbf{A} = \frac{1}{\rho}\frac{\partial}{\partial \rho}(\rho A_\rho) + \frac{1}{\rho}\frac{\partial A_\phi}{\partial \phi} + \frac{\partial A_z}{\partial z} = \frac{1}{\rho}\frac{\partial}{\partial \rho}(\rho) = \frac{1}{\rho}.$$

Thus,

$$\int_{\mathcal{V}} \nabla \bullet \mathbf{A}\, d\mathcal{V} = \int_0^1 \int_0^{2\pi} \int_0^1 \frac{1}{\rho}\, \rho\, d\rho\, d\phi\, dz = 2\pi.$$

◁

Curl

The curl of a vector **A** is represented by $\nabla \times \mathbf{A}$. Mathematically, the curl can be described by

$$\nabla \times \mathbf{A} = \lim_{\Delta s \to 0} \frac{\oint_{\mathcal{L}} \mathbf{A} \bullet d\mathbf{l}}{\Delta s} \hat{\mathbf{n}}. \tag{1.82}$$

Equation (1.82) indicates that the curl of a vector **A** at a point P is the circulation of **A** per unit area as the area shrinks to zero. The magnitude of the curl is a measure of how much a vector **A** rotates about the point P, and its direction is normal to the area and can be determined from the right-hand rule.

In Cartesian, cylindrical, and spherical coordinates, respectively, the curl is written as

$$\nabla \times \mathbf{A} = \left(\frac{\partial A_z}{\partial y} - \frac{\partial A_y}{\partial z}\right)\hat{\mathbf{x}} + \left(\frac{\partial A_x}{\partial z} - \frac{\partial A_z}{\partial x}\right)\hat{\mathbf{y}} + \left(\frac{\partial A_y}{\partial x} - \frac{\partial A_x}{\partial y}\right)\hat{\mathbf{z}}, \tag{1.83}$$

$$\nabla \times \mathbf{A} = \left(\frac{1}{\rho}\frac{\partial A_z}{\partial \phi} - \frac{\partial A_\phi}{\partial z}\right)\hat{\boldsymbol{\rho}} + \left(\frac{\partial A_\rho}{\partial z} - \frac{\partial A_z}{\partial \rho}\right)\hat{\boldsymbol{\phi}} + \frac{1}{\rho}\left(\frac{\partial(\rho A_\phi)}{\partial \rho} - \frac{\partial A_\rho}{\partial \phi}\right)\hat{\mathbf{z}}, \tag{1.84}$$

and

$$\nabla \times \mathbf{A} = \frac{1}{r \sin\theta}\left(\frac{\partial(A_\phi \sin\theta)}{\partial\theta} - \frac{\partial A_\theta}{\partial\phi}\right)\hat{\mathbf{r}} + \frac{1}{r}\left(\frac{1}{\sin\theta}\frac{\partial A_r}{\partial\phi} - \frac{\partial(rA_\phi)}{\partial r}\right)\hat{\boldsymbol{\theta}} + \frac{1}{r}\left(\frac{\partial(rA_\theta)}{\partial r} - \frac{\partial A_r}{\partial\theta}\right)\hat{\boldsymbol{\phi}}. \tag{1.85}$$

Stokes theorem: Stokes theorem states that the circulation of a vector **A** around a closed path $\mathcal{L}$ is equal to the surface integral of $\nabla \times \mathbf{A}$ over the surface $\mathcal{S}$ formed by $\mathcal{L}$. Mathematically,

$$\oint_{\mathcal{L}} \mathbf{A} \bullet d\mathbf{l} = \int_{\mathcal{S}} (\nabla \times \mathbf{A}) \bullet d\mathbf{S}. \tag{1.86}$$

Example 1.20 Verify Stokes theorem using Example 1.12.

Solution. We found that $\oint_{\mathcal{L}} \mathbf{A} \bullet d\mathbf{l} = 0$. Now,

$$\nabla \times \mathbf{A} = \left(\frac{\partial(e^x)}{\partial y} - \frac{\partial(-yz)}{\partial z}\right)\hat{\mathbf{x}} + \left(\frac{\partial x^2(1-z)}{\partial z} - \frac{\partial(e^x)}{\partial x}\right)\hat{\mathbf{y}} + \left(\frac{\partial(-yz)}{\partial x^2(1-z)} - \frac{\partial x^2}{\partial y}\right)\hat{\mathbf{z}} = y\,\hat{\mathbf{x}} - (e^x - x^2)\,\hat{\mathbf{y}}.$$

Thus,

$$\int_{\mathcal{S}} (\nabla \times \mathbf{A}) \bullet d\mathbf{S} = \int_{-1}^{1}\int_{-1}^{1} [y\,\hat{\mathbf{x}} - (e^x - x^2)\,\hat{\mathbf{y}}] \bullet dxdy\,\hat{\mathbf{z}}\Big|_{z=0} = 0.$$

Notice that $\hat{\mathbf{n}} = \hat{\mathbf{z}}$ because the orientation of the curve is counter clockwise. If clockwise orientation was chosen, it would have been $\hat{\mathbf{n}} = -\hat{\mathbf{z}}$. ◁

Laplacian

The Laplacian operator (∇^2) can operate on a scalar V or on a vector **A**. For a scalar, the Laplacian is given by

$$\nabla^2 V = \nabla \bullet \nabla V, \tag{1.87}$$

whereas for a vector, it is given by

$$\nabla^2 \mathbf{A} = \nabla(\nabla \bullet \mathbf{A}) - \nabla \times (\nabla \times \mathbf{A}). \tag{1.88}$$

In Cartesian, cylindrical, and spherical coordinates, respectively, the Laplacian of a scalar V is written as

$$\nabla^2 V = \frac{\partial^2 V}{\partial x^2} + \frac{\partial^2 V}{\partial y^2} + \frac{\partial^2 V}{\partial z^2}, \tag{1.89}$$

$$\nabla^2 V = \frac{\partial^2 V}{\partial \rho^2} + \frac{1}{\rho}\frac{\partial V}{\partial \rho} + \frac{1}{\rho^2}\frac{\partial^2 V}{\partial \phi^2} + \frac{\partial^2 V}{\partial z^2}, \tag{1.90}$$

and

$$\nabla^2 V = \frac{\partial^2 V}{\partial r^2} + \frac{2}{r}\frac{\partial V}{\partial r} + \frac{1}{r^2}\frac{\partial^2 V}{\partial \theta^2} + \frac{\cot\theta}{r^2}\frac{\partial V}{\partial \theta} + \frac{1}{r^2 \sin^2\theta}\frac{\partial^2 V}{\partial \phi^2}. \tag{1.91}$$

In Cartesian coordinates, and after noting that $\hat{\mathbf{x}}$, $\hat{\mathbf{y}}$, and $\hat{\mathbf{z}}$ are constant vectors, the Laplacian of a vector **A** can be found easily as

$$\nabla^2 \mathbf{A} = \nabla^2 A_x \,\hat{\mathbf{x}} + \nabla^2 A_y \,\hat{\mathbf{y}} + \nabla^2 A_z \,\hat{\mathbf{z}}. \tag{1.92}$$

Example 1.21 Obtain the Laplacian of (a) $V = e^x \sin y \cos z$, and for (b) $\mathbf{A} = xy\hat{\mathbf{x}} + x^2\hat{\mathbf{y}} + e^x \sin y\hat{\mathbf{z}}$.

Solution.
(a)

$$\begin{aligned}
\nabla^2 V &= \frac{\partial^2 V}{\partial x^2} + \frac{\partial^2 V}{\partial y^2} + \frac{\partial^2 V}{\partial z^2} \\
&= \frac{\partial^2 (e^x \sin y \cos z)}{\partial x^2} + \frac{\partial^2 (e^x \sin y \cos z)}{\partial y^2} + \frac{\partial^2 (e^x \sin y \cos z)}{\partial z^2} \\
&= e^x \sin y \cos z - e^x \sin y \cos z - e^x \sin y \cos z = -e^x \sin y \cos z.
\end{aligned}$$

(b) We note that $A_x = xy$, $A_y = x^2$, and $A_z = e^x \sin y$. Then, $\nabla^2 A_x = 0$, $\nabla^2 A_y = 2$, and $\nabla^2 A_z = e^x \sin y - e^x \sin y = 0$. Hence, $\nabla^2 \mathbf{A} = 2\,\hat{\mathbf{y}}$. ◁

1.3.4 Classification of Vectors

A vector is solenoidal (or divergenceless) if

$$\nabla \bullet \mathbf{A} = 0. \tag{1.93}$$

In general, one can show that

$$\nabla \bullet (\nabla \times \mathbf{F}) = 0 \tag{1.94}$$

holds true for any vector $\mathbf{F}$. Hence, a solenoidal vector $\mathbf{A}$ can be expressed in terms of another vector $\mathbf{F}$ by

$$\mathbf{A} = \nabla \times \mathbf{F}. \tag{1.95}$$

A vector is conservative (or irrotational) if

$$\nabla \times \mathbf{A} = \mathbf{0}. \tag{1.96}$$

In general, one can show that

$$\nabla \times (\nabla V) = \mathbf{0} \tag{1.97}$$

holds true for any scalar. Hence, a conservative vector $\mathbf{A}$ can be expressed in terms of a scalar V by

$$\mathbf{A} = \nabla V. \tag{1.98}$$

It is to be noted that a vector can be determined, provided that its curl and its divergence are specified. This is known as Helmholtz theorem.

1.4 Vector Calculus in a General Curvilinear Coordinate System

In Sect. 1.3, we discussed vector calculus in the three common coordinate systems. In this section, we generalize to a general curvilinear orthogonal coordinate system.

Starting from Cartesian variables $\{x, y, z\}$, curvilinear variables $\{v_1, v_2, v_3\}$ can be introduced from Cartesian coordinates using

$$\left.\begin{aligned} x &= f(v_1, v_2, v_3) \\ y &= g(v_1, v_2, v_3) \\ z &= w(v_1, v_2, v_3) \end{aligned}\right\}, \tag{1.99}$$

where $f(\bullet)$, $g(\bullet)$, and $w(\bullet)$ are functions of the curvilinear variables. Then, we define

$$\left.\begin{aligned} h_1 &= \sqrt{\left(\frac{\partial x}{\partial v_1}\right)^2 + \left(\frac{\partial y}{\partial v_1}\right)^2 + \left(\frac{\partial z}{\partial v_1}\right)^2} = \sqrt{\left(\frac{\partial f}{\partial v_1}\right)^2 + \left(\frac{\partial g}{\partial v_1}\right)^2 + \left(\frac{\partial w}{\partial v_1}\right)^2} \\ h_2 &= \sqrt{\left(\frac{\partial x}{\partial v_2}\right)^2 + \left(\frac{\partial y}{\partial v_2}\right)^2 + \left(\frac{\partial z}{\partial v_2}\right)^2} = \sqrt{\left(\frac{\partial f}{\partial v_2}\right)^2 + \left(\frac{\partial g}{\partial v_2}\right)^2 + \left(\frac{\partial w}{\partial v_2}\right)^2} \\ h_3 &= \sqrt{\left(\frac{\partial x}{\partial v_3}\right)^2 + \left(\frac{\partial y}{\partial v_3}\right)^2 + \left(\frac{\partial z}{\partial v_3}\right)^2} = \sqrt{\left(\frac{\partial f}{\partial v_3}\right)^2 + \left(\frac{\partial g}{\partial v_3}\right)^2 + \left(\frac{\partial w}{\partial v_3}\right)^2} \end{aligned}\right\} \tag{1.100}$$

as the metric coefficients.

Associated with the curvilinear variables $\{v_1, v_2, v_3\}$, we define the corresponding curvilinear bases $\{\hat{\mathbf{v}}_1, \hat{\mathbf{v}}_2, \hat{\mathbf{v}}_3\}$. Then, using the metric coefficients, the following differential elements of a general curvilinear coordinate system can be defined.

$$d\mathbf{l} = h_1 dv_1\, \hat{\mathbf{v}}_1 + h_2 dv_2\, \hat{\mathbf{v}}_2 + h_3 dv_3\, \hat{\mathbf{v}}_3, \tag{1.101}$$

$$d\mathbf{S}\Big|_{v_1=v_o} = h_2 h_3 dv_2 dv_3\, \hat{\mathbf{v}}_1 \quad d\mathbf{S}\Big|_{v_2=v_o} = h_1 h_3 dv_1 dv_3\, \hat{\mathbf{v}}_2 \quad d\mathbf{S}\Big|_{v_3=v_o} = h_1 h_2 dv_1 dv_2\, \hat{\mathbf{v}}_3 \Big\}, \tag{1.102}$$

where v_o is a constant, and

$$d\mathcal{V} = h_1 h_2 h_3 dv_1 dv_2 dv_3. \tag{1.103}$$

Given a scalar $V(\mathbf{r})$ and a vector $\mathbf{A}(\mathbf{r}) = A_{v_1}(\mathbf{r})\,\hat{\mathbf{v}}_1 + A_{v_2}(\mathbf{r})\,\hat{\mathbf{v}}_2 + A_{v_3}(\mathbf{r})\,\hat{\mathbf{v}}_3$, the gradient, divergence, curl, and Laplacian in a curvilinear coordinate system can be written as follows.

$$\nabla V = \frac{1}{h_1}\frac{\partial V}{\partial v_1}\hat{\mathbf{v}}_1 + \frac{1}{h_2}\frac{\partial V}{\partial v_2}\hat{\mathbf{v}}_2 + \frac{1}{h_3}\frac{\partial V}{\partial v_3}\hat{\mathbf{v}}_3, \tag{1.104}$$

$$\nabla \bullet \mathbf{A} = \frac{1}{h_1 h_2 h_3}\left[\frac{\partial}{\partial v_1}\left(h_2 h_3 \frac{\partial A_{v_1}}{\partial v_1}\right) + \frac{\partial}{\partial v_2}\left(h_1 h_3 \frac{\partial A_{v_2}}{\partial v_2}\right) + \frac{\partial}{\partial v_3}\left(h_1 h_2 \frac{\partial A_{v_3}}{\partial v_3}\right)\right], \tag{1.105}$$

$$\nabla \times \mathbf{A} = \begin{vmatrix} h_1\hat{\mathbf{v}}_1 & h_2\hat{\mathbf{v}}_2 & h_3\hat{\mathbf{v}}_3 \\ \dfrac{\partial}{\partial v_1} & \dfrac{\partial}{\partial v_2} & \dfrac{\partial}{\partial v_3} \\ h_1 A_{v_1} & h_2 A_{v_2} & h_3 A_{v_3} \end{vmatrix}, \tag{1.106}$$

$$\nabla^2 V = \frac{1}{h_1 h_2 h_3}\left[\frac{\partial}{\partial v_1}\left(\frac{h_2 h_3}{h_1}\frac{\partial V}{\partial v_1}\right) + \frac{\partial}{\partial v_2}\left(\frac{h_1 h_3}{h_2}\frac{\partial V}{\partial v_2}\right) + \frac{\partial}{\partial v_3}\left(\frac{h_1 h_2}{h_3}\frac{\partial V}{\partial v_3}\right)\right], \tag{1.107}$$

and

$$\nabla^2 \mathbf{A} = \nabla(\nabla \bullet \mathbf{A}) - \nabla \times (\nabla \times \mathbf{A}). \tag{1.108}$$

Cartesian bases $\{\hat{\mathbf{x}}, \hat{\mathbf{y}}, \hat{\mathbf{z}}\}$ can be related to curvilinear bases $\{\hat{\mathbf{v}}_1, \hat{\mathbf{v}}_2, \hat{\mathbf{v}}_3\}$ through

$$\left.\hat{\mathbf{x}} = \frac{\nabla f}{|\nabla f|} \quad \hat{\mathbf{y}} = \frac{\nabla g}{|\nabla g|} \quad \hat{\mathbf{z}} = \frac{\nabla w}{|\nabla w|} \quad\right\}, \tag{1.109}$$

where the gradient operators on the right side of Eq. (1.109) are performed in the chosen curvilinear coordinates. Equation (1.109) can be written in matrix form as

$$\begin{pmatrix} \hat{\mathbf{x}} \\ \hat{\mathbf{y}} \\ \hat{\mathbf{z}} \end{pmatrix} = \begin{pmatrix} \dfrac{\nabla f}{|\nabla f|} \\ \dfrac{\nabla g}{|\nabla g|} \\ \dfrac{\nabla w}{|\nabla w|} \end{pmatrix} \begin{pmatrix} \hat{\mathbf{v}}_1 \\ \hat{\mathbf{v}}_2 \\ \hat{\mathbf{v}}_3 \end{pmatrix}. \tag{1.110}$$

Inverse relation can be obtaining by simply taking the transpose of the square matrix resulting in Eq. (1.110).

As an example, elliptical cylindrical coordinates are related to the Cartesian coordinates through

$$\left.x = a\cosh\mu\cos\phi \quad y = a\sinh\mu\sin\phi \quad z = z\right\}, \tag{1.111}$$

where a is a constant, $\mu \in [0, \infty)$, $\phi \in [0, 2\pi)$, and $z \in (-\infty, \infty)$. With these, one can find the associated metric coefficients as

$$\left.h_1 = h_2 = a\sqrt{\sinh^2\mu + \sin^2\phi} \quad h_3 = 1\right\}. \tag{1.112}$$

These metric coefficients can be used, for instance, to find the gradient

$$\nabla V = \frac{1}{a\sqrt{\sinh^2\mu + \sin^2\phi}}\left(\frac{\partial V}{\partial \mu}\hat{\boldsymbol{\mu}} + \frac{\partial V}{\partial \phi}\hat{\boldsymbol{\phi}}\right) + \frac{\partial V}{\partial z}\hat{\mathbf{z}}. \tag{1.113}$$

Using the gradient, conversion from Cartesian bases to elliptical cylindrical bases can then be done by

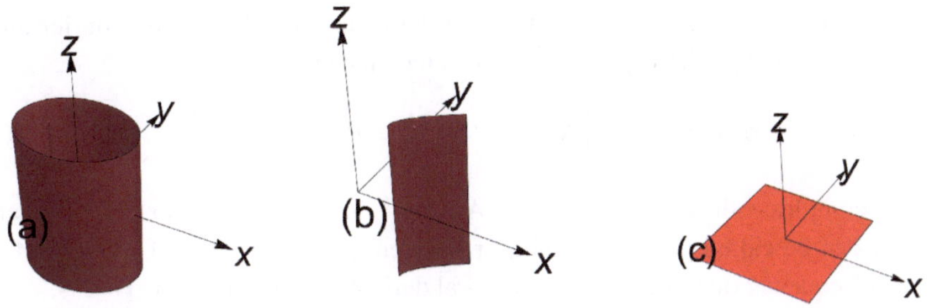

Fig. 1.13 **a** $\mu = 1$, **b** $\phi = \pi/6$, and **c** $z = 0$

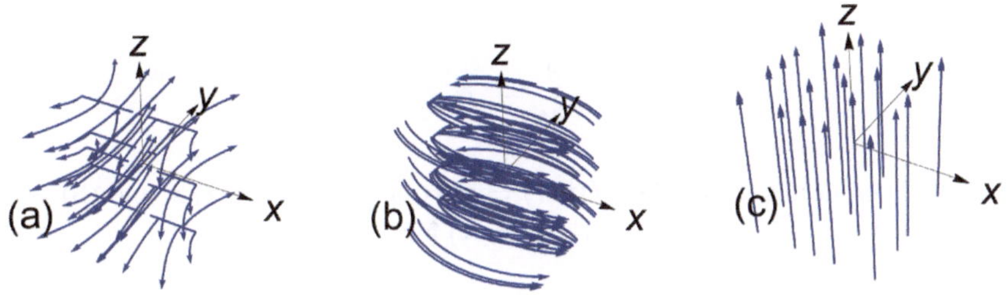

Fig. 1.14 **a** $\hat{\boldsymbol{\mu}}$, **b** $\hat{\boldsymbol{\phi}}$, and **c** $\hat{\mathbf{z}}$

$$\begin{pmatrix} \hat{\mathbf{x}} \\ \hat{\mathbf{y}} \\ \hat{\mathbf{z}} \end{pmatrix} = \begin{pmatrix} \frac{\sqrt{2}\sinh\mu\cos\phi}{\sqrt{\cosh 2\mu - \cos 2\phi}} & -\frac{\sqrt{2}\cosh\mu\sin\phi}{\sqrt{\cosh 2\mu - \cos 2\phi}} & 0 \\ \frac{\sqrt{2}\cosh\mu\sin\phi}{\sqrt{\cosh 2\mu - \cos 2\phi}} & \frac{\sqrt{2}\sinh\mu\cos\phi}{\sqrt{\cosh 2\mu - \cos 2\phi}} & 0 \\ 0 & 0 & 1 \end{pmatrix} \begin{pmatrix} \hat{\boldsymbol{\mu}} \\ \hat{\boldsymbol{\phi}} \\ \hat{\mathbf{z}} \end{pmatrix}. \tag{1.114}$$

Figures 1.13 and 1.14, respectively, show elliptical cylindrical surfaces and bases when $a = 4$.

1.5 Time-Harmonic Vectors

Given a vector $\mathscr{A}(\mathbf{r}, t)$ dependent upon spatial variables $\mathbf{r}$, as well as upon a temporal variable t, without loss of generality, we can assume that the temporal variation of the vector is sinusoidal with an angular frequency $\omega = 2\pi f$ (in $rad\ s^{-1}$), where f is a linear frequency (in Hz). That is

$$\mathscr{A}(\mathbf{r}, t) = \mathbf{A}_o(\mathbf{r}) \cos(\omega t + \theta), \tag{1.115}$$

where $\mathbf{A}_o(\mathbf{r})$ is the amplitude, and θ is the angle (or phase). Such a vector can be called a time-harmonic vector. Generality is not lost when dealing with time-harmonic vectors because non-sinusoidal temporal variation can be dealt with the use of Fourier analysis. Since $\cos x = \mathrm{Re}\{e^{jx}\}$, where $j = \sqrt{-1}$, we can deduce that

$$\mathscr{A}(\mathbf{r}, t) = \mathrm{Re}\left\{\mathbf{A}_o(\mathbf{r})e^{j\theta}e^{j\omega t}\right\} = \mathrm{Re}\left\{\mathbf{A}(\mathbf{r})e^{j\omega t}\right\}, \tag{1.116}$$

where the complex-valued vector $\mathbf{A}(\mathbf{r}) = \mathbf{A}_o(\mathbf{r})e^{j\theta}$ is called a phasor. The domain that involves ω can be called phasor domain, or frequency domain. One of the advantages of working in frequency domain is that a temporal derivative is eliminated. That is,

$$\frac{\partial \mathscr{A}(\mathbf{r}, t)}{\partial t} = \frac{\partial}{\partial t}\mathrm{Re}\left\{\mathbf{A}(\mathbf{r})e^{j\omega t}\right\} = \mathrm{Re}\left\{j\omega\mathbf{A}(\mathbf{r})e^{j\omega t}\right\}. \tag{1.117}$$

Hence,

$$\frac{\partial}{\partial t} \to j\omega. \tag{1.118}$$

Example 1.22 Obtain $\mathbf{A}(\mathbf{r})$ for (a) $\mathscr{A}(\mathbf{r}, t) = \cos(\omega t - 10x + \pi/2)\,\hat{\mathbf{z}}$, and (b) $\mathscr{A}(\mathbf{r}, t) = 4\cos(\omega t + 2y + \pi/2)\,\hat{\mathbf{x}} + 2\cos(\omega t + x - \pi/4)\,\hat{\mathbf{y}}$.

Solution. (a) $\mathbf{A}(\mathbf{r}) = e^{j(-10x+\pi/2)}\hat{\mathbf{z}}$, and (b) $\mathbf{A}(\mathbf{r}) = 4e^{j(2y+\pi/2)}\hat{\mathbf{x}} + 2\,e^{j(x-\pi/4)}\hat{\mathbf{y}}$. ◁

Example 1.23 Obtain $\mathscr{A}(\mathbf{r}, t)$ for (a) $\mathbf{A}(\mathbf{r}) = j\hat{\mathbf{x}} + e^{j2\pi z/2}\hat{\mathbf{y}}$, and (b) $\mathbf{A}(\mathbf{r}) = \hat{\mathbf{z}}e^{-jy}\sin\pi x$.

Solution.
(a)

$$\begin{aligned}\mathscr{A}(\mathbf{r}, t) &= \mathrm{Re}\left\{[j\hat{\mathbf{x}} + e^{j2\pi z/2}\hat{\mathbf{y}}]e^{j\omega t}\right\} = \mathrm{Re}\left\{[e^{j\pi/2}\hat{\mathbf{x}} + e^{j2\pi z/2}\hat{\mathbf{y}}]e^{j\omega t}\right\} = \mathrm{Re}\left\{e^{j(\omega t+\pi/2)}\hat{\mathbf{x}} + e^{j(\omega+2\pi z/2)}\hat{\mathbf{y}}\right\}\\ &= \cos(\omega t + \pi/2)\hat{\mathbf{x}} + \cos\left(\omega t + \frac{2\pi z}{2}\right)\hat{\mathbf{y}} = -\sin\omega t\,\hat{\mathbf{x}} + \cos\left(\omega t + \frac{2\pi z}{2}\right)\hat{\mathbf{y}}.\end{aligned}$$

(b)

$$\mathscr{A}(\mathbf{r}, t) = \mathrm{Re}\left\{\hat{\mathbf{z}}e^{-jy}\sin\pi x e^{j\omega t}\right\} = (\hat{\mathbf{z}}\sin\pi x)\,\mathrm{Re}\left\{e^{j(\omega t-y)}\right\} = \sin\pi x\cos(\omega t - y)\hat{\mathbf{z}}.$$

◁

Although not evident from the previous examples, a phasor $\mathbf{A}(\mathbf{r})$ of a vector $\mathscr{A}(\mathbf{r}, t)$ can be generally frequency dependent. Therefore, it is customary to include the frequency dependence into its argument [i.e., the phasor form of $\mathscr{A}(\mathbf{r}, t)$ is $\mathbf{A}(\mathbf{r}, \omega)$].

Problems

1.1 Given $\mathbf{A} = -\hat{\mathbf{x}} + \hat{\mathbf{y}} + \hat{\mathbf{z}}$ and $\mathbf{B} = \hat{\mathbf{x}} - \hat{\mathbf{y}} + \hat{\mathbf{z}}$, find (a) a unit vector normal to both **A** and **B**, and (b) the angle between **A** and **B**.

1.2 Given $\mathbf{A} = 3\hat{\mathbf{x}} + 2\hat{\mathbf{y}} - \hat{\mathbf{z}}$, $\mathbf{B} = 3\hat{\mathbf{x}} - 4\hat{\mathbf{y}} - 5\hat{\mathbf{z}}$, and $\mathbf{C} = \hat{\mathbf{x}} - \hat{\mathbf{y}} + \hat{\mathbf{z}}$, find (a) $\mathbf{A} \cdot (\mathbf{B} \times \mathbf{C})$, (b) $\mathbf{A} \times (\mathbf{B} \times \mathbf{C})$, and (c) the angle between **A** and **B**, as well as the angle between **A** and **C**.

1.3 Find the area of a disk of radius a using (a) Cartesian coordinates, and (b) using cylindrical coordinates. **Hint:** $\int \sqrt{a^2 - x^2}dx = \frac{1}{2}[x\sqrt{a^2 - x^2} + a^2 \sin^{-1}(x/a)]$.

1.4 Show that

$$(\mathbf{A} \times \mathbf{B}) \cdot \mathbf{C} = (\mathbf{B} \times \mathbf{C}) \cdot \mathbf{A}.$$

1.5 Find the gradient of (a) $V = xz + x^3$, and (b) $V = r \cos\theta + (1/r^2) \sin\phi$.

1.6 Find the divergence of (a) $\mathbf{A} = (xy^2z^3)(\hat{\mathbf{x}} + \hat{\mathbf{y}} + \hat{\mathbf{z}})$, and (b) $\mathbf{A} = \rho \cos\phi \hat{\boldsymbol{\rho}} + [(z/\rho) \sin\phi]\hat{\mathbf{z}}$.

1.7 Show that $\nabla \times (\nabla V) = \mathbf{0}$ and that $\nabla \cdot (\nabla \times \mathbf{A}) = 0$.

1.8 Show that

$$\nabla \cdot [V(\mathbf{r})\mathbf{A}(\mathbf{r})] = V(\mathbf{r})[\nabla \cdot \mathbf{A}(\mathbf{r})] + \mathbf{A}(\mathbf{r}) \cdot [\nabla V(\mathbf{r})].$$

1.9 Verify the divergence theorem using the function

$$\mathbf{A}(\mathbf{r}) = y^2 \,\hat{\mathbf{x}} + (2xy + z^2)\,\hat{\mathbf{y}} + 2yz\,\hat{\mathbf{z}},$$

for a cube of side length of unity.

1.10 Verify the divergence theorem using the function

$$\mathbf{A}(\mathbf{r}) = r \cos\theta \,\hat{\mathbf{r}} + r \sin\theta \,\hat{\boldsymbol{\theta}} + r \sin\theta \cos\phi \,\hat{\boldsymbol{\phi}},$$

for the hemisphere $\{r = a, \theta \in [0, \pi/2]\}$.

1.11 Show that

$$\nabla \left(\frac{1}{R}\right) = -\frac{\mathbf{R}}{R^3},$$

and that

$$\nabla' \left(\frac{1}{R}\right) = \frac{\mathbf{R}}{R^3},$$

where ∇ being the del operator operating on the unprimed variables, and ∇′ being the del operator operating on the primed variables.

Appendix

The following computer program can be used to produce Figs. 1.9a and 1.10a. Extension to other cases can be readily made.

```
v1v2v3 = CoordinateTransform["Cartesian" -> "Spherical", {x, y, z}];
v1[x_, y_, z_] := v1v2v3[[1]];
v2[x_, y_, z_] := v1v2v3[[2]];
v3[x_, y_, z_] := v1v2v3[[3]];

a = 4;
Show[
 Graphics3D@{Arrow[{{0, 0, 0}, {2 a, 0, 0}}]},
 Graphics3D@{Arrow[{{0, 0, 0}, {0, 2 a, 0}}]},
 Graphics3D@{Arrow[{{0, 0, 0}, {0, 0, 2 a}}]},
 Graphics3D@{Text[x, {2 a + 0.5, 0, 0}]},
 Graphics3D@{Text[y, {0, 2 a + 0.5, 0}]},
 Graphics3D@{Text[z, {0, 0, 2 a + 0.5}]},
 ContourPlot3D[v1[x, y, z] == 2, {x, -a, a}, {y, -a, a}, {z, -a, a},
  ContourStyle -> Red, Mesh -> 0, ImageSize -> 72*9, AspectRatio -> 1,
  LabelStyle -> Directive[20, Bold]], BaseStyle -> FontSize -> 40,
 Boxed -> False]

vhat[x_, y_, z_] = Grad[v1[x, y, z], {x, y, z}, "Cartesian"]/
  Norm[Grad[v1[x, y, z], {x, y, z}, "Cartesian"]];
vx[x_, y_, z_] := vhat[x, y, z] . {1, 0, 0};
vy[x_, y_, z_] := vhat[x, y, z] . {0, 1, 0};
vz[x_, y_, z_] := vhat[x, y, z] . {0, 0, 1};
data1 = Table[{{x, y, z}, {vx[x, y, z], vy[x, y, z],
    vz[x, y, z]}}, {x, -2.1, 2, 0.2}, {y, -2.1, 2, 0.2}, {z, -2.1, 2,
    0.2}];

Show[
 Graphics3D@{Arrow[{{0, 0, 0}, {2 a, 0, 0}}]},
 Graphics3D@{Arrow[{{0, 0, 0}, {0, 2 a, 0}}]},
 Graphics3D@{Arrow[{{0, 0, 0}, {0, 0, 2 a}}]},
 Graphics3D@{Text[x, {2 a + 0.5, 0, 0}]},
 Graphics3D@{Text[y, {0, 2 a + 0.5, 0}]},
 Graphics3D@{Text[z, {0, 0, 2 a + 0.5}]}, ListStreamPlot3D[data1,
  ImageSize -> 72*9, AspectRatio -> 1, StreamColorFunction -> None,
  StreamStyle -> {Blue}, StreamMarkers -> "Arrow", StreamScale -> Full,
  StreamPoints -> 50], BaseStyle -> FontSize -> 40,
 Boxed -> False]
```

The following computer program can be used to compute the gradient, divergence, curl, Laplacian of a scalar, and Laplacian of a vector in various coordinate systems.

```
Grad[x y z, {x, y, z}, "Cartesian"]
Div[{1, 0, 1}, {\[Rho], \[Phi], z}, "Cylindrical"]
Curl[{1/r Cos[\[Theta]], -(1/r) Cos[\[Theta]],
  0}, {r, \[Theta], \[Phi]}, "Spherical"]
Laplacian[\[Xi] \[Eta] \[Phi], {\[Xi], \[Eta], \[Phi]}, "Toroidal"]
Laplacian[{\[Mu], \[Mu] z, \[Phi]}, {\[Mu], \[Phi],
  z}, "EllipticCylindrical"]
```

Maxwell Equations and the Structure of Electromagnetics

2

This chapter serves as a bedrock of electromagnetics. Starting from elementary concepts of linear systems in Sect. 2.1, and after introducing Maxwell equations in Sect. 2.2, we treat these equations as a set of governing equations relating inputs to outputs. Then, the main target becomes finding these outputs in relation to their inputs. Finally, we see in Sect. 2.3 that electrostatics and magnetostatics arise as approximations to the general theory.

2.1 Linear Systems and the Green Function

In one-dimensional linear systems, given an input $f(x)$ existing in a domain $x \in [a, b]$, the output $V(x)$ can be related to the input through

$$\mathcal{L}V(x) = f(x), \tag{2.1}$$

where $\mathcal{L}$ is a differential, integral, or integro-differential operator. The formal solution can be written as

$$V(x) = V_c(x) + \int_a^b G(x, x')f(x')\,dx', \tag{2.2}$$

where $V_c(x)$ is the solution of the homogeneous equation $\mathcal{L}V(x) = 0$, (sometimes called the complementary solution) and $G(x, x')$ is the Green function [1–3]. Notice that inside the integral, the input's argument is designated by a primed letter, while the output's argument is designated by an unprimed letter. The Green function acts as a transfer function that converts the input to the output. Once it is found, the problem is formally solved.

H. M. Alkhoori, *Concise Introduction to Electromagnetic Fields*, Synthesis Lectures on Electromagnetics, https://doi.org/10.1007/978-3-031-60331-0_2

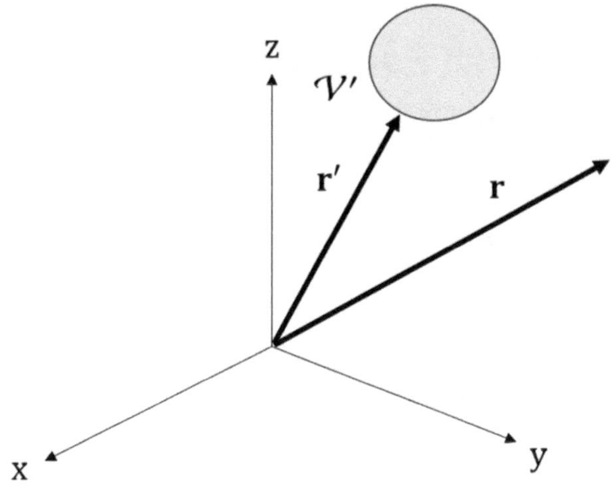

Fig. 2.1 Source point $\mathbf{r}'$ and field point $\mathbf{r}$

In three-dimensional linear systems, given an input $f(\mathbf{r})$ existing in a region $\mathbf{r} \in \mathcal{V}'$ (see Fig. 2.1), the output $V(\mathbf{r})$ can be related to the input through

$$\mathcal{L}V(\mathbf{r}) = f(\mathbf{r}). \tag{2.3}$$

Then, the solution in an unbounded region is

$$V(\mathbf{r}) = V_c(\mathbf{r}) + \int_{\mathcal{V}'} G(\mathbf{r}, \mathbf{r}') f(\mathbf{r}')\, d\mathcal{V}', \tag{2.4}$$

where $V_c(\mathbf{r})$ is the solution of the homogeneous equation $\mathcal{L}V_c(\mathbf{r}) = 0$.

For vectors, given an input $\mathbf{f}(\mathbf{r})$ existing in a region $\mathbf{r} \in \mathcal{V}'$, the output $\mathbf{A}(\mathbf{r})$ can be related to the input through

$$\mathcal{L}\mathbf{A}(\mathbf{r}) = \mathbf{f}(\mathbf{r}). \tag{2.5}$$

For vectors, the output is not necessarily parallel to the input. Therefore, the Green function in this case has to have the ability to change the direction of the input. Such can be done using dyadics (i.e., second-rank tensors) [4]. Therefore, the solution is

$$\mathbf{A}(\mathbf{r}) = \mathbf{A}_c(\mathbf{r}) + \int_{\mathcal{V}'} \underline{\underline{G}}(\mathbf{r}, \mathbf{r}') \cdot \mathbf{f}(\mathbf{r}')\, d\mathcal{V}', \tag{2.6}$$

where $\mathbf{A}_c(\mathbf{r})$ is the solution of the homogeneous equations $\mathcal{L}\mathbf{A}_c(\mathbf{r}) = \mathbf{0}$, and $\underline{\underline{G}}(\mathbf{r}, \mathbf{r}')$ is the dyadic Green function [1, 5].

2.2 Electromagnetics

2.2.1 Maxwell Equations

In any medium, Maxwell equations, the governing equations of electromagnetic phenomenon, are written as

$$\left.\begin{aligned}\nabla \times \mathscr{E}(\mathbf{r}, t) + \frac{\partial}{\partial t}\mathscr{B}(\mathbf{r}, t) &= \mathbf{0}\\ \nabla \times \mathscr{B}(\mathbf{r}, t) - \mu_0\varepsilon_0\frac{\partial}{\partial t}\mathscr{E}(\mathbf{r}, t) &= \mu_0\mathscr{J}_{\text{tot}}(\mathbf{r}, t)\\ \nabla \bullet \mathscr{E}(\mathbf{r}, t) &= \frac{1}{\varepsilon_0}\rho_{\text{tot}}(\mathbf{r}, t)\\ \nabla \bullet \mathscr{B}(\mathbf{r}, t) &= 0\end{aligned}\right\}, \tag{2.7}$$

where $\mathscr{E}(\mathbf{r}, t)$ is the electric field (in Vm^{-1}), $\mathscr{B}(\mathbf{r}, t)$ is the magnetic flux density (in T), $\mathscr{J}_{\text{tot}}(\mathbf{r}, t)$ is the electric volume current density (in Am^{-2}), $\rho_{\text{tot}}(\mathbf{r}, t)$ is the electric volume charge density (in cm^{-3}), and $\varepsilon_0 = 10^{-9}/36\pi\ \text{Fm}^{-1}$ and $\mu_0 = 4\pi \times 10^{-7}\ \text{Hm}^{-1}$ are the permittivity and permeability of free space, respectively. When time-harmonic assumption on the temporal part of all quantities is employed, the frequency-domain Maxwell equations are obtained as

$$\left.\begin{aligned}\nabla \times \mathbf{E}(\mathbf{r}, \omega) + j\omega\mathbf{B}(\mathbf{r}, \omega) &= \mathbf{0}\\ \nabla \times \mathbf{B}(\mathbf{r}, \omega) - j\omega\mu_0\varepsilon_0\mathbf{E}(\mathbf{r}, \omega) &= \mu_0\mathbf{J}_{\text{tot}}(\mathbf{r}, \omega)\\ \nabla \bullet \mathbf{E}(\mathbf{r}, \omega) &= \frac{1}{\varepsilon_0}\rho_{\text{tot}}(\mathbf{r}, \omega)\\ \nabla \bullet \mathbf{B}(\mathbf{r}, \omega) &= 0\end{aligned}\right\}, \tag{2.8}$$

In Eqs. (2.8), $\mathbf{E}(\mathbf{r}, \omega)$ is the electric field phasor, $\mathbf{B}(\mathbf{r}, \omega)$ is the magnetic flux density phasor, $\mathbf{J}_{\text{tot}}(\mathbf{r}, \omega)$ is the electric volume current density phasor, and $\rho_{\text{tot}}(\mathbf{r}, \omega)$ is the electric volume charge density phasor. Henceforth, we omit the ω dependence.

To account for the interaction between the electromagnetic fields and any medium, the electric volume charge density phasor and the electric volume current density phasor are partitioned, respectively, as

$$\left.\begin{aligned}\rho_{\text{tot}}(\mathbf{r}) &= \rho_{\text{ext}}(\mathbf{r}) + \rho_{\text{int}}(\mathbf{r})\\ \mathbf{J}_{\text{tot}}(\mathbf{r}) &= \mathbf{J}_{\text{ext}}(\mathbf{r}) + \mathbf{J}_{\text{int}}(\mathbf{r})\end{aligned}\right\}, \tag{2.9}$$

where $\rho_{\text{ext}}(\mathbf{r})$ is the external electric volume charge density phasor, $\rho_{\text{int}}(\mathbf{r})$ is the internal electric volume charge density phasor arising from the interaction, $\mathbf{J}_{\text{ext}}(\mathbf{r})$ is the external electric volume current density phasor, and $\mathbf{J}_{\text{int}}(\mathbf{r})$ is the internal electric volume current density phasor arising from the interaction. It is customary to represent $\rho_{\text{int}}(\mathbf{r})$ as

$$\rho_{\text{int}}(\mathbf{r}) = -\nabla \bullet \mathbf{P}(\mathbf{r}), \tag{2.10}$$

where $\mathbf{P}(\mathbf{r})$ is the electric polarization phasor (i.e., electric dipole moment $\mathbf{p}$ per unit volume in Cm^{-2}), and to represent $\mathbf{J}_{\text{int}}(\mathbf{r})$ as

$$\mathbf{J}_{\text{int}}(\mathbf{r}) = j\omega\mathbf{P}(\mathbf{r}) + \frac{1}{\mu_0}\nabla \times \mathbf{M}(\mathbf{r}), \tag{2.11}$$

where $\mathbf{M}(\mathbf{r})$ is the magnetization phasor (or magnetic polarization phasor) (i.e., magnetic dipole moment $\mathbf{m}$ per unit volume in Am^{-1}). Therefore, Eq. (2.9) become

$$\left.\begin{aligned} \rho_{\text{tot}}(\mathbf{r}) &= \rho_{\text{ext}}(\mathbf{r}) - \nabla \bullet \mathbf{P}(\mathbf{r}) \\ \mathbf{J}_{\text{tot}}(\mathbf{r}) &= \mathbf{J}_{\text{ext}}(\mathbf{r}) + j\omega\mathbf{P}(\mathbf{r}) + \frac{1}{\mu_0}\nabla \times \mathbf{M}(\mathbf{r}) \end{aligned}\right\}. \tag{2.12}$$

The electric polarization vector and the magnetization vector will be discussed in more details, respectively, in Chaps. 3 and 5. After substituting Eqs. (2.12) into (2.8), equations containing both the external and internal source densities emerge. The phasors

$$\left.\begin{aligned} \mathbf{D}(\mathbf{r}) &= \varepsilon_0\mathbf{E}(\mathbf{r}) + \mathbf{P}(\mathbf{r}) \\ \mathbf{H}(\mathbf{r}) &= \frac{1}{\mu_0}\mathbf{B}(\mathbf{r}) - \mathbf{M}(\mathbf{r}) \end{aligned}\right\} \tag{2.13}$$

are defined for convenience, where $\mathbf{D}(\mathbf{r})$ is the electric flux density phasor (in Cm^{-2}) and $\mathbf{H}(\mathbf{r})$ is the magnetic field phasor (in Am^{-1}). Since the Lorentz force is calculated using $\mathbf{E}(\mathbf{r})$ and $\mathbf{B}(\mathbf{r})$, they are regarded as the primitive fields. The fields $\mathbf{D}(\mathbf{r})$ and $\mathbf{H}(\mathbf{r})$ emerge due to the interaction between the electromagnetic fields and the particles that constitute a certain medium. Therefore, they are regarded as the induction fields. Hence, the frequency-domain Maxwell equations can be written as

$$\left.\begin{aligned} \nabla \times \mathbf{E}(\mathbf{r}) + j\omega\mathbf{B}(\mathbf{r}) &= \mathbf{0} \\ \nabla \times \mathbf{H}(\mathbf{r}) - j\omega\mathbf{D}(\mathbf{r}) &= \mathbf{J}_{\text{ext}}(\mathbf{r}) \\ \nabla \bullet \mathbf{D}(\mathbf{r}) &= \rho_{\text{ext}}(\mathbf{r}) \\ \nabla \bullet \mathbf{B}(\mathbf{r}) &= 0 \end{aligned}\right\}. \tag{2.14}$$

Henceforth, the subscript 'ext' in the source terms in Eqs. (2.14) is replaced by the subscript 'e' which indicates the electric nature of the sources. For later convenience, we introduce the magnetic volume charge density phasor $\rho_m(\mathbf{r})$ (in Tm^{-3}) and the magnetic volume current density phasor $\mathbf{J}_m(\mathbf{r})$ (in Vm^{-2}) into Eqs. (2.14). Although magnetic charges and currents have not been proved to exist, they are used in solving electromagnetics problems. Therefore, Eqs. (2.14) can be written as

$$\left.\begin{aligned} \nabla \times \mathbf{E}(\mathbf{r}) + j\omega\mathbf{B}(\mathbf{r}) &= -\mathbf{J}_m(\mathbf{r}) \\ \nabla \times \mathbf{H}(\mathbf{r}) - j\omega\mathbf{D}(\mathbf{r}) &= \mathbf{J}_e(\mathbf{r}) \\ \nabla \bullet \mathbf{D}(\mathbf{r}) &= \rho_e(\mathbf{r}) \\ \nabla \bullet \mathbf{B}(\mathbf{r}) &= \rho_m(\mathbf{r}) \end{aligned}\right\}. \tag{2.15}$$

Apparently, there are four source terms in Maxwell equations; these are ρ_e, $\mathbf{J}_e$, ρ_m, and $\mathbf{J}_m$. After all, only two are independent. This can be shown as follows. On taking the divergence of both sides of Eq. $(2.15)_2$, and then substituting Eq. $(2.15)_3$, we get the electric current continuity equation (in frequency domain)

$$\nabla \bullet \mathbf{J}_e(\mathbf{r}) = -j\omega\rho_e(\mathbf{r}). \tag{2.16}$$

Similarly, on taking the divergence of both sides of Eq. $(2.15)_1$, and then substituting Eq. $(2.15)_4$, we get the magnetic current continuity equation (in frequency domain)

$$\nabla \bullet \mathbf{J}_m(\mathbf{r}) = -j\omega\rho_m(\mathbf{r}). \tag{2.17}$$

This shows that the ultimate sources of the fields can be though of as either the electric and magnetic charges, or the electric and magnetic currents.

2.2.2 Maxwell Equations in Isotropic Dielectric-Magnetic Medium

An isotropic dielectric-magnetic medium is the simplest medium possible in electromagnetics. It can be incorporated into Maxwell equations mathematically by substitution of

$$\left.\begin{aligned} \mathbf{D}(\mathbf{r}) &= \varepsilon\mathbf{E}(\mathbf{r}) \\ \mathbf{B}(\mathbf{r}) &= \mu\mathbf{H}(\mathbf{r}) \end{aligned}\right\} \tag{2.18}$$

into Maxwell equations. In Eqs. (2.18), $\varepsilon = \varepsilon' - j\varepsilon''$ is the permittivity scalar, where ε' is the real part of the permittivity scalar and ε'' is the imaginary part of the permittivity scalar, and $\mu = \mu' - j\mu''$ is the permeability scalar, where μ' is the real part of the permeability scalar and μ'' is the imaginary part of the permeability scalar. It is to be noted that both ε and μ are generally frequency-dependent functions. In such a case, the medium is said to be dispersive. If it happens that both of ε and μ are extremely-weakly dependent upon the frequency, then such a medium is said to be non dispersive. Also, the terms ε'' and μ'' account for the dissipation (or losses) associated with the medium. In the event that the frequency becomes zero, both the permittivity and permeability scalars become purely real; that is, $\varepsilon = \varepsilon' - \varepsilon_0\varepsilon_r$ with ε_r being the relative permittivity, and $\mu = \mu' = \mu_0\mu_r$ with μ_r being the relative permeability. Equations (2.18) are known as the constitutive relations of an isotropic dielectric magnetic medium. Therefore, Maxwell equations in an isotropic dielectric-magnetic medium become

$$\left.\begin{aligned}\nabla\times\mathbf{E}(\mathbf{r})+j\omega\mu\mathbf{H}(\mathbf{r})&=-\mathbf{J}_m(\mathbf{r})\\ \nabla\times\mathbf{H}(\mathbf{r})-j\omega\varepsilon\mathbf{E}(\mathbf{r})&=\mathbf{J}_e(\mathbf{r})\\ \nabla\bullet\mathbf{E}(\mathbf{r})&=\frac{1}{\varepsilon}\rho_e(\mathbf{r})\\ \nabla\bullet\mathbf{H}(\mathbf{r})&=\frac{1}{\mu}\rho_m(\mathbf{r})\end{aligned}\right\}. \tag{2.19}$$

Notice that for each of **E** and **H**, there is one curl equation and one divergence equation. By virtue of Helmholtz theorem, this ensures that **E** and **H** can be found. Also, if either **E** or **H** is known, the other can found using Maxwell equations, as to be shown next.

Example 2.1 In a source-free region, the electric field is given by

$$\mathscr{E}(\mathbf{r},t)=E_o\cos(\omega t+kx)\,\hat{\mathbf{y}},$$

where k is a constant. Obtain the magnetic field $\mathscr{H}(\mathbf{r},t)$.

Solution. After conversion to frequency domain, we find $\mathbf{H}(\mathbf{r})$ using $\nabla\times\mathbf{E}(\mathbf{r})=-j\omega\mu\,\mathbf{H}(\mathbf{r})$. Then we find $\mathscr{H}(\mathbf{r},t)$ from $\mathbf{H}(\mathbf{r})$ upon conversion to time domain. That is,

$$\left.\begin{aligned}&\mathbf{E}(\mathbf{r})=E_o\,e^{jkx}\hat{\mathbf{y}}\\ &\mathbf{H}(\mathbf{r})=\frac{j}{\omega\mu}\nabla\times\mathbf{E}(\mathbf{r})=\frac{jE_o}{\omega\mu}\nabla\times e^{jkx}\hat{\mathbf{y}}=\frac{jE_o}{\omega\mu}\begin{vmatrix}\hat{\mathbf{x}}&\hat{\mathbf{y}}&\hat{\mathbf{z}}\\ \frac{\partial}{\partial x}&\frac{\partial}{\partial y}&\frac{\partial}{\partial z}\\ 0&e^{jkx}&0\end{vmatrix}=\frac{jE_o}{\omega\mu}jke^{jkx}\,\hat{\mathbf{z}}=-\frac{kE_o}{\omega\mu}e^{jkx}\,\hat{\mathbf{z}}\\ &\mathscr{H}(\mathbf{r},t)=-\frac{kE_o}{\omega\mu}\cos(\omega t+kx)\,\hat{\mathbf{z}}\end{aligned}\right\}.$$

◁

Example 2.2 In a source-free region, the magnetic field phasor is given by

$$\mathbf{H}(\mathbf{r})=H_o(\hat{\mathbf{x}}-j\hat{\mathbf{y}})e^{-jkz}.$$

Obtain the electric field phasor $\mathbf{E}(\mathbf{r})$.

Solution. We find $\mathbf{E}(\mathbf{r})$ using $\nabla\times\mathbf{H}(\mathbf{r})=j\omega\varepsilon\mathbf{E}(\mathbf{r})$.

$$\mathbf{E}(\mathbf{r})=-\frac{j}{\omega\varepsilon}\nabla\times\mathbf{H}(\mathbf{r})=-\frac{jH_o}{\omega\varepsilon}\nabla\times[(\hat{\mathbf{x}}-j\hat{\mathbf{y}})e^{-jkz}]=-\frac{jH_o}{\omega\varepsilon}\begin{vmatrix}\hat{\mathbf{x}}&\hat{\mathbf{y}}&\hat{\mathbf{z}}\\ \frac{\partial}{\partial x}&\frac{\partial}{\partial y}&\frac{\partial}{\partial z}\\ e^{-jkz}&-je^{-jkz}&0\end{vmatrix}=-\frac{jkH_o}{\omega\varepsilon}(\hat{\mathbf{x}}-j\hat{\mathbf{y}})e^{-jkz}.$$

◁

2.2.3 Input/Output Relation

As stated earlier, only two of the four sources ρ_e, ρ_m, $\mathbf{J}_e$, and $\mathbf{J}_m$ are independent. In electromagnetics literature, it is more common to regard the current, instead of charges, as the sources. Using the continuity equations, we see that

$$\left.\begin{aligned}\rho_e &= -\frac{1}{j\omega}\nabla\bullet\mathbf{J}_e\\ \rho_m &= -\frac{1}{j\omega}\nabla\bullet\mathbf{J}_m\end{aligned}\right\}. \tag{2.20}$$

Therefore, frequency-domain Maxwell equations become

$$\left.\begin{aligned}\nabla\times\mathbf{E}(\mathbf{r})+j\omega\mu\mathbf{H}(\mathbf{r}) &= -\mathbf{J}_m(\mathbf{r})\\ \nabla\times\mathbf{H}(\mathbf{r})-j\omega\varepsilon\mathbf{E}(\mathbf{r}) &= \mathbf{J}_e(\mathbf{r})\\ \nabla\bullet\mathbf{E}(\mathbf{r}) &= -\frac{1}{j\omega\varepsilon}\nabla\bullet\mathbf{J}_e(\mathbf{r})\\ \nabla\bullet\mathbf{H}(\mathbf{r}) &= -\frac{1}{j\omega\mu}\nabla\bullet\mathbf{J}_m(\mathbf{r})\end{aligned}\right\}. \tag{2.21}$$

Equation (2.21) clearly show that the inputs are $\mathbf{J}_e$ and $\mathbf{J}_m$, while the outputs are $\mathbf{E}$ and $\mathbf{H}$. Symbolically, Eq. (2.21) can be recast into

$$\mathcal{L}\begin{Bmatrix}\mathbf{E}\\ \mathbf{H}\end{Bmatrix}=\begin{Bmatrix}\mathbf{J}_e\\ \mathbf{J}_m\end{Bmatrix}. \tag{2.22}$$

Thus, our ultimate target is to find $\mathbf{E}$ and $\mathbf{H}$, which can be written symbolically as

$$\begin{Bmatrix}\mathbf{E}\\ \mathbf{H}\end{Bmatrix}=\mathcal{L}^{-1}\begin{Bmatrix}\mathbf{J}_e\\ \mathbf{J}_m\end{Bmatrix}. \tag{2.23}$$

It is important to keep in mind that obtaining the inverse operator $\mathcal{L}^{-1}$ relies on the environment around the sources. In other words, if a boundary exists, boundary conditions, in addition to Eq. (2.21), must be supplemented [6, 7]. In the absence of a boundary, then the so-called radiation condition, and finiteness condition must be imposed [6, 8, 9].

Before leaving this section, it is to be noted that the coupling between $\mathbf{E}$ and $\mathbf{H}$ is attributed to the nonzero angular frequency ω. As to be shown in Sect. 2.3, in the zero-frequency limit, $\mathbf{E}$ and $\mathbf{H}$ decouple. Moreover, notice that $\mathbf{E}$ and $\mathbf{H}$ coexist. That is, we don't regard that $\mathbf{E}$ is caused by $\mathbf{H}$ or vice versa.

2.3 Special Cases

Before considering special cases, let us suppress the magnetic charge ρ_m and magnetic current $\mathbf{J}_m$. Also, let us make the changes $\rho_e \to \rho$ and $\mathbf{J}_e \to \mathbf{J}$. Then, Eqs. (2.19) become

$$\left.\begin{aligned} \nabla \times \mathbf{E}(\mathbf{r}) + j\omega\mu\mathbf{H}(\mathbf{r}) &= \mathbf{0} \\ \nabla \times \mathbf{H}(\mathbf{r}) - j\omega\varepsilon\mathbf{E}(\mathbf{r}) &= \mathbf{J}(\mathbf{r}) \\ \nabla \bullet \mathbf{E}(\mathbf{r}) &= \frac{1}{\varepsilon}\rho(\mathbf{r}) \\ \nabla \bullet \mathbf{H}(\mathbf{r}) &= 0 \end{aligned}\right\}. \tag{2.24}$$

As a special case, when $\omega = 0$, we see immediately that Eqs. (2.24) become

$$\left.\begin{aligned} \nabla \times \mathbf{E}(\mathbf{r}) &= \mathbf{0} \\ \nabla \times \mathbf{H}(\mathbf{r}) &= \mathbf{J}(\mathbf{r}) \\ \nabla \bullet \mathbf{E}(\mathbf{r}) &= \frac{1}{\varepsilon}\rho(\mathbf{r}) \\ \nabla \bullet \mathbf{H}(\mathbf{r}) &= 0 \end{aligned}\right\}. \tag{2.25}$$

A direct consequence is that the electric field decouple from the magnetic field. That is, **E** satisfies

$$\left.\begin{aligned} \nabla \times \mathbf{E}(\mathbf{r}) &= \mathbf{0} \\ \nabla \bullet \mathbf{E}(\mathbf{r}) &= \frac{1}{\varepsilon}\rho(\mathbf{r}) \end{aligned}\right\}, \tag{2.26}$$

and **H** satisfies

$$\left.\begin{aligned} \nabla \times \mathbf{H}(\mathbf{r}) &= \mathbf{J}(\mathbf{r}) \\ \nabla \bullet \mathbf{H}(\mathbf{r}) &= 0 \end{aligned}\right\}. \tag{2.27}$$

Equations (2.26) constitute the subject electrostatics in an isotropic dielectric medium, whereas Eqs. (2.27) constitute the subject magnetostatics in an isotropic magnetic medium. Thus, we see that electrostatics and magnetostatics emerge in the zero frequency limit. In a general dielectric medium, Eqs. (2.26) generalize to

$$\left.\begin{aligned} \nabla \times \mathbf{E}(\mathbf{r}) &= \mathbf{0} \\ \nabla \bullet \mathbf{D}(\mathbf{r}) &= \rho(\mathbf{r}) \end{aligned}\right\}, \tag{2.28}$$

whereas in a general magnetic medium, Eqs. (2.27) generalize to

$$\left.\begin{aligned} \nabla \times \mathbf{H}(\mathbf{r}) &= \mathbf{J}(\mathbf{r}) \\ \nabla \bullet \mathbf{B}(\mathbf{r}) &= 0 \end{aligned}\right\}. \tag{2.29}$$

In time domain, zero-frequency limit implies time independence. Hence, we regard electrostatics and magnetostatics as time-independent electromagnetics. Such a subject is covered in Part II. Time-dependent electromagnetics is covered in Parts III and IV.

Problems

2.1 Given $\mathscr{E} = 4\cos(\omega t + 2y + x + \pi/2)\,\hat{\mathbf{x}}$, obtain $\mathbf{E}$.

2.2 Given $\mathbf{E} = e^{-jkz}\,\hat{\mathbf{x}}$, obtain $\mathscr{H}$.

2.3 An inhomogeneous dielectric-magnetic medium [10] is described by

$$\left.\begin{aligned} \mathbf{D} &= \varepsilon(\mathbf{r})\mathbf{E} \\ \mathbf{B} &= \mu(\mathbf{r})\mathbf{H} \end{aligned}\right\},$$

where $\varepsilon(\mathbf{r})$ and $\mu(\mathbf{r})$ are spatially dependent quantities. Starting from Eqs. (2.28), derive the corresponding frequency-domain, source-free (i.e., sources are set to zero) Maxwell equations that involve the fields $\mathbf{E}$ and $\mathbf{H}$ only.

2.4 A chiral medium [11] is described by

$$\left.\begin{aligned} \mathbf{D} &= \varepsilon\mathbf{E} - j\chi\mathbf{H} \\ \mathbf{B} &= \mu\mathbf{H} + j\chi\mathbf{E} \end{aligned}\right\},$$

where χ is called the chirality parameter. Starting from Eqs. (2.28), derive the corresponding frequency-domain, source-free (i.e., sources are set to zero) Maxwell equations that involve the fields $\mathbf{E}$ and $\mathbf{H}$ only.

References

1. M. Faryad and A. Lakhtakia, *Infinite-Space Dyadic Green Functions in Electromagnetism* (Morgan & Claypool, 2018)
2. D. G. Zill, *Advanced Engineering Mathematics* (Jones & Bartlett, 2018)
3. G. Barton, *Elements of Green's Functions and Propagation* (Clarendon, 1989)
4. H. C. Chen, *Theory of Electromagnetic Waves: A Coordinate-free Approach* (McGraw-Hill, 1985)
5. C. T. Tai, *Dyadic Green Functions in Electromagnetic Theory* (IEEE Press, 1994)
6. E. J. Rothwell and M. J. Cloud, *Electromagnetics* 3rd edn. (CRC Press, 2018)
7. J. A. Stratton, *Electromagnetic Theory* (McGraw–Hill, 1941)
8. P. M. Morse and H. Feshbach, *Methods of Theoretical Physics* (McGraw-Hill, 1953)
9. J. Van Bladel, *Electromagnetic Fields* (John Wiley, 2007)
10. J. Jin, *Theory and Computation of Electromagnetic Fields* 2nd edn. (Wiley, 2015)
11. A. Lakhtakia, *Beltrami Fields in Chiral Media* (World Scientific, 1994)

Part II
Time-Independent Electromagnetics

In time independent regime, Maxwell equations separate into two sets, one belonging to electrostatics, and the other belonging to magnetostatics. The second part of the book is devoted to those aforementioned topics. It consists of three chapters. Chapter 3 is about electrostatics. In particular, solution of Maxwell equations in such a regime is discussed. Chapter 4 is about currents and resistance. Finally, Chap. 5 discusses magnetostatics and solution of Maxwell equations in such a regime.

Electrostatics

3

We saw in Sect. 2.3 that electrostatics is the subject where a source ρ gives rise to a field $\mathbf{E}$. These are governed by

$$\left.\begin{aligned} \nabla \times \mathbf{E}(\mathbf{r}) &= \mathbf{0} \\ \nabla \bullet \mathbf{D}(\mathbf{r}) &= \rho(\mathbf{r}) \end{aligned}\right\}. \tag{3.1}$$

This chapter is devoted to electrostatics. In Sect. 3.1, we learn how to compute the electrostatic field in an isotropic dielectric medium using (i) the so-called Coulomb law, (ii) Gauss law, and (iii) the method of potentials. We then discuss energy in Sect. 3.2, electric dipoles in Sect. 3.3, electrostatic perturbation by a dielectric object in Sect. 3.4, boundary conditions of fields existing in two mediums in Sect. 3.5, and conductors and capacitance in Sect. 3.6. A useful computer program is given in the appendix at the end of the chapter.

3.1 Electric Field Computation

3.1.1 Solution of Maxwell Equations

In an isotropic dielectric medium (i.e., $\mathbf{D} = \varepsilon \mathbf{E}$), Maxwell equations become

$$\left.\begin{aligned} \nabla \times \mathbf{E}(\mathbf{r}) &= \mathbf{0} \\ \nabla \bullet \mathbf{E}(\mathbf{r}) &= \frac{\rho(\mathbf{r})}{\varepsilon} \end{aligned}\right\}. \tag{3.2}$$

From Eq. $(3.2)_1$, we have

$$\mathbf{E}(\mathbf{r}) = -\nabla V(\mathbf{r}), \tag{3.3}$$

where $V(\mathbf{r})$ (in V) is called the electric scalar potential, and the minus sign is for the sake of convenience (to be explained shortly). On substituting Eq. (3.3) into Eq. $(3.2)_2$, one gets

H. M. Alkhoori, *Concise Introduction to Electromagnetic Fields*, Synthesis Lectures on Electromagnetics, https://doi.org/10.1007/978-3-031-60331-0_3

$$\nabla^2 V(\mathbf{r}) = -\frac{\rho(\mathbf{r})}{\varepsilon}. \tag{3.4}$$

Equation (3.4) is an inhomogeneous Laplace equation. If ρ is distributed in a region $\mathcal{V}'$, solution to Eq. (3.4) in an unbounded space is [1]

$$V(\mathbf{r}) = \int_{\mathcal{V}'} G(\mathbf{r}, \mathbf{r}') \left[-\frac{\rho(\mathbf{r}')}{\varepsilon} \right] d\mathcal{V}', \tag{3.5}$$

where $G(\mathbf{r}, \mathbf{r}')$ is the Green function given by

$$G(\mathbf{r}, \mathbf{r}') = -\frac{1}{4\pi R}, \qquad R = |\mathbf{r} - \mathbf{r}'|. \tag{3.6}$$

Substituting Eqs. (3.6) into (3.5) gives

$$V(\mathbf{r}) = \frac{1}{4\pi\varepsilon} \int_{\mathcal{V}'} \frac{\rho(\mathbf{r}')}{R} d\mathcal{V}'. \tag{3.7}$$

Finally, applying $\mathbf{E} = -\nabla V$ and noting that

$$\nabla \left(\frac{1}{R} \right) = -\frac{\mathbf{R}}{R^3} \tag{3.8}$$

give

$$\mathbf{E}(\mathbf{r}) = \frac{1}{4\pi\varepsilon} \int_{\mathcal{V}'} \rho(\mathbf{r}') \frac{\mathbf{R}}{R^3} d\mathcal{V}'. \tag{3.9}$$

Equation (3.9) is known as Coulomb law. To avoid any possible ambiguity, let us make the volume charge density $\rho \to \rho_v$.

So, for a volume charge ρ_v, the field is found from

$$\mathbf{E}(\mathbf{r}) = \frac{1}{4\pi\varepsilon} \int_{\mathcal{V}'} \rho_v(\mathbf{r}') \frac{\mathbf{R}}{R^3} d\mathcal{V}'. \tag{3.10}$$

Suppose that the charge is distributed over a surface $\mathcal{S}'$. Then, it is convenient to replace the volume charge density ρ_v by a surface charge density ρ_s (in Cm^{-2}), and to replace the volume integral by a surface integral. Then,

$$\mathbf{E}(\mathbf{r}) = \frac{1}{4\pi\varepsilon} \int_{\mathcal{S}'} \rho_s(\mathbf{r}') \frac{\mathbf{R}}{R^3} d\mathcal{S}', \tag{3.11}$$

Next, suppose that the charge is distributed over a line $\mathcal{L}'$. Then, it is convenient to replace the volume charge density ρ_v by a line charge density ρ_l (in Cm^{-1}), and to replace the volume integral by a line integral. Then,

$$\mathbf{E}(\mathbf{r}) = \frac{1}{4\pi\varepsilon} \int_{\mathcal{L}'} \rho_l(\mathbf{r}') \frac{\mathbf{R}}{R^3} d\mathcal{L}', \tag{3.12}$$

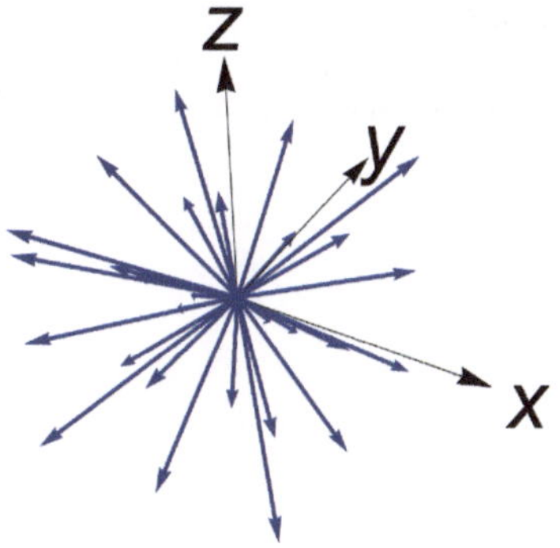

Fig. 3.1 Streamline plots of $\mathbf{E}$ when $Q > 0$. The source is located at the origin

Finally, suppose that the charge is simply a point located at point $\mathbf{r}'$. Then, it is convenient to replace the volume charge density ρ_v by a charge Q (in C), and to remove the volume integral. Then,

$$\mathbf{E}(\mathbf{r}) = \frac{Q}{4\pi\varepsilon}\frac{\mathbf{R}}{R^3}. \tag{3.13}$$

Figure 3.1 shows streamline plots of the electric field for a point charge located at the origin with $Q > 0$. Notice that the electric field lines diverge from the source because $Q > 0$. If $Q < 0$, then the electric field lines will converge to the source. If $\mathbf{E} = \nabla V$ were adopted instead of $\mathbf{E} = -\nabla V$, then the converse would be true. That is the reason for the minus sign appearing in $\mathbf{E} = -\nabla V$.

Example 3.1 The line $\{x = 0, y = 0, z \in (-l, l)\}$ carries a uniform line charge ρ_l. (a) Obtain $\mathbf{E}$ at $(x, 0, 0)$. (b) Obtain $\mathbf{E}$ at $(x, 0, 0)$ when $l \to \infty$. **Hint:** $\int \frac{du}{(u^2 + a^2)^{3/2}} = \frac{u/a^2}{\sqrt{u^2 + a^2}}$, and $\int \frac{u\,du}{(u^2 + a^2)^{3/2}} = -\frac{1}{\sqrt{u^2 + a^2}}$.

Solution. (a) For a line charge,

$$\mathbf{E} = \frac{1}{4\pi\varepsilon}\int_{\mathcal{L}'} \rho_l(\mathbf{r}')\frac{\mathbf{R}}{R^3}\,d\mathcal{L}' = \frac{\rho_l}{4\pi\varepsilon}\int_{\mathcal{L}'}\frac{\mathbf{R}}{R^3}\,d\mathcal{L}'.$$

Now, $d\mathbf{l}' = dz'\hat{\mathbf{z}}, d\mathcal{L}' = dz', \mathbf{r} = x\hat{\mathbf{x}}, \mathbf{r}' = z'\hat{\mathbf{z}}, \mathbf{R} = x\hat{\mathbf{x}} - z'\hat{\mathbf{z}}$, and $R = (x^2 + z'^2)^{3/2}$. Therefore,

$$\begin{aligned}\mathbf{E} &= \frac{\rho_l}{4\pi\varepsilon}\int_{-l}^{l}\frac{x\hat{\mathbf{x}} - z'\hat{\mathbf{z}}}{(x^2 + z'^2)^{3/2}}\bigg|_{x'=0,y'-0} dz' = \frac{\rho_l}{4\pi\varepsilon}\left(x\hat{\mathbf{x}}\int_{-l}^{l}\frac{dz'}{(x^2 + z'^2)^{3/2}} \quad \hat{\mathbf{z}}\int_{-l}^{l}\frac{z'\,dz'}{(x^2 + z'^2)^{3/2}}\right)\\ &= \frac{\rho_l}{2\pi\varepsilon}\frac{l}{x\sqrt{x^2 + l^2}}\hat{\mathbf{x}}\end{aligned}$$

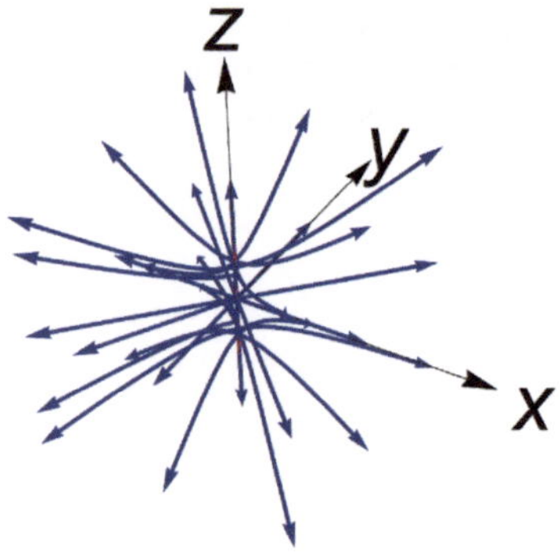

Fig. 3.2 Streamline plots of **E** for a line when $\rho_l > 0$. The source is represented by a red line

Figure 3.2 shows streamline plot of the electric field everywhere when $\rho_l > 0$. Notice that the electric field lines diverge from the source. Also, notice that the electric field has an $\hat{\mathbf{x}}$ component only along the x axis, which agrees with the result we obtained.

(b) Letting $l \to \infty$ gives the field of an infinite line. Thus,

$$\mathbf{E} = \lim_{l\to\infty} \frac{\rho_l}{2\pi\varepsilon} \frac{l}{x\sqrt{x^2+l^2}} \hat{\mathbf{x}} = \frac{\rho_l}{2\pi\varepsilon x} \hat{\mathbf{x}}.$$

◁

In general, for an infinite line along the z axis carrying a uniform line charge ρ_l, the electric field is given by

$$\mathbf{E} = \frac{\rho_l}{2\pi\varepsilon\rho} \hat{\boldsymbol{\rho}}. \tag{3.14}$$

Example 3.2 Obtain **E** at $(0, 0, z)$ for the circular loop $\{\rho = a, z = 0\}$ that carries a uniform line charge density ρ_l.

Solution.

$$\mathbf{E} = \frac{1}{4\pi\varepsilon} \int_{\mathcal{L}'} \rho_l(\mathbf{r}') \frac{\mathbf{R}}{R^3}\, d\mathcal{L}' = \frac{\rho_l}{4\pi\varepsilon} \int_{\mathcal{L}'} \frac{\mathbf{R}}{R^3} d\mathcal{L}'.$$

We have $d\mathbf{l}' = \rho' d\phi' \hat{\boldsymbol{\phi}}'$, $d\mathcal{L}' = \rho' d\phi'$, $\mathbf{r} = z\hat{\mathbf{z}}$, $\mathbf{r}' = \rho'\hat{\boldsymbol{\rho}}'$, $\mathbf{R} = -\rho'\hat{\boldsymbol{\rho}}' + z\hat{\mathbf{z}}$, and $R = (\rho'^2 + z^2)^{3/2}$. Then,

$$\mathbf{E} = \frac{\rho_l}{4\pi\varepsilon} \int_0^{2\pi} \frac{-\rho'\hat{\boldsymbol{\rho}}' + z\hat{\mathbf{z}}}{(\rho'^2+z^2)^{3/2}} \rho' d\phi' \Big|_{\rho'=a, z'=0} = \frac{\rho_l}{4\pi\varepsilon} \int_0^{2\pi} \frac{-a\hat{\boldsymbol{\rho}}' + z\hat{\mathbf{z}}}{(a^2+z^2)^{3/2}} a d\phi' = \frac{\rho_l}{2\varepsilon} \frac{za}{(a^2+z^2)^{3/2}} \hat{\mathbf{z}}.$$

Figure 3.3 shows streamline plot of the electric field everywhere when $\rho_l > 0$. ◁

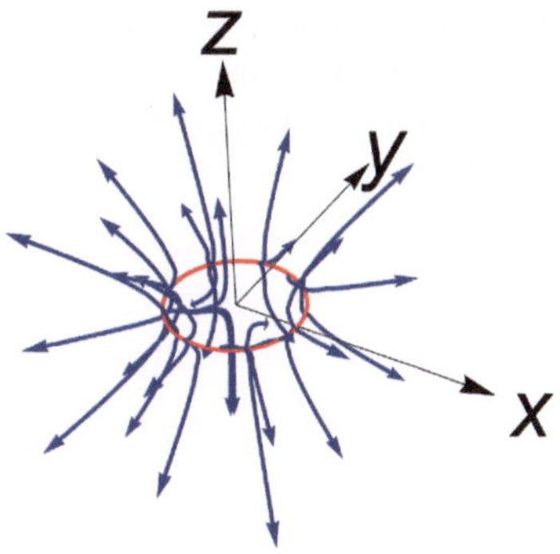

Fig. 3.3 Streamline plots of **E** for a loop when $\rho_l > 0$. The source is represented by a red line

Example 3.3 Obtain **E** at $(0, 0, z)$ for the plane $z = 0$ that carries a uniform surface charge density ρ_s. **Hint:** $\int_{-\infty}^{\infty}\int_{-\infty}^{\infty} \frac{x\,dxdy}{(x^2+y^2+z^2)^{3/2}} = \int_{-\infty}^{\infty}\int_{-\infty}^{\infty} \frac{y\,dxdy}{(x^2+y^2+z^2)^{3/2}} = 0$ and $\int_{-\infty}^{\infty}\int_{-\infty}^{\infty} \frac{dxdy}{(x^2+y^2+z^2)^{3/2}} = \frac{2\pi}{z}$.

Solution.

$$\mathbf{E} = \frac{1}{4\pi\varepsilon}\int_{S'} \rho_s(\mathbf{r}')\frac{\mathbf{R}}{R^3}\,dS' = \frac{\rho_s}{4\pi\varepsilon}\int_{S'} \frac{\mathbf{R}}{R^3}dS'.$$

We have $d\mathbf{S}' = dx'dy'\,\hat{\mathbf{z}}, dS' = dx'dy', \mathbf{r} = z\hat{\mathbf{z}}, \mathbf{r}' = x'\hat{\mathbf{x}} + y'\hat{\mathbf{y}}, \mathbf{R} = -x'\hat{\mathbf{x}} - y'\hat{\mathbf{y}} + z\hat{\mathbf{z}}$, and $R = (x'^2 + y'^2 + z^2)^{3/2}$. Then,

$$\mathbf{E} = \frac{\rho_s}{4\pi\varepsilon}\int_{-\infty}^{\infty}\int_{-\infty}^{\infty} \frac{-x'\hat{\mathbf{x}} - y'\hat{\mathbf{y}} + z\hat{\mathbf{z}}}{(x'^2 + y'^2 + z^2)^{3/2}}dx'dy'\bigg|_{z'=0} = \frac{\rho_s z}{4\pi\varepsilon}\frac{2\pi}{z}\hat{\mathbf{z}} = \frac{\rho_s}{2\varepsilon}\hat{\mathbf{z}}.$$

Figure 3.4 shows streamline plot of the electric field everywhere when $\rho_s > 0$ for the finite plane $\{x \in [-1, 1], y \in [-1, 1], z = 0\}$. Notice that the electric field has a $\hat{\mathbf{z}}$ component only along the z axis, which agrees with the result we obtained. ◁

In general, for an infinite plane with a unit normal $\hat{\mathbf{n}}$ carrying a uniform surface charge ρ_s, the electric field is given by

$$\mathbf{E} = \frac{\rho_s}{2\varepsilon}\hat{\mathbf{n}}. \tag{3.15}$$

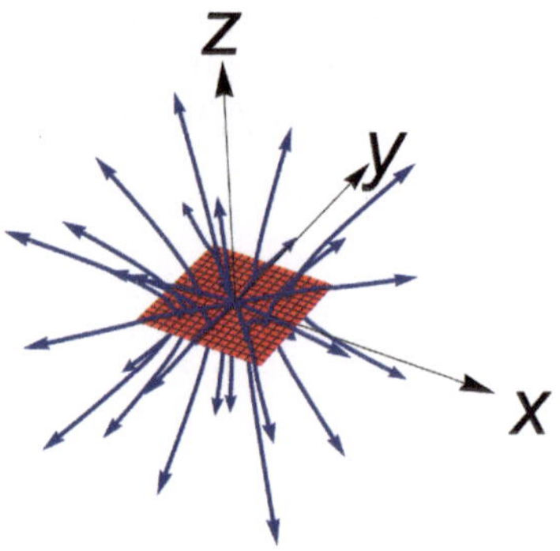

Fig. 3.4 Streamline plots of $\mathbf{E}$ for a finite plane when $\rho_S > 0$. The source is represented by a red surface

3.1.2 Gauss Law

Equation $(3.2)_2$, namely,

$$\nabla \bullet \mathbf{E}(\mathbf{r}) = \frac{1}{\varepsilon}\rho(\mathbf{r}), \tag{3.16}$$

is called Gauss law in differential form. An integral form can be obtained as follows. We integrate both sides of Eq. (3.16) over a volume $\mathcal{V}$. This gives

$$\int_{\mathcal{V}} \nabla \bullet \mathbf{E}(\mathbf{r})\, d\mathcal{V} = \frac{1}{\varepsilon}\int_{\mathcal{V}} \rho(\mathbf{r})\, d\mathcal{V}. \tag{3.17}$$

After using the divergence theorem on the left side of Eq. (3.17), and defining

$$\int_{\mathcal{V}} \rho(\mathbf{r})\, d\mathcal{V} = \mathcal{Q}_{\text{enc}}, \tag{3.18}$$

we get

$$\oint_{\mathcal{S}} \mathbf{E}(\mathbf{r}) \bullet d\mathbf{S} = \frac{1}{\varepsilon}\mathcal{Q}_{\text{enc}}, \tag{3.19}$$

where $\mathcal{S}$ is a closed surface (often called a gaussian surface) bounding $\mathcal{V}$. Notice that $\mathcal{Q}_{\text{enc}}$ is the amount of charge enclosed by $\mathcal{S}$. Equation (3.19) is the the integral form of Gauss law.

Gauss law is true for any source configuration; however, its usefulness is limited. Remember that our ultimate goal is to obtain the electric field of a given source configuration. In Eq. (3.19), we see three quantities: $\mathbf{E}(\mathbf{r})$ (unknown), $\mathcal{Q}_{\text{enc}}$ (known), and $\mathcal{S}$ (free to choose). The issue is $\mathbf{E}(\mathbf{r})$ is buried under the integral sign. Yet, Eq. (3.19) can be made useful only if $\mathbf{E}(\mathbf{r})$ can be taken outside the integral. This can be done by properly choosing $\mathcal{S}$ in very special configurations of sources. Such problems, that involve these very special configurations of sources, are said to possess a symmetry. Only the following problems can be solved using Gauss law in the three coordinate systems that we considered. These are

- an infinite line with uniform ρ_l,
- an infinite plane with uniform ρ_s,
- an infinite cylinder with uniform ρ_v, nonuniform ρ_v that is a function of ρ only, or uniform ρ_s, and
- a sphere with uniform ρ_v, nonuniform ρ_v that is a function of r only, or uniform ρ_s.

Summary of Gauss law:

1. Determine the independent variables and the direction of **E**; for a cylindrical source, $\mathbf{E} = E(\rho)\hat{\boldsymbol{\rho}}$, and for a spherical source, $\mathbf{E} = E(r)\hat{\mathbf{r}}$.
2. From the direction of **E**, choose your gaussian surface $\mathcal{S}$ accordingly. If the direction is $\hat{\boldsymbol{\rho}}$, then $\mathcal{S}$ will be a finite cylinder of radius ρ and height l. If the direction is $\hat{\mathbf{r}}$, then $\mathcal{S}$ will be a sphere of radius r.
3. Evaluate $\oint_{\mathcal{S}} \mathbf{E} \cdot d\mathbf{S}$; for a cylindrical source, it will be $2\pi\rho l E(\rho)$, and for a spherical source, it will be $4\pi r^2 E(r)$.
4. Evaluate Q_{enc}.
5. Find **E** from $\oint_{\mathcal{S}} \mathbf{E} \cdot d\mathbf{S} = Q_{\text{enc}}/\varepsilon$.

Example 3.4 Obtain **E** for an infinite line along the z axis carrying a uniform line charge density ρ_l using Gauss law.

Solution. By intuition, **E** must be along $\hat{\boldsymbol{\rho}}$, and thus, the gaussian surface here will be a finite cylinder of radius ρ and height l. Hence, $\oint_{\mathcal{S}} \mathbf{E} \cdot d\mathbf{S} = 2\pi\rho_l l E$. Also, $Q_{\text{enc}} = \int_{\mathcal{L}} \rho_l \, d\mathcal{L} = \int_0^l \rho_l \, dz = \rho_l l$. Therefore,

$$2\pi\rho l E = \rho_l l/\varepsilon \Rightarrow E = \frac{\rho_l}{2\pi\varepsilon\rho} \Rightarrow \mathbf{E} = \frac{\rho_l}{2\pi\varepsilon\rho}\hat{\boldsymbol{\rho}}.$$

◁

Example 3.5 For an infinite cylinder of radius a carrying a uniform volume charge density ρ_v, obtain **E** inside ($\rho < a$) and outside ($\rho > a$) the cylinder using Gauss law.

Solution. By intuition, **E** must be along $\hat{\boldsymbol{\rho}}$, and thus, the gaussian surface here will be a finite cylinder of radius ρ and height l. Therefore, $\oint_{\mathcal{S}} \mathbf{E} \cdot d\mathbf{S} = 2\pi\rho l E$. When $\rho < a$, we have $Q_{\text{enc}} = \int_{\mathcal{V}} \rho_v \, d\mathcal{V} = \int_0^l \int_0^{2\pi} \int_0^{\rho} \rho_v \, \rho d\rho d\phi dz = \rho_v \pi \rho^2 l$. Hence,

$$2\pi\rho l E = \rho_v \pi \rho^2 l/\varepsilon \Rightarrow E = \frac{\rho_v \rho}{2\varepsilon} \Rightarrow \mathbf{E} = \frac{\rho_v \rho}{2\varepsilon}\hat{\boldsymbol{\rho}}, \ \rho < a.$$

When $\rho > a$, we have $Q_{\text{enc}} = \int_{\mathcal{V}} \rho_v \, d\mathcal{V} = \int_0^l \int_0^{2\pi} \int_0^a \rho_v \, \rho d\rho d\phi dz + \int_0^l \int_0^{2\pi} \int_a^\rho 0 \, \rho d\rho d\phi dz =$ $\rho_v \pi a^2 l$. Hence,

$$2\pi\rho l E = \rho_v \pi a^2 l/\varepsilon \Rightarrow E = \frac{\rho_v a^2}{2\varepsilon\rho} \Rightarrow \mathbf{E} = \frac{\rho_v a^2}{2\varepsilon\rho}\hat{\boldsymbol{\rho}}, \; \rho > a.$$

Therefore,

$$\mathbf{E} = \begin{cases} \dfrac{\rho_v \rho}{2\varepsilon}\hat{\boldsymbol{\rho}}, & 0 < \rho < a, \\ \dfrac{\rho_v a^2}{2\varepsilon\rho}\hat{\boldsymbol{\rho}}, & a < \rho < \infty. \end{cases}$$

◁

Example 3.6 For an infinite cylinder of radius a carrying a uniform surface charge density ρ_s, obtain **E** inside ($\rho < a$) and outside ($\rho > a$) the cylinder using Gauss law.

Solution. By intuition, **E** must be along $\hat{\boldsymbol{\rho}}$, and thus, the gaussian surface here will be a finite cylinder of radius ρ and height l. Hence, $\oint_S \mathbf{E} \cdot d\mathbf{S} = 2\pi\rho l E$. In $\rho < a$, no charge is enclosed. Hence, $Q_{\text{enc}} = 0$, and thus, $\mathbf{E} = \mathbf{0}$ $\rho < a$. In $\rho > a$, we have $Q_{\text{enc}} = \int_S \rho_s \, dS =$ $\int_0^l \int_0^{2\pi} \rho_s \, a d\phi dz = 2\pi a l \rho_s$. Thus,

$$2\pi\rho l E = \rho_s 2\pi a l \rho_s/\varepsilon \Rightarrow E = \frac{\rho_s a}{\varepsilon\rho} \Rightarrow \mathbf{E} = \frac{\rho_s a}{\varepsilon\rho}\hat{\boldsymbol{\rho}}, \; \rho > a.$$

Therefore,

$$\mathbf{E} = \begin{cases} \mathbf{0}, & 0 < \rho < a, \\ \dfrac{\rho_s a}{\varepsilon\rho}\hat{\boldsymbol{\rho}}, & a < \rho < \infty. \end{cases}$$

◁

Example 3.7 For a sphere $r = a$ carrying a uniform volume charge density ρ_v, obtain **E** inside ($r < a$) and outside ($r > a$) the sphere using Gauss law.

Solution. By intuition, **E** must be along $\hat{\mathbf{r}}$, and thus, the gaussian surface here will be sphere of radius a. Hence, $\oint_S \mathbf{E} \cdot d\mathbf{S} = 4\pi r^2 E$. In $r < a$, we have $Q_{\text{enc}} = \int_{\mathcal{V}} \rho_v \, d\mathcal{V} =$ $\int_0^{2\pi} \int_0^{\pi} \int_0^r \rho_v \, r^2 \sin\theta dr d\theta d\phi = \rho_v 4\pi r^3/3$. Thus,

$$4\pi r^2 E = \rho_v 4\pi r^3/3\varepsilon \Rightarrow E = \frac{\rho_v r}{3\varepsilon} \Rightarrow \mathbf{E} = \frac{\rho_v r}{3\varepsilon}\hat{\mathbf{r}}, \; r < a.$$

In $r > a$, we have $Q_{\text{enc}} = \int_{\mathcal{V}} \rho_v \, d\mathcal{V} = \int_0^{2\pi}\int_0^{\pi}\int_0^{a} \rho_v \, r^2 \sin\theta dr d\theta d\phi + \int_0^{2\pi}\int_0^{\pi}\int_a^{r} 0 \, r^2 \sin\theta dr d\theta d\phi = \rho_v 4\pi a^3/3$. Thus,

$$4\pi r^2 E = \rho_v 4\pi a^3/3\varepsilon \Rightarrow E = \frac{\rho_v a^3}{3\varepsilon r^2} \Rightarrow \mathbf{E} = \frac{\rho_v a^3}{3\varepsilon r^2}\hat{\mathbf{r}}, \; r > a.$$

Therefore,

$$\mathbf{E} = \begin{cases} \dfrac{\rho_v r}{3\varepsilon}\hat{\mathbf{r}}, & 0 < r < a, \\ \dfrac{\rho_v a^3}{3\varepsilon r^2}\hat{\mathbf{r}}, & a < r < \infty. \end{cases}$$

◁

Example 3.8 For a sphere $r = a$ carrying a uniform surface charge density ρ_s, obtain **E** inside ($r < a$) and outside ($r > a$) the sphere using Gauss law.

Solution. By intuition, **E** must be along $\hat{\mathbf{r}}$, and thus, the gaussian surface here will be sphere of radius a. Hence, $\oint_S \mathbf{E} \cdot d\mathbf{S} = 4\pi r^2 E$. In $r < a$, we have $Q_{\text{enc}} = 0$. So, $\mathbf{E} = \mathbf{0} \; r < a$. In $r > a$, we have $Q_{\text{enc}} = \int_S \rho_s \, dS = \int_0^{2\pi}\int_0^{\pi} \rho_s \, a^2 \sin\theta d\theta d\phi = \rho_s 4\pi a^2$. Thus,

$$4\pi r^2 E = \rho_s 4\pi a^2/\varepsilon \Rightarrow E = \frac{\rho_s a^2}{\varepsilon r^2} \Rightarrow \mathbf{E} = \frac{\rho_s a^2}{\varepsilon r^2}\hat{\mathbf{r}}, \; r > a.$$

Therefore,

$$\mathbf{E} = \begin{cases} \mathbf{0}, & 0 < r < a, \\ \dfrac{\rho_s a^2}{\varepsilon r^2}\hat{\mathbf{r}}, & a < r < \infty. \end{cases}$$

◁

3.1.3 The Method of Potentials

In Sect. 3.1.1, we computed **E** directly from the source by carrying out a vector integral. However, we already saw that **E** can be found from the scalar V using $\mathbf{E} = -\nabla V$, and that the scalar V can be found from the source using a scalar integral [see Eq. (3.7)]. Such an approach can be called the method of potentials, and it is adopted here. Just like electric field formulas, the following can be used to compute the potential of the various sources.

- Point charge:

$$V(\mathbf{r}) = \frac{1}{4\pi\varepsilon}\frac{Q}{R}. \tag{3.20}$$

- Line charge:

$$V(\mathbf{r}) = \frac{1}{4\pi\varepsilon}\int_{\mathcal{L}'}\frac{\rho_l(\mathbf{r}')}{R}\,d\mathcal{L}'. \tag{3.21}$$

- Surface charge:

$$V(\mathbf{r}) = \frac{1}{4\pi\varepsilon}\int_{\mathcal{S}'}\frac{\rho_s(\mathbf{r}')}{R}\,d\mathcal{S}'. \tag{3.22}$$

- Volume charge:

$$V(\mathbf{r}) = \frac{1}{4\pi\varepsilon}\int_{\mathcal{V}'}\frac{\rho_v(\mathbf{r}')}{R}\,d\mathcal{V}'. \tag{3.23}$$

Example 3.9 Obtain $\mathbf{E}$ in Example 3.1 using the method of potentials. **Hint:** $\int \frac{du}{(u^2+a^2)^{1/2}} = \ln\left(\sqrt{u^2+a^2}+u\right)$.

Solution.

$$V = \frac{1}{4\pi\varepsilon}\int_{\mathcal{L}'}\frac{\rho_l(\mathbf{r}')}{R}\,d\mathcal{L}' = \frac{\rho_l}{4\pi\varepsilon}\int_{\mathcal{L}'}\frac{1}{R}d\mathcal{L}'.$$

We have $d\mathbf{l}' = dz'\hat{\mathbf{z}}$, $d\mathcal{L}' = dz'$, $\mathbf{r} = x\hat{\mathbf{x}}$, $\mathbf{r}' = z'$, $\mathbf{R} = x\hat{\mathbf{x}} - z'\hat{\mathbf{z}}$, and $R = (x^2+z'^2)^{1/2}$. Therefore,

$$V = \frac{\rho_l}{4\pi\varepsilon}\int_{-l}^{l}\frac{1}{(x^2+z'^2)^{1/2}}dz' = \frac{\rho_l}{4\pi\varepsilon}\left[\ln(\sqrt{x^2+z'^2}+z')\Big|_{z'=-l}^{z'=l}\right] = \frac{\rho_l}{4\pi\varepsilon}\left[\ln(\sqrt{x^2+l^2}+l) - \ln(\sqrt{x^2+l^2}-l)\right].$$

Finally,

$$\mathbf{E} = -\nabla V = -\hat{\mathbf{x}}\frac{\partial V}{\partial x} = \frac{\rho_l}{2\pi\varepsilon}\frac{l}{x\sqrt{x^2+l^2}}\hat{\mathbf{x}}.$$

◁

Obtaining *V* from E

Given the electric field $\mathbf{E}$, the potential V can be found as follows.

$$\left.\begin{aligned} \mathbf{E} &= -\nabla V \\ \mathbf{E}\bullet d\mathbf{l} &= -\nabla V\bullet d\mathbf{l} \\ \int_{\mathcal{L}}\mathbf{E}\bullet d\mathbf{l} &= -\int_{\mathcal{L}}\nabla V\bullet d\mathbf{l} \\ &= -\int_{\mathcal{L}} dV = -V\Big|_{\mathcal{L}} \end{aligned}\right\}, \tag{3.24}$$

where $\mathcal{L}$ is a curve with end points $\mathbf{r}$ and O. This can be rewritten as

$$\int_{O}^{\mathbf{r}} \mathbf{E} \bullet d\mathbf{l} = -V\Big|_{O}^{\mathbf{r}} = -\left[V(\mathbf{r}) - V(O)\right], \tag{3.25}$$

Thus,

$$V(\mathbf{r}) - V(O) = -\int_{O}^{\mathbf{r}} \mathbf{E} \bullet d\mathbf{l}. \tag{3.26}$$

This gives the potential difference between two points, or, the potential at point $\mathbf{r}$ relative to the reference point O. In electrostatics, it is natural to choose a reference point such that $V(O) = 0$. For sources with a finite extent, that point turns out to be the point at infinity. The point at infinity implies that we are very far away from the source of the electric field. Therefore, the potential at a point $\mathbf{r}$ relative to infinity is

$$V(\mathbf{r}) = -\int_{\infty}^{\mathbf{r}} \mathbf{E} \bullet d\mathbf{l}. \tag{3.27}$$

Example 3.10 The electric field of a point charge is given by

$$\mathbf{E} = \frac{Q}{4\pi \varepsilon r^2}\hat{\mathbf{r}}. \tag{3.28}$$

Find the potential at r relative to infinity.

Solution.

$$V(r) - V(\infty) = V(r) = -\int_{\infty}^{r} \mathbf{E} \bullet d\mathbf{l} = -\int_{\infty}^{r} \frac{Q}{4\pi \varepsilon r^2}\hat{\mathbf{r}} \bullet \hat{\mathbf{r}} dr = \frac{Q}{4\pi \varepsilon r}.$$

◁

Example 3.11 Plane $z = 0$ carries $-\rho_s$ and plane $z = l$ carries ρ_s. The electric field between the two planes is

$$\mathbf{E} = -\frac{\rho_s}{\varepsilon}\hat{\mathbf{z}}. \tag{3.29}$$

Find the potential at $z = l$ when the reference is $z = 0$.

Solution.

$$V(l) - V(0) = -\int_{0}^{l} \mathbf{E} \bullet d\mathbf{l} = \int_{0}^{l} \frac{\rho_s}{\varepsilon}\hat{\mathbf{z}} \bullet \hat{\mathbf{z}} dz = \frac{\rho_s l}{\varepsilon}.$$

◁

3.2 Electrostatic Energy

In this section, we derive electrostatic energy formula from basic physical principles. To facilitate the concepts, let us first discuss energy in free space (i.e., $\varepsilon = \varepsilon_0$). Then, extension to a general medium is readily made.

3.2.1 Energy in Free Space

In the presence of an electric field **E**, if we move a point charge Q from a reference point O to a point **r**, we need to oppose the force resulting from that field. The minimum force needed is $\mathbf{F} = -Q\mathbf{E}$. Exerting a force over a specific path yields work, or energy W (in J). This energy is given by

$$W = \int_O^{\mathbf{r}} \mathbf{F} \cdot d\mathbf{l} = -Q \int_O^{\mathbf{r}} \mathbf{E} \cdot d\mathbf{l} = Q[V(\mathbf{r}) - V(O)]. \tag{3.30}$$

Now, suppose we want to bring the point charge q from the point at infinity to the point **r**. The energy then will be equal to

$$W = QV(\mathbf{r}). \tag{3.31}$$

This is the energy needed to form a single charge configuration. With this in mind, suppose that we want to assemble a collection of point charges Q_1, Q_2, and Q_3. How much work should that take? We start by bringing Q_1 to $\mathbf{r}_1$. At the beginning, there is no field for Q_1 to fight against. Hence,

$$W_1 = Q_1 \times 0 = 0. \tag{3.32}$$

After Q_1 is placed at $\mathbf{r}_1$, there will be a field with an associated potential $V_1(\mathbf{r})$ given by

$$V_1(r) = \frac{1}{4\pi\varepsilon_0} \frac{Q_1}{|\mathbf{r} - \mathbf{r}_1|}. \tag{3.33}$$

Now, when we bring charge Q_2 to point $\mathbf{r}_2$, there is a field to fight against, and thus, the work is

$$W_2 = Q_2 V_1(\mathbf{r}_2) = Q_2 \frac{1}{4\pi\varepsilon_0} \frac{Q_1}{|\mathbf{r}_2 - \mathbf{r}_1|} = \frac{1}{4\pi\varepsilon_0} \frac{Q_1 Q_2}{R_{12}}, \tag{3.34}$$

where $R_{12} = |\mathbf{r}_2 - \mathbf{r}_1|$. To bring charge Q_3, there are two fields to fight against. Hence, the work is

$$W_3 = Q_3 V_1(\mathbf{r}_3) + Q_3 V_2(\mathbf{r}_3) = \frac{1}{4\pi\varepsilon_0} \left(\frac{Q_1 Q_3}{R_{13}} + \frac{Q_2 Q_3}{R_{23}} \right), \tag{3.35}$$

where $R_{13} = |\mathbf{r}_3 - \mathbf{r}_1|$ and $R_{23} = |\mathbf{r}_3 - \mathbf{r}_2|$. Therefore, the total work required to assemble this collection of point charges is

$$W = W_1 + W_2 + W_3 = \frac{1}{4\pi\varepsilon_0}\left(\frac{Q_1Q_2}{R_{12}} + \frac{Q_1Q_3}{R_{13}} + \frac{Q_2Q_3}{R_{23}}\right). \tag{3.36}$$

By the same token, the work required to form a collection of n charges is

$$W = \frac{1}{4\pi\varepsilon_0}\sum_{i=1}^{n}\sum_{j>i}^{n}\frac{Q_iQ_j}{R_{ij}}, \tag{3.37}$$

where $R_{ij} = |\mathbf{r}_j - \mathbf{r}_i|$. The condition $j > i$ guarantees that we don't count each pair twice. To get rid of it, we count each pair twice and then divide by 2. Therefore,

$$W = \frac{1}{8\pi\varepsilon_0}\sum_{i=1}^{n}\sum_{j\neq i}^{n}\frac{Q_iQ_j}{R_{ij}} \tag{3.38}$$

is the total work required to assemble a collection of n point charges. Equation (3.38) can be written as

$$W = \frac{1}{2}\sum_{i=1}^{n} Q_i V(\mathbf{r}_i), \tag{3.39}$$

where

$$V(\mathbf{r}_i) = \frac{1}{4\pi\varepsilon_0}\sum_{j\neq i}^{n}\frac{Q_j}{R_{ij}}. \tag{3.40}$$

This is the amount of energy stored in the total field associated with that collection of point charges.

For a line charge distribution, the energy is obtained by replacing Eq. (3.39) with

$$W = \frac{1}{2}\int_{\mathcal{L}} \rho_l(\mathbf{r})V(\mathbf{r})\,d\mathcal{L}. \tag{3.41}$$

For a surface charge distribution, the energy is

$$W = \frac{1}{2}\int_{\mathcal{S}} \rho_s(\mathbf{r})V(\mathbf{r})\,d\mathcal{S}. \tag{3.42}$$

Finally, for a volume charge distribution, the energy is

$$W = \frac{1}{2}\int_{\mathcal{V}} \rho_v(\mathbf{r})V(\mathbf{r})\,d\mathcal{V}. \tag{3.43}$$

Note that the regions $\mathcal{L}$, $\mathcal{S}$, and $\mathcal{V}$ are not necessarily the source regions; these are any regions that contain the source region.

Equations (3.41)–(3.43) require the knowledge of the potential, as well as the charge density. Another formula that requires the knowledge of the field only can be obtained from Eq. (3.43) as follows. We replace ρ_v in Eq. (3.43) by $\varepsilon_0 \nabla \cdot \mathbf{E}$. Then, Eq. (3.43) can be written as

$$W = \frac{\varepsilon_0}{2} \int_{\mathcal{V}} [\nabla \cdot \mathbf{E}(\mathbf{r})] V(\mathbf{r}) \, d\mathcal{V}. \tag{3.44}$$

Using

$$\nabla \cdot (V\mathbf{A}) = \mathbf{A} \cdot \nabla V + V \nabla \cdot \mathbf{A}, \tag{3.45}$$

Equation (3.44) becomes

$$W = \frac{\varepsilon_0}{2} \left[-\int_{\mathcal{V}} \mathbf{E} \cdot \nabla V \, d\mathcal{V} + \int_{\mathcal{V}} \nabla \cdot (V\mathbf{E}) \, d\mathcal{V} \right]. \tag{3.46}$$

Using $\mathbf{E} = -\nabla V$ in the first integral and the divergence theorem in the second integral, we get

$$W = \frac{\varepsilon_0}{2} \left[\int_{\mathcal{V}} E^2 \, d\mathcal{V} + \oint_{\mathcal{S}} V\mathbf{E} \cdot d\mathbf{S} \right]. \tag{3.47}$$

Remember that the region $\mathcal{V}$ bounded by the surface $\mathcal{S}$ is not necessarily the source region $\mathcal{V}'$, but any region that contains $\mathcal{V}'$. In other words, we are free to choose our region $\mathcal{V}$ as well as the shape of its bounding surface $\mathcal{S}$ so long as it encloses $\mathcal{V}'$. With this argument, we can make our region $\mathcal{V}$ be the entire space $\mathcal{V}_\infty$ bounded by a spherical surface $\mathcal{S}_\infty$ of radius of infinity. Because $E \propto \frac{1}{r^2}$, $V \propto \frac{1}{r}$, and $dS \propto r^2$, the product $V\mathbf{E} \cdot d\mathbf{S}$ will be proportional to $\frac{1}{r}$. When integrated over $\mathcal{S}_\infty$, it will vanish. Hence, Eq. (3.47) becomes

$$W = \frac{\varepsilon_0}{2} \int_{\mathcal{V}_\infty} E^2 \, d\mathcal{V}. \tag{3.48}$$

Example 3.12 The sphere $r = a$ carries a uniform surface charge density ρ_s. The field is given by

$$\mathbf{E} = \begin{cases} \mathbf{0}, & 0 < r < a, \\ \dfrac{\rho_s a^2}{\varepsilon_0 r^2} \hat{\mathbf{r}}, & a < r < \infty. \end{cases}$$

Obtain the energy.

Solution.

$$W = \frac{\varepsilon_0}{2}\int_{\mathcal{V}_\infty} E^2\,d\mathcal{V} = \frac{\varepsilon_0}{2}\int_0^{2\pi}\int_0^{\pi}\int_0^{\infty} E^2 r^2 \sin\theta dr d\theta d\phi$$

$$= \frac{\varepsilon_0}{2}\int_0^{2\pi}\int_0^{\pi}\left[\int_0^a 0 + \int_a^{\infty}\left(\frac{\rho_s a^2}{\varepsilon_0 r^2}\right)^2\right] r^2 \sin\theta dr d\theta d\phi = \frac{\rho_s^2 a^4}{2\varepsilon_0}\int_0^{2\pi}\int_0^{\pi}\int_a^{\infty}\frac{1}{r^2}\sin\theta dr d\theta d\phi = \frac{2\pi\rho_s^2 a^3}{\varepsilon_0}.$$

◁

3.2.2 Energy in a General Medium

In free space, the electrostatic energy is found using Eq. (3.48). In an isotropic dielectric medium with permittivity ε, we simply make $\varepsilon_0 \to \varepsilon$. Then, the energy becomes

$$W = \frac{\varepsilon}{2}\int_{\mathcal{V}_\infty} E^2\,d\mathcal{V}, \tag{3.49}$$

which can be rewritten as

$$W = \frac{1}{2}\int_{\mathcal{V}_\infty} \mathbf{D}\cdot\mathbf{E}\,d\mathcal{V}. \tag{3.50}$$

Notice that, although Eq. (3.50) was derived for an isotropic dielectric medium, it is valid for any dielectric medium [2, 3]. Note that Eq. (3.50) represents the total energy (i.e., the energy of the external sources W_{ext} and the energy of the internal sources W_{int}). To obtain the energy of the external sources only, we use free-space electric field to get

$$W_{\text{ext}} = \frac{\varepsilon_0}{2}\int_{\mathcal{V}_\infty} \mathbf{E}\cdot\mathbf{E}\,d\mathcal{V}. \tag{3.51}$$

Example 3.13 A sphere of radius a composed of an isotropic dielectric medium with permittivity ε carries a uniform volume charge density ρ_v. Find (a) the total energy W, and (b) the external-source energy W_{ext}.

Solution. We first find the **D** and **E** fields as

$$\mathbf{D} = \begin{cases} \dfrac{\rho_v r}{3}, & 0 < r < a, \\ \dfrac{\rho_v a^3}{3r^2}\hat{\mathbf{r}}, & a < r < \infty. \end{cases} \qquad \mathbf{E} = \begin{cases} \dfrac{\rho_v r}{3\varepsilon}, & 0 < r < a, \\ \dfrac{\rho_v a^3}{3\varepsilon_0 r^2}\hat{\mathbf{r}}, & a < r < \infty. \end{cases}$$

(a)

$$W = \frac{1}{2}\int_{\mathcal{V}_\infty} \mathbf{D}\cdot\mathbf{E}\, d\mathcal{V} = 2\pi\left(\int_0^a \mathbf{D}\cdot\mathbf{E}\, r^2 dr + \int_a^\infty \mathbf{D}\cdot\mathbf{E}\, r^2 dr\right)$$

$$= 2\pi\left(\int_0^a \frac{1}{\varepsilon}\left(\frac{\rho_v r}{3}\right)^2 r^2 dr + \int_a^\infty \frac{1}{\varepsilon_0}\left(\frac{\rho_v a^3}{3r^2}\right)^2 r^2 dr\right) = \frac{2\pi\rho_v^2 a^5}{9}\left(\frac{1}{5\varepsilon} + \frac{1}{\varepsilon_0}\right).$$

(b) It should make sense that the energy in free space can be obtained from part (a) by letting $\varepsilon \to \varepsilon_0$. Then,

$$W_{\text{ext}} = \frac{2\pi\rho_v^2 a^5}{9}\left(\frac{1}{5\varepsilon_0} + \frac{1}{\varepsilon_0}\right) = \frac{4\pi\rho_v^2 a^5}{15\varepsilon_0}.$$

◁

3.3 Electric Dipole

For a source ρ_v occupying a region $\mathcal{V}'$, the electric scalar potential is given by

$$V = \frac{1}{4\pi\varepsilon}\int_{\mathcal{V}'} \frac{\rho_v(\mathbf{r}')}{R}\, d\mathcal{V}'. \tag{3.52}$$

With the assumption that $\mathbf{r}$ lies along the z axis, the term $1/R$ can be expanded as [4, 5]

$$\frac{1}{R} = \frac{1}{\sqrt{r^2 - 2rr'\cos\theta' + r'^2}} = \frac{1}{r\sqrt{1 - 2\frac{r'}{r}\cos\theta' + \left(\frac{r'}{r}\right)^2}} = \frac{1}{r}\sum_{n=0}^{\infty} P_n(\cos\theta')\left(\frac{r'}{r}\right)^n, \tag{3.53}$$

where $P_n(x)$ is the Legendre polynomial of the first kind of order n. It is known that $P_0(x) = 1$, $P_1(x) = x$, etc. If $r >> r'$ (i.e., either the field point is very far away from the source, or the source is very small), then, we may retrieve the first two terms. Thus,

$$\frac{1}{R} \approx \frac{1}{r}\left(P_0(\cos\theta') + P_1(\cos\theta')\frac{r'}{r}\right) = \frac{1}{r}\left(1 + \cos\theta'\frac{r'}{r}\right). \tag{3.54}$$

Therefore,

$$\begin{aligned} V &\approx \frac{1}{4\pi\varepsilon r}\int_{\mathcal{V}'} \rho_v(\mathbf{r}')\, d\mathcal{V}' + \frac{1}{4\pi\varepsilon r^2}\int_{\mathcal{V}'} [r'\cos\theta' \rho_v(\mathbf{r}')]\, d\mathcal{V}' \\ &= \frac{1}{4\pi\varepsilon r}\int_{\mathcal{V}'} \rho_v(\mathbf{r}')\, d\mathcal{V}' + \frac{1}{4\pi\varepsilon r^2}\hat{\mathbf{r}}\cdot\int_{\mathcal{V}'} [\mathbf{r}'\rho_v(\mathbf{r}')]\, d\mathcal{V}'. \end{aligned} \tag{3.55}$$

The first term is called the electric monopole term, and the second term the electric dipole term. We note that the quantity $\int_{\mathcal{V}'} \rho_v(\mathbf{r}')\, d\mathcal{V}'$ in the first integral is nothing but the total charge Q, which we call here a monopole. That is,

$$Q = \int_{\mathcal{V}'} \rho_v(\mathbf{r}')\, d\mathcal{V}'. \tag{3.56}$$

In the second integral, we call the quantity $\int_{\mathcal{V}'} \mathbf{r}' \rho_v(\mathbf{r}')\, d\mathcal{V}'$ the dipole moment vector $\mathbf{p}$ (in Cm). That is,

$$\mathbf{p} = \int_{\mathcal{V}'} \mathbf{r}' \rho_v(\mathbf{r}')\, d\mathcal{V}'. \tag{3.57}$$

With these definitions, Eq. (3.55) becomes

$$V \approx V_{\text{mono}} + V_{\text{dip}}, \tag{3.58}$$

where

$$V_{\text{mono}} = \frac{1}{4\pi\varepsilon} \frac{Q}{r}, \tag{3.59}$$

and

$$V_{\text{dip}} = \frac{1}{4\pi\varepsilon} \frac{\mathbf{p} \bullet \hat{\mathbf{r}}}{r^2}. \tag{3.60}$$

Although we assumed that $\mathbf{r}$ lies along the z axis, the same result holds for any $\mathbf{r}$.

The Electric Field of an Electric Dipole

The potential of a dipole when $\mathbf{p}$ is oriented along the z axis (i.e., $\mathbf{p} = p\,\hat{\mathbf{z}}$) is

$$V_{\text{dip}} = \frac{p}{4\pi\varepsilon} \frac{\cos\theta}{r^2}. \tag{3.61}$$

The electric field in this case is

$$\mathbf{E}_{\text{dip}} = -\nabla V_{\text{dip}}, \tag{3.62}$$

which becomes

$$\mathbf{E}_{\text{dip}} = \frac{p}{4\pi\varepsilon r^3} (2\cos\theta\, \hat{\mathbf{r}} + \sin\theta\, \hat{\boldsymbol{\theta}}). \tag{3.63}$$

Notice that, whereas the potential of a monopole term is proportional to $\frac{1}{r}$, the potential of a dipole term is proportional to $\frac{1}{r^2}$. Accordingly, the field of a monopole term is proportional to $\frac{1}{r^2}$ and the field of a dipole term is proportional to $\frac{1}{r^3}$.

Dipoles can be used as an atom model. As a simplest picture, an atom can be thought of as two point charges, one of charge Q and the other of charge $-Q$, that are displaced a distance $\mathbf{d}$ apart (the vector $\mathbf{d}$ is the distance vector from the positive charge to the negative charge). For such a configuration, the dipole moment becomes [2]

$$\mathbf{p} = Q\mathbf{d}. \tag{3.64}$$

A streamline plot of the electric field of a dipole located at the origin and oriented along the z axis is shown in Fig. 3.5.

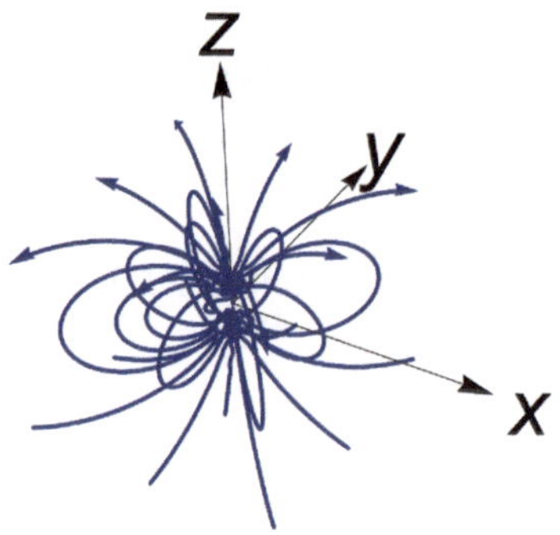

Fig. 3.5 Streamline plots of $\mathbf{E}$ for a dipole $\mathbf{p} = 4\pi\varepsilon_0\hat{\mathbf{z}}$ located at the origin

Example 3.14 Let point charge $2\,C$ be situated at point $(1, 2, 3)$ and point charge $-2\,C$ be situated at point $(4, 5, 6)$. Determine $\mathbf{p}$.

Solution.

$$\mathbf{p} = 2[(1, 2, 3) - (4, 5, 6)] = -6(\hat{\mathbf{x}} + \hat{\mathbf{y}} + \hat{\mathbf{z}}).$$

◁

3.4 Electrostatic Perturbation by a Dielectric Object

Consider the situation where an electrostatic field exists in free space in the presence of a dielectric object. How will the dielectric object interact with the field? In order to answer that question, we first model the dielectric object using electric dipoles. Then, after understanding the role of the dielectric object, we mention the overall interaction that takes place.

3.4.1 A Dielectric Object

An atom can be thought of as a cloud of electrons surrounding a nucleus. You may think of this cloud of electrons as an effective negative point charge and the nucleus as an effective positive charge. Assuming both particles having the same charge, an atom can be considered as a dipole with dipole moment $\mathbf{p}$.

A dielectric object can be thought of as a group of atoms. Unlike a single atom, a dielectric object is not described by $\mathbf{p}$, but rather by a polarization vector $\mathbf{P}$. Suppose that inside a dielectric object, there exists a region ΔV with a total dipole moment $\Delta\mathbf{p}$. Then, the polarization vector is defined as

$$\mathbf{P} = \lim_{\Delta \to 0} \frac{\Delta \mathbf{p}}{\Delta V} = \frac{d\mathbf{p}}{d\mathcal{V}}. \tag{3.65}$$

Consider a dielectric object occupying a region $\mathcal{V}'$ bounded by a surface $\mathcal{S}'$ and is characterized by a polarization vector **P**. Since **P** is nothing but a dipole density, and since a single dipole gives an electric field, then, a dielectric object will give an electric field as well. How to compute that electric field from the polarization vector **P**? This can be done more easily by deriving an expression for the potential V due to the polarization **P**. Once V is found, the electric field can be found from $\mathbf{E} = -\nabla V$. The differential potential due to a single dipole moment in $\mathcal{V}'$ is given by

$$dV = \frac{1}{4\pi\varepsilon_0} \frac{d\mathbf{p} \bullet \mathbf{R}}{R^3}. \tag{3.66}$$

Using Eq. (3.65), the potential becomes

$$V = \frac{1}{4\pi\varepsilon_0} \int_{\mathcal{V}'} \frac{\mathbf{P}(\mathbf{r}') \bullet \mathbf{R}}{R^3} d\mathcal{V}'. \tag{3.67}$$

Then, $\mathbf{E} = -\nabla V$.

Internal Charges

As another point of view, using the identity

$$\frac{\mathbf{R}}{R^3} = \nabla' \left(\frac{1}{R} \right), \tag{3.68}$$

with ∇' operating on the primed variables, and the identity

$$\nabla \bullet (V\mathbf{A}) = \mathbf{A} \bullet \nabla V + V \nabla \bullet \mathbf{A}, \tag{3.69}$$

the potential can be rewritten as

$$V = \frac{1}{4\pi\varepsilon_0} \oint_{\mathcal{S}'} \frac{\mathbf{P}(\mathbf{r}') \bullet \hat{\mathbf{n}}'}{R} d\mathcal{S}' + \frac{1}{4\pi\varepsilon_0} \int_{\mathcal{V}'} \frac{-\nabla' \bullet \mathbf{P}(\mathbf{r}')}{R} d\mathcal{V}'. \tag{3.70}$$

If we define

$$\rho_{\text{int}}^{s}(\mathbf{r}) = \mathbf{P}(\mathbf{r}) \bullet \hat{\mathbf{n}}, \tag{3.71}$$

as the internal surface charge density, and

$$\rho_{\text{int}}^{v}(\mathbf{r}) = -\nabla \bullet \mathbf{P}(\mathbf{r}), \tag{3.72}$$

as the internal volume charge density, Eq. (3.70) can be written as

$$V = \frac{1}{4\pi\varepsilon_0}\oint_{S'}\frac{\rho^s_{\text{int}}(\mathbf{r}')}{R}dS' + \frac{1}{4\pi\varepsilon_0}\int_{\mathcal{V}'}\frac{\rho^v_{\text{int}}(\mathbf{r}')}{R}d\mathcal{V}'. \tag{3.73}$$

Equation (3.73) shows that the potential (and hence the field) due to a dielectric object is equivalent to a potential due to a surface charge density ρ^s_{int}, and to a potential due to a volume charge density ρ^v_{int}.

In a situation where an external source is immersed in a dielectric medium, the internal charges can be separated from the external charges once a relation between **P** and **E**, where **E** is the total field due to both external and internal charges, is given. In an isotropic dielectric medium, such a relation is given by

$$\mathbf{P} = \mathbf{D} - \varepsilon_0\mathbf{E} = (\varepsilon - \varepsilon_0)\mathbf{E} = \varepsilon_0\chi_e\mathbf{E}, \tag{3.74}$$

where $\chi_e = (\varepsilon_r - 1)$ is the electric susceptibility scalar.

Example 3.15 An infinite cylinder of radius a is composed of an isotropic dielectric medium with permittivity ε and is carrying a uniform volume charge density ρ_v. Given that

$$\mathbf{E} = \begin{cases} \dfrac{\rho_v\rho}{2\varepsilon}\hat{\boldsymbol{\rho}}, & 0 < \rho < a, \\ \dfrac{\rho_v a^2}{2\rho\varepsilon_0}\hat{\boldsymbol{\rho}}, & a < \rho < \infty. \end{cases}$$

obtain the internal charges.

Solution. We first find the polarization as

$$\mathbf{P} = \varepsilon_0\chi_e\mathbf{E} = \frac{\varepsilon_r - 1}{\varepsilon_r}\frac{\rho_v\rho}{2}\hat{\boldsymbol{\rho}}.$$

Then,

$$\rho^v_{\text{int}} = -\nabla\bullet\mathbf{P} = -\frac{\varepsilon_r - 1}{\varepsilon_r}\rho_v,$$

and

$$\rho^s_{\text{int}} = \mathbf{P}\bullet\hat{\boldsymbol{\rho}}\Big|_{\rho=a} = \frac{\varepsilon_r - 1}{\varepsilon_r}\frac{\rho_v a}{2}.$$

◁

Example 3.16 A sphere of radius a carries a uniform surface charge density ρ_s is immersed in a sphere of radius b ($b > a$) composed of an isotropic dielectric medium with permittivity ε. (a) Find **E** everywhere. (b) Find the internal charges.

Solution. (a) We use Gauss law as follows.

$$\oint_S \mathbf{E} \cdot d\mathbf{S} = 4\pi r^2 E.$$

$$r < a : Q_{\text{enc}} = \int_S \rho_s \, dS = 0 \Rightarrow \mathbf{E} = \mathbf{0}.$$

$$a < r < b : Q_{\text{enc}} = \int_S \rho_s \, dS = \rho_s 4\pi a^2 \Rightarrow \mathbf{E} = \frac{\rho_s a^2}{\varepsilon r^2}\hat{\mathbf{r}}.$$

$$r > b : Q_{\text{enc}} = \int_S \rho_s \, dS = \rho_s 4\pi a^2 \Rightarrow \mathbf{E} = \frac{\rho_s a^2}{\varepsilon_0 r^2}\hat{\mathbf{r}}.$$

$$\Rightarrow \mathbf{E} = \begin{cases} \mathbf{0}, & 0 < r < a, \\ \dfrac{\rho_s a^2}{\varepsilon r^2}\hat{\mathbf{r}}, & a < r < b, \\ \dfrac{\rho_s a^2}{\varepsilon_0 r^2}\hat{\mathbf{r}}, & b < r < \infty. \end{cases}$$

(b) As **E** is given, we find **P**. Then, from **P**, the bound charges are found. So,

$$\mathbf{P} = \varepsilon_0 \chi_e \mathbf{E} = \frac{\varepsilon_r - 1}{\varepsilon_r}\frac{\rho_s a^2}{r^2}\hat{\mathbf{r}}.$$

$$\rho^v_{\text{int}} = -\nabla \cdot \mathbf{P} = 0.$$

$$\rho^s_{\text{int}} = \begin{cases} \mathbf{P} \cdot (-\hat{\mathbf{r}}), & r = a, \\ \mathbf{P} \cdot \hat{\mathbf{r}}, & r = b. \end{cases} = \begin{cases} -\dfrac{\varepsilon_r - 1}{\varepsilon_r}\rho_s, & r = a, \\ \dfrac{\varepsilon_r - 1}{\varepsilon_r}\dfrac{\rho_s a^2}{b^2}, & r = b. \end{cases}$$

◁

3.4.2 Interaction Between Electrostatic Field and a Dielectric Object

Consider a dielectric object suspended in free space. Suppose that initially, $\mathbf{P} = \mathbf{0}$. When a source electric field $\mathbf{E}_{\text{source}}$ is applied, the dipoles inside the object align themselves in a certain configuration with respect to $\mathbf{E}_{\text{source}}$. These dipole are said to be polarized. Due to their new arrangement, there will be a net **P**, which gives rise to a perturbation electric field $\mathbf{E}_{\text{pert}}$ outside the object, and an internal electric field $\mathbf{E}_{\text{int}}$ inside the object. Those fields must depend not only on $\mathbf{E}_{\text{source}}$, but also on the composition of the medium including (i) the force due to the mass of the electron, (ii) the binding force between the electron and the proton, and (iii) the damping force. Computation of the perturbation electric field is possible once a relation between **P** and its cause $\mathbf{E}_{\text{source}}$ is found.

Figure 3.6a shows streamline plots of a uniform source electric field $\mathbf{E}_{\text{source}} = \hat{\mathbf{z}}$ in free space on the $z = 0$ plane. When a sphere of radius a made of an isotropic dielectric medium of

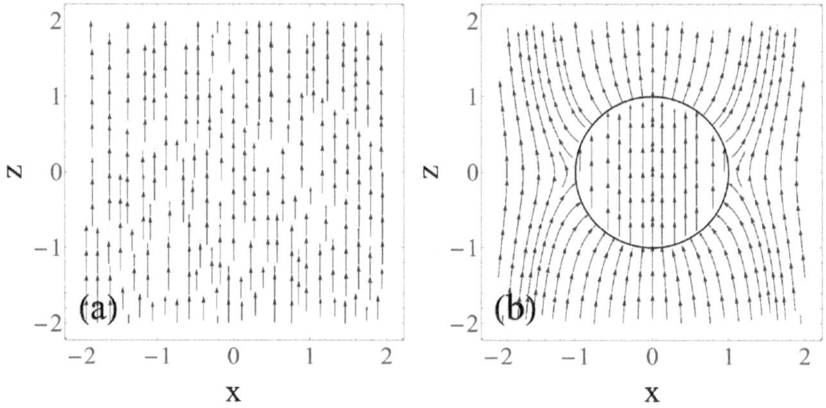

Fig. 3.6 Streamline plots of **E** on the $z = 0$ plane **a** in the absence of the dielectric sphere, and **b** in the presence of the dielectric sphere

permittivity ε is immersed, internal electric field $\mathbf{E}_{\text{int}}$ inside the sphere, as well as perturbation electric field $\mathbf{E}_{\text{pert}}$ outside the sphere are induced. The total field outside the sphere is, therefore, $\mathbf{E}_{\text{source}} + \mathbf{E}_{\text{pert}}$, whereas inside the sphere, it is $\mathbf{E}_{\text{int}}$. Such a scenario is depicted in Fig. 3.6b.

3.5 Boundary Conditions

When two or more mediums are adjacent to each other, the electric fields of these mediums can be related to each other at the boundaries of these mediums. Such relations can be obtained from the boundary conditions among these mediums.

Consider region 1 to be composed of an isotropic dielectric medium of ε_1 and region 2 to be composed of an isotropic dielectric medium of ε_2. A surface S (a boundary) is separating the two mediums. The field in region 1, $\mathbf{E}_1$, can be decomposed into a component tangential to the surface $\mathbf{E}_1^{\parallel}$ and a component normal to the surface $\mathbf{E}_1^{\perp}$. That is,

$$\mathbf{E}_1 = \mathbf{E}_1^{\parallel} + \mathbf{E}_1^{\perp}. \tag{3.75}$$

Likewise, the field in region 2, $\mathbf{E}_2$, can be decomposed as

$$\mathbf{E}_2 = \mathbf{E}_2^{\parallel} + \mathbf{E}_2^{\perp}. \tag{3.76}$$

On the surface S, we want to find a relation between $\mathbf{E}_1^{\parallel}$ and $\mathbf{E}_2^{\parallel}$, and between $\mathbf{E}_1^{\perp}$ and $\mathbf{E}_2^{\perp}$. This becomes useful if one field is known and the other is unknown. Our starting point is the integral form of Maxwell equations in Eqs. (3.2), which can be written as

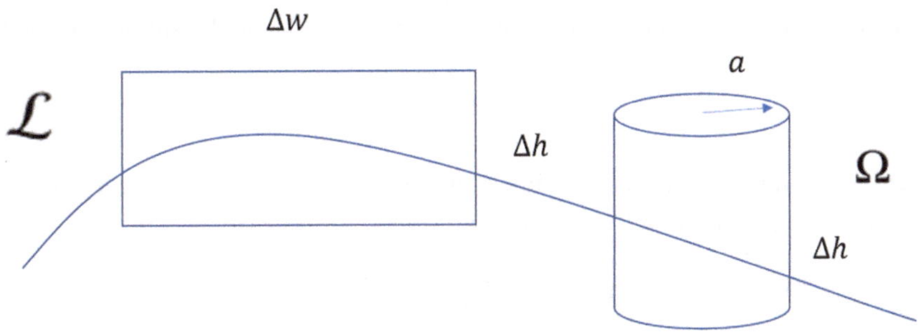

Fig. 3.7 Integration curve $\mathcal{L}$ and surface Ω

$$\left.\begin{aligned} \oint_{\mathcal{L}} \mathbf{E} \cdot d\mathbf{l} = 0 \\ \oint_{S} \mathbf{D} \cdot d\mathbf{S} = Q_{\text{enc}} \end{aligned}\right\}. \tag{3.77}$$

Equation $(3.77)_1$ will be used to find a relation between the tangential components, whereas Eq. $(3.77)_2$ will be used to find a relation between the normal components.

Tangential components: To find a relation between $\mathbf{E}_1^{\|}$ and $\mathbf{E}_2^{\|}$ on the surface S, we apply

$$\oint_{\mathcal{L}} \mathbf{E} \cdot d\mathbf{l} = 0, \tag{3.78}$$

where $\mathcal{L}$ is shown in Fig. 3.7.

With the assumption that the dimensions Δw and Δh are very small, we can take the integrand out of the integral. Therefore, Eq. (3.78) becomes

$$E_1^{\|}\Delta w - E_1^{\perp}\frac{\Delta h}{2} - E_2^{\perp}\frac{\Delta h}{2} - E_2^{\|}\Delta w + E_2^{\perp}\frac{\Delta h}{2} + E_1^{\perp}\frac{\Delta h}{2} = 0, \tag{3.79}$$

which reduces to

$$(E_1^{\|} - E_2^{\|})\Delta w = 0. \tag{3.80}$$

To make $\mathcal{L}$ intersect with S, we must let $\Delta h \to 0$. Therefore,

$$E_1^{\|} = E_2^{\|}, \tag{3.81}$$

which can be written in vector form as

$$\hat{\mathbf{n}} \times (\mathbf{E}_1 - \mathbf{E}_2) = \mathbf{0}, \tag{3.82}$$

where $\hat{\mathbf{n}}$ is a unit normal on S that points into region 1. Therefore, the tangential component of the electric field is said to be continuous across S. Keep in your mind that this boundary condition is valid only on the surface S.

Normal components: To find a relation between $\mathbf{E}_1^{\perp}$ and $\mathbf{E}_2^{\perp}$ on the surface $\mathcal{S}$, we apply

$$\oint_{\Omega} \mathbf{D} \bullet d\mathbf{S} = Q_{\text{enc}}, \tag{3.83}$$

where Ω is a cylindrical surface of radius a as shown in Fig. 3.7. With the assumption that the dimensions ΔS and Δh are very small, Eq. (3.83) becomes

$$D_1^{\perp}\Delta S + D_1^{\parallel}(2\pi a\Delta h) - D_2^{\perp}\Delta S + D_2^{\parallel}(2\pi a\Delta h) = \rho_s \Delta S. \tag{3.84}$$

Upon letting $\Delta h \to 0$, Eq. (3.84) becomes

$$D_1^{\perp} - D_2^{\perp} = \rho_s, \tag{3.85}$$

or, in vector form,

$$\hat{\mathbf{n}} \bullet (\mathbf{D}_1 - \mathbf{D}_2) = \rho_s. \tag{3.86}$$

Again, this boundary condition is valid only on the surface $\mathcal{S}$.

In summary, the boundary conditions are

$$\left.\begin{array}{l} \mathbf{E}_1^{\parallel} = \mathbf{E}_2^{\parallel} \\ D_1^{\perp} - D_2^{\perp} = \rho_s \end{array}\right\}. \tag{3.87}$$

Assume that both $\mathbf{E}_1$ and $\mathbf{E}_2$ are constant vectors, if $\mathbf{E}_1$ in known, then $\mathbf{E}_2$ can be found as

$$\mathbf{E}_2 = \mathbf{E}_1^{\parallel} + \frac{1}{\varepsilon_2}\left(\varepsilon_1 E_1^{\perp} - \rho_s\right)\hat{\mathbf{n}}. \tag{3.88}$$

Example 3.17 Free space $\varepsilon_1 = \varepsilon_0$ and a medium with $\varepsilon_2 = 4\varepsilon_0$ are separated by the plane $x = 0$. Free space occupies $x > 0$ and the medium occupies $x < 0$. Given $\mathbf{E}_1 = 12\hat{\mathbf{x}} - 10\hat{\mathbf{y}} + 4\hat{\mathbf{z}}$, find $\mathbf{E}_2$.

Solution. After noting that $\hat{\mathbf{n}} = \hat{\mathbf{x}}$, we see that

$$\mathbf{E}_1^{\perp} = (\mathbf{E}_1 \bullet \hat{\mathbf{n}})\hat{\mathbf{n}} = 12\hat{\mathbf{x}} \Rightarrow \mathbf{E}_2^{\perp} = \frac{\varepsilon_1}{\varepsilon_2}\mathbf{E}_1^{\perp} = 3\hat{\mathbf{x}},$$

and

$$\mathbf{E}_1^{\parallel} = \mathbf{E}_1 - \mathbf{E}_1^{\perp} = -10\hat{\mathbf{y}} + 4\hat{\mathbf{z}} \Rightarrow \mathbf{E}_2^{\parallel} = \mathbf{E}_1^{\parallel} = -10\hat{\mathbf{y}} + 4\hat{\mathbf{z}}.$$

Therefore, $\mathbf{E}_2 = 3\hat{\mathbf{x}} - 10\hat{\mathbf{y}} + 4\hat{\mathbf{z}}$. ◁

Example 3.18 Two mediums with $\varepsilon_1 = 4\varepsilon_0$ and $\varepsilon_2 = 3\varepsilon_0$ are separated by the plane $z = 1$. Medium 1 occupies $z > 1$ and medium 2 occupies $z < 1$. Given $\mathbf{E}_1 = 5\hat{\mathbf{x}} - 2\hat{\mathbf{y}} + 3\hat{\mathbf{z}}$, find $\mathbf{E}_2$.

Solution. After noting that $\hat{\mathbf{n}} = \hat{\mathbf{z}}$, we see that

$$\mathbf{E}_1^{\perp} = (\mathbf{E}_1 \bullet \hat{\mathbf{n}})\hat{\mathbf{n}} = 3\hat{\mathbf{z}} \Rightarrow \mathbf{E}_2^{\perp} = \frac{\varepsilon_1}{\varepsilon_2}\mathbf{E}_1^{\perp} = 4\hat{\mathbf{z}},$$

and

$$\mathbf{E}_1^{\parallel} = \mathbf{E}_1 - \mathbf{E}_1^{\perp} = 5\hat{\mathbf{x}} - 2\hat{\mathbf{y}} \Rightarrow \mathbf{E}_2^{\parallel} = \mathbf{E}_1^{\parallel} = 5\hat{\mathbf{x}} - 2\hat{\mathbf{y}}.$$

Therefore, $\mathbf{E}_2 = 5\hat{\mathbf{x}} - 2\hat{\mathbf{y}} + 4\hat{\mathbf{z}}$. ◁

Example 3.19 Repeat Example 3.18 when a surface charge $\rho_s = \varepsilon_0$ exists on the surface.

Solution. Due to the surface charge, $\mathbf{E}_2^{\perp}$ becomes

$$\mathbf{E}_2^{\perp} = \frac{1}{\varepsilon_2}(\varepsilon_1 E_1^{\perp} - \rho_s)\hat{\mathbf{z}} = \frac{1}{3}(12 - 1)\hat{\mathbf{z}} = \frac{11}{3}\hat{\mathbf{z}}.$$

Therefor, $\mathbf{E}_2 = 5\hat{\mathbf{x}} - 2\hat{\mathbf{y}} + \frac{11}{3}\hat{\mathbf{z}}$. ◁

3.6 Conductors and Capacitance

In a dielectric medium, each atom contains electrons and protons that are attached to each other. In a conductor, by contrast, one or more electrons are free to move (these are called free electrons). One consequence of this freedom is that a conductor can conduct a current (to be discussed in Chap. 4). The mathematical consequence of this freedom is apparent in the form of the electric susceptibility χ_e of a conductor. In a perfect conductor [known also as an electric perfect conductor (PEC)], the number of these free electrons is infinite. The electric susceptibility χ_e is directly proportional to the number of electrons. Because the number of electrons if infinite, it follows that $\chi_e \to \infty$. From $\mathbf{P} = \varepsilon_0 \chi_e \mathbf{E}$, because $\mathbf{P}$ has to be finite, we conclude that $\mathbf{E} = \mathbf{0}$ inside a conductor. Physically, when a source field exists in a conductor, the electrons respond by generating an internal field that opposes the source field. Because the number of electrons is infinite, the cancellation of the source field is perfect and the total field (source + internal) becomes zero. A PEC may not exist in real life, but metals display characters very similar to that of a PEC. The fact that $\mathbf{E} = \mathbf{0}$ inside a PEC has the following consequences:

1. The volume charge inside a PEC is zero (follows from Gauss law).
2. Consequently, the charges reside on the surface, constituting a surface charge density.
3. The potential difference between any two points on the PEC surface is zero; hence, a PEC is an equipotential surface.
4. The electric field outside a PEC is perpendicular to its surface (follows from boundary conditions).

3.6.1 Capacitance

Consider two PECs, one with a positive charge Q and the other with a negative charge $-Q$ being separated by a certain distance. The potential difference (V_o) between these two PECs is given by

$$V_0 = -\int_{-}^{+} \mathbf{E} \cdot d\mathbf{l}, \tag{3.89}$$

where $\mathbf{E}$ is the electric field that exists in the space surrounding the two PECs, and the plus (minus) sign indicates the conductor with positive (negative) charge. Note that $V_0 \propto \mathbf{E}$. From $\mathbf{E} = \dfrac{1}{4\pi\varepsilon}\displaystyle\int_{S'} \rho_s(\mathbf{r}')\frac{\mathbf{R}}{R^3}\,dS'$ and $Q = \displaystyle\int_{S'} \rho_s\,dS'$, one notices that $\mathbf{E} \propto \rho_s$ and $\rho_s \propto Q$. Hence, it follows that $\mathbf{E} \propto Q$, and thus, $V_0 \propto Q$. This proportionality is described by the capacitance C (in F) through

$$Q = C V_0, \tag{3.90}$$

or

$$C = \frac{Q}{V_0}. \tag{3.91}$$

Capacitance is a quantity that depends on the shapes of the PECs, as well as on the medium surrounding them. We assume that the shapes of the two PECs are the same. Hence, a capacitor is merely two PECs of the same shape, and the same magnitude of charge but with an opposite sign. It is used to store charges. Higher capacitance means that more charges can be stored into the capacitor, and vice versa.

As another approach for finding capacitance, the energy is given by $W = \dfrac{1}{2}\displaystyle\int_{\mathcal{V}_\infty} \mathbf{D} \cdot \mathbf{E}\,d\mathcal{V}$, which we can write as $W = \dfrac{1}{2} V_0 Q$. Hence,

$$C = \frac{Q^2}{2\,W}. \tag{3.92}$$

Hence, after finding the energy, capacitance is readily found.

Example 3.20 Consider the plane $z = 0$ and the plane $z = l$. Plane $z = 0$ carries $\rho_s = -Q/A$ and plane $z = l$ carries $\rho_s = Q/A$. The region in between is filled by an isotropic dielectric medium of permittivity ε. Find the capacitance for an area A of this configuration.

Solution. After finding

$$\mathbf{D} = \begin{cases} \mathbf{0}, & -\infty < z < 0, \\ -\dfrac{Q}{A}\hat{\mathbf{z}}, & 0 < z < l \\ \mathbf{0}, & l < z < \infty. \end{cases} \Rightarrow \mathbf{E} = \begin{cases} \mathbf{0}, & -\infty < z < 0, \\ -\dfrac{Q}{\varepsilon A}\hat{\mathbf{z}}, & 0 < z < l \\ \mathbf{0}, & l < z < \infty. \end{cases}$$

we find energy as

$$W = \frac{\varepsilon}{2}\int_{\mathcal{V}_\infty} E^2\, d\mathcal{V} = \frac{Q^2 l}{2\varepsilon A}.$$

Hence, the capacitance is

$$C = \frac{\varepsilon A}{l}.$$

◁

Example 3.21 Two coaxial cylinders of radii a and b ($b > a$) carry $\rho_s = Q/2\pi la$ and $\rho_s = -Q/2\pi lb$, respectively. The region in between is filled by an isotropic dielectric medium of permittivity ε. Find the capacitance for a length l of this configuration.

Solution. After finding

$$\mathbf{D} = \begin{cases} \mathbf{0}, & 0 < \rho < a, \\ \dfrac{Q}{2\pi l\rho}\hat{\boldsymbol{\rho}}, & a < \rho < b, \\ \mathbf{0}, & b < \rho < \infty. \end{cases} \Rightarrow \mathbf{E} = \begin{cases} \mathbf{0}, & 0 < \rho < a, \\ \dfrac{Q}{2\pi l\varepsilon\rho}\hat{\boldsymbol{\rho}}, & a < \rho < b, \\ \mathbf{0}, & b < \rho < \infty. \end{cases}$$

we find the energy as

$$W = \frac{\varepsilon}{2}\int_{\mathcal{V}_\infty} E^2\, d\mathcal{V} = \varepsilon\pi l\int_a^b \left(\frac{Q}{2\pi l\varepsilon\rho}\right)^2 \rho\, d\rho = \frac{Q^2}{4\pi l\varepsilon}\ln(b/a).$$

Therefore, the capacitance is

$$C = \frac{2\pi\varepsilon l}{\ln(b/a)}.$$

◁

Example 3.22 Two concentric spheres of radii a and b ($b > a$) carry $\rho_s = Q/4\pi a^2$ and $\rho_s = -Q/4\pi b^2$, respectively. The region in between is filled by an isotropic dielectric medium of permittivity ε. Find the capacitance of this configuration.

Solution. After finding

$$\mathbf{D} = \begin{cases} \mathbf{0}, & 0 < r < a, \\ \dfrac{Q}{4\pi r^2}\hat{\mathbf{r}}, & a < r < b, \\ \mathbf{0}, & b < r < \infty. \end{cases} \Rightarrow \mathbf{E} = \begin{cases} \mathbf{0}, & 0 < r < a \\ \dfrac{Q}{4\varepsilon\pi r^2}\hat{\mathbf{r}}, & a < r < b, \\ \mathbf{0}, & b < r < \infty, \end{cases}$$

we find the energy as

$$W = \frac{\varepsilon}{2}\int_{\mathcal{V}_\infty} E^2\, d\mathcal{V} = \varepsilon 2\pi \int_a^b \left(\frac{Q}{4\varepsilon\pi r^2}\right)^2 r^2 dr = \frac{Q^2}{8\pi\varepsilon}\left(\frac{1}{a} - \frac{1}{b}\right).$$

Therefore, the capacitance is

$$C = \frac{4\pi\varepsilon}{\dfrac{1}{a} - \dfrac{1}{b}}.$$

◁

Problems

3.1 The finite plane $\{x \in [-a, a], y \in [-b, b], z = 0\}$ carries a uniform surface charge density ρ_s. Find E_z at $(0, 0, z)$ using Coulomb law. **Hint:** $\displaystyle\int_{-a}^{a} \frac{dx'}{(x'^2 + y'^2 + z^2)^{3/2}} = \frac{2a}{(y'^2 + z^2)\sqrt{a^2 + y'^2 + z^2}}$, and $\displaystyle\int_{-b}^{b} \frac{dy'}{(y'^2 + z^2)\sqrt{a^2 + y'^2 + z^2}} = \frac{2b}{z}\tan^{-1}\left(\frac{ab}{z\sqrt{a^2 + b^2 + z^2}}\right)$.

3.2 The circular loop $\{\rho = a, z = 0\}$ carries a line charge density $\rho_l(\phi)$ given by

$$\rho_l(\phi) = \begin{cases} \rho_0, & 0 < \phi < \pi, \\ -\rho_0, & \pi < \phi < 2\pi, \end{cases}$$

where ρ_o is a constant. Find $\mathbf{E}$ at $(0, 0, z)$ using Coulomb law.

3.3 The disk $\{\rho \le a, z = 0\}$ carries a uniform surface charge density ρ_s. Find $\mathbf{E}$ at $(0, 0, z)$ using Coulomb law. **Hint:** $\displaystyle\int \frac{u\,du}{(u^2 + a^2)^{3/2}} = -\frac{1}{\sqrt{u^2 + a^2}}$.

3.4 Two spheres with radii a and b $(a < b)$ are concentric. A volume charge density $\rho_v(r)$ is distributed in the region $a \leq r \leq b$, and is given by

$$\rho_v(r) = \frac{1}{r^2}, \qquad a < r < b.$$

Find **E** using Gauss law in: (a) $r < a$, (b) $a \leq r \leq b$, and (c) $r > b$.

3.5 Two cylinders with radii a and b $(b > a)$ are coaxial. A volume charge density $\rho_{v1}(\rho)$ is distributed in the region $a > \rho$, and is given by

$$\rho_{v1}(\rho) = \alpha\rho, \qquad a > \rho,$$

where α is a constant. A volume charge density $\rho_{v2}(\rho)$ is distributed in the region $a \leq \rho \leq b$, and is given by

$$\rho_{v2}(\rho) = \frac{\beta}{\rho^2}, \qquad a < \rho < b,$$

where β is a constant. A surface charge density ρ_s is distributed on the surface $\rho = b$. Find **E** using Gauss law in: (a) $\rho < a$, (b) $a < \rho < b$, and (c) $\rho > b$.

3.6 A sphere of radius a carries a non-uniform surface charge density $\rho_s(\theta) = \rho_o \cos\theta$, where ρ_o is a constant. Find **E** at $(0, 0, z)$ in $r > a$ using the method of potentials. **Hint:**
$$\int \frac{\sin\theta\cos\theta}{\sqrt{z^2 + a^2 - 2za\cos\theta}}\, d\theta = \frac{\sqrt{z^2 + a^2 - 2za\cos\theta}\,(z^2 + a^2 + za\cos\theta)}{3z^2a^2}.$$

3.7 Obtain the potential V at point $(x, 0, 0)$ for the line $\{x = 0,\ y = 0,\ z \in (0, l)\}$ carrying a non-uniform charge $\rho_l = 4\pi\varepsilon_0 z$. **Hint:** $\displaystyle\int \frac{u\,du}{\sqrt{u^2 + a^2}} = \sqrt{u^2 + a^2}$.

3.8 Obtain the energy W for a charged sphere of radius 1 when the field is give by

$$\mathbf{E} = \begin{cases} \dfrac{r}{\varepsilon_0}\hat{\mathbf{r}}, & 0 < r < 1, \\ \dfrac{1}{\varepsilon_0 r^2}\hat{\mathbf{r}}, & 1 < r < \infty. \end{cases}$$

3.9 The potential for the dipole $\mathbf{p} = p_o\,\hat{\mathbf{y}}$ is

$$V = \frac{p_o \sin\theta \sin\phi}{4\pi\varepsilon r^2}.$$

Obtain **E**.

3.10 Two cylinders with radii a and c ($a < c$) are coaxial. The cylinder of radius a carries ρ_s while the cylinder of radius c carries $-\rho_s$. The medium in $a < \rho < b$ has ε_1 and the medium in $b < \rho < c$ has ε_2. Obtain the capacitance of this configuration.

Appendix

The following computer program can be used to produce streamline plots of Example 3.1.

```
\[Epsilon]o = (10^-9)/(36 \[Pi]);
l = 1;
\[Rho]l[z_] := 4 \[Pi] \[Epsilon]o;
Gf[x_, y_, z_, xp_, yp_, zp_] :=
  1/(4 \[Pi] \[Epsilon]o) 1/((x - xp)^2 + (y - yp)^2 + (z - zp)^2)^(
   1/2);
V[x_, y_, z_] :=
  NIntegrate[\[Rho]l[zp] Gf[x, y, z, 0, 0, zp], {zp, -l, l}];

Efield[x_, y_, z_] =-Grad[V[x, y, z], {x, y, z}];
Ex[x_, y_, z_] := Efield[x, y, z] . {1, 0, 0};
Ey[x_, y_, z_] := Efield[x, y, z] . {0, 1, 0};
Ez[x_, y_, z_] := Efield[x, y, z] . {0, 0, 1};
data = Table[{{x, y, z}, {Ex[x, y, z], Ey[x, y, z],
      Ez[x, y, z]}}, {x, -2.1, 2.1, 0.2}, {y, -2.1, 2.1,
    0.2}, {z, -2.1, 2.1, 0.2}];
a = 2;
Show[
 Graphics3D@{Arrow[{{0, 0, 0}, {2 a, 0, 0}}]},
 Graphics3D@{Arrow[{{0, 0, 0}, {0, 2 a, 0}}]},
 Graphics3D@{Arrow[{{0, 0, 0}, {0, 0, 2 a}}]},
 Graphics3D@{Text[x, {2 a + 0.5, 0, 0}]},
 Graphics3D@{Text[y, {0, 2 a + 0.5, 0}]},
 Graphics3D@{Text[z, {0, 0, 2 a + 0.5}]}, ListStreamPlot3D[data,
  ImageSize -> 72*9, AspectRatio -> 1, StreamColorFunction -> None,
  StreamStyle -> {Blue}, StreamMarkers -> "Arrow",
  StreamScale -> Full, StreamPoints -> 30,
  Epilog -> {Inset[Style["", FontSize -> 60], Scaled[{0.1, 0.1}]]}],
 ParametricPlot3D[{0, 0, t}, {t, -l, l},
  PlotStyle -> {FontFamily -> "Times New Roman", 50, Red}],
 Boxed -> False, BaseStyle -> FontSize -> 40]
```

References

1. P.M. Morse, H. Feshbach, *Methods of Theoretical Physics* (McGraw-Hill, 1953)

2. J.D. Jackson, *Classical Electrodynamics* (Wiley, 2007)
3. A. Zangwill, *Modern Electrodynamics* (Cambridge University Press, 2013)
4. M.D. Greenberg, *Advanced Engineering Mathematics* (Prentice Hall, 1988)
5. G.B. Arfken, H.J. Weber, F.E. Harris, *Mathematical Methods for Physicists* (Elsevier, 2013)

4 Currents and Resistance

In Chap. 3, we computed electric fields due to several static charge distributions. We next want to consider the situation where charges are moving with a uniform velocity. Such a situation is called a steady current, and fields that arise due to steady currents are static magnetic fields (to be discussed in Chap. 5). In this chapter, we first discuss types of currents in Sect. 4.1 and the continuity equation in Sect. 4.2. Then, we discuss currents that arise in conductors in Sect. 4.3. Due to collisions among charges that occur inside a conductor, the resistance of a conductor is introduced as a quantity that represents these collisions. We compute the resistance for several configurations of conductors in Sect. 4.4.

4.1 Currents

Line Current

If a line charge density ρ_l moves with a uniform velocity $\mathbf{v}$ (in ms^{-1}), then the line current $\mathbf{I}$ (in A) is defined as

$$\mathbf{I} = \rho_l \mathbf{v}. \tag{4.1}$$

Note that a line current is a vector quantity. However, when it flows in a wire, its direction is dictated by the wire. Hence, we can regard a line current as a scalar quantity $I = |\mathbf{I}|$.

H. M. Alkhoori, *Concise Introduction to Electromagnetic Fields*, Synthesis Lectures on Electromagnetics, https://doi.org/10.1007/978-3-031-60331-0_4

Surface Current

If a surface charge density ρ_s moves with a uniform velocity $\mathbf{v}$, then the surface current density $\mathbf{K}$ in Am^{-1} is defined as

$$\mathbf{K} = \rho_s \mathbf{v}. \tag{4.2}$$

Given a surface current $\mathbf{K}$ and a line $\mathcal{L}$, the total current I flowing in $\mathcal{L}$ is given by [1]

$$I = \int_{\mathcal{L}} \mathbf{K} \bullet (\hat{\mathbf{n}} \times d\mathbf{l}), \tag{4.3}$$

where $\hat{\mathbf{n}}$ is a unit vector normal to both $\mathbf{K}$ and $d\mathbf{l}$.

Example 4.1 A uniform surface current density $\mathbf{K} = K_0\hat{\mathbf{z}}$ passes through a loop of radius a. Find the total current I.

Solution.

$$I = \int_0^{2\pi} K_0\hat{\mathbf{z}} \bullet (\hat{\rho} \times a d\phi\, \hat{\boldsymbol{\phi}}) = K_0 2\pi a.$$

◁

Volume Current

If a volume charge density ρ_v moves with a uniform velocity $\mathbf{v}$, then the volume current density $\mathbf{J}$ is defined as

$$\mathbf{J} = \rho_v \mathbf{v}. \tag{4.4}$$

Given a volume current $\mathbf{J}$ and a surface $\mathcal{S}$, the total current I flowing in $\mathcal{S}$ is given by

$$I = \int_{\mathcal{S}} \mathbf{J} \bullet d\mathbf{S}. \tag{4.5}$$

In this perspective, the current I can be thought of as the flux of the current density $\mathbf{J}$ passing through a surface $\mathcal{S}$.

Example 4.2 A uniform volume current density $\mathbf{J} = J_0\hat{\mathbf{z}}$ passes through a disk of radius a. Find the total current I.

Solution.

$$I = \int_S \mathbf{J} \cdot d\mathbf{S} = \int_0^{2\pi} \int_0^{a} J_0 \hat{\mathbf{z}} \cdot \hat{\mathbf{z}} \rho d\rho d\phi = J_0 \pi a^2.$$

◁

4.2 The Continuity Equation

We already saw from Sect. 2.2.1 that

$$\nabla \cdot \mathbf{J}(\mathbf{r}) = -j\omega\rho_v(\mathbf{r}). \tag{4.6}$$

Converting to time domain, we get

$$\nabla \cdot \boldsymbol{\mathcal{J}}(\mathbf{r}, t) = -\frac{\partial \rho_v(\mathbf{r}, t)}{\partial t}. \tag{4.7}$$

This is the continuity equation in differential form. To make it in integral form, we integrate both sides of Eq. (4.7) over a region $\mathcal{V}$ to get

$$\int_{\mathcal{V}} \nabla \cdot \boldsymbol{\mathcal{J}}(\mathbf{r}, t)\, d\mathcal{V} = -\int_{\mathcal{V}} \frac{\partial \rho_v(\mathbf{r}, t)}{\partial t}\, d\mathcal{V}. \tag{4.8}$$

Making use of the divergence theorem on the left side gives

$$\oint_S \boldsymbol{\mathcal{J}}(\mathbf{r}, t) \cdot d\mathbf{S} = -\int_{\mathcal{V}} \frac{\partial \rho_v(\mathbf{r}, t)}{\partial t}\, d\mathcal{V}, \tag{4.9}$$

where S is a surface enclosing $\mathcal{V}$. Finally, assuming that $\mathcal{V}$ is stationary, we get

$$\oint_S \boldsymbol{\mathcal{J}}(\mathbf{r}, t) \cdot d\mathbf{S} = -\frac{d}{dt} \int_{\mathcal{V}} \rho_v(\mathbf{r}, t)\, d\mathcal{V}, \tag{4.10}$$

which is the continuity equation in integral form. Yet, we can obtain the circuit version of the continuity equation after noting that

$$I(t) = \int_S \boldsymbol{\mathcal{J}}(\mathbf{r}, t) \cdot d\mathbf{S}, \tag{4.11}$$

and that

$$Q(t) = \int_{\mathcal{V}} \rho_v(\mathbf{r}, t)\, d\mathcal{V}. \tag{4.12}$$

Then, we get

$$I(t) = -\frac{dQ(t)}{dt}, \tag{4.13}$$

which is the continuity equation in circuit form. It is a mathematical statement of the principle of conservation of charge. It says that the amount of decrease of charge in a certain region is equal to the amount of flow of charge out of that region.

Suppose that the current phasor $\mathbf{J}$ enters a region $\mathcal{V}$. If there is no accumulation of charge inside $\mathcal{V}$ (i.e., $\partial \rho_v / \partial t = 0$), then when $\mathbf{J}$ exits that region, its value should not change. Mathematically, this is stated as

$$\nabla \bullet \mathbf{J} = 0. \tag{4.14}$$

Such a current is called a steady current.

4.3 Conduction Current

Recall that a conductor contains a number of free charges. Producing a current in a conductor requires a force. The volume current density in the conductor must be proportional to that force by

$$\mathbf{J} = \sigma \mathbf{f}, \tag{4.15}$$

where σ (in Sm^{-1}) is the proportionality factor, and $\mathbf{f}$ is the force per unit charge required to make the current. The force $\mathbf{f}$ can be any type of force (i.e., gravitational, nuclear, electromagnetic, etc.). We will confine ourselves to an electromagnetic type of force. The electromagnetic force per unit charge is described by Lorentz force law as

$$\mathbf{f} = \mathbf{E} + \mathbf{v} \times \mathbf{B}, \tag{4.16}$$

where $\mathbf{v}$ is the velocity of the charges. Therefore, the volume current density inside a conductor is given, by

$$\mathbf{J} = \sigma(\mathbf{E} + \mathbf{v} \times \mathbf{B}). \tag{4.17}$$

In situations where $|\mathbf{E}| >> |\mathbf{v} \times \mathbf{B}|$, Eq. (4.17) can be approximated as

$$\mathbf{J} \approx \sigma \mathbf{E}. \tag{4.18}$$

Equation (4.18) is called Ohm law. Such a current is called a conduction current because it is associated with conductors. An example where the condition $|\mathbf{E}| >> |\mathbf{v} \times \mathbf{B}|$ occurs is when the charges move in free space with a speed much smaller than the speed of light (3×10^8 ms^{-1}).

The quantity σ can be obtained by modeling the conductor. As a simple model, the force equation for a single charge inside the conductor is given by

$$\mathbf{F} = q\mathbf{E}. \tag{4.19}$$

The only force that affects the charge is the force resulting from collisions. If the charge has a mass m (in g), and τ (in s) is the average time interval between collisions, then $\mathbf{F} = m\dfrac{d\mathbf{v}}{dt} \approx \dfrac{m\mathbf{v}}{\tau}$. Hence, Eq. (4.19) becomes

$$\frac{m\mathbf{v}}{\tau} = q\mathbf{E}, \tag{4.20}$$

which gives

$$\mathbf{v} = \frac{q\tau}{m}\mathbf{E}. \tag{4.21}$$

But

$$\mathbf{J} = \rho_v \mathbf{v} = nq\mathbf{v}, \tag{4.22}$$

where n (in m^{-3}) is the number of charges per unit volume. Therefore, on substituting Eqs. (4.21) into (4.22), one obtains

$$\mathbf{J} = \frac{nq^2\tau}{m}\mathbf{E}. \tag{4.23}$$

On comparing Eqs. (4.23) with (4.18), we conclude that

$$\sigma = \frac{nq^2\tau}{m}. \tag{4.24}$$

The quantity σ is called the conductivity of the medium.

Remarks

- Since $\mathbf{v} \propto q$, positive charge (protons) move parallel to $\mathbf{E}$, while negative charges (electrons) move antiparallel to $\mathbf{E}$.
- The mass of the electrons m_e is much smaller than the mass of the protons m_p. Since $\mathbf{v} \propto \dfrac{1}{m}$, this implies that the speed of the electrons v_e is much larger than the speed of the protons v_p, that we can assume $v_p \approx 0$.
- Since σ is always positive, the current $\mathbf{J}$ is always parallel to the field $\mathbf{E}$, regardless of the type of charge.

4.4 Resistance

Consider two PECs bounding a conducting medium characterized by a conductivity σ. An electric field $\mathbf{E}$ existing in the medium gives a conduction current $\mathbf{J}$. The input/output relation can be written as

$$\mathbf{J} = \sigma\mathbf{E}. \tag{4.25}$$

A circuit-version of this input/output relation can be obtained as follows. Suppose that one PEC is held at $V = V_0$ and the other PEC at $V = 0$. Since V_0 gives **E**, **E** gives **J**, and **J** gives I, then I inside the conductor has to be proportional to V_0. This is described by

$$I = \frac{1}{R} V_0, \tag{4.26}$$

where R (in Ω) is called the resistance. Therefore, resistance is given by

$$R = \frac{V_0}{I} = \frac{V_0}{\int_S \sigma \mathbf{E} \cdot d\mathbf{S}} = -\frac{V_0}{\int_S \sigma \nabla V \cdot d\mathbf{S}} \tag{4.27}$$

Resistance is a quantity that depends on the shape of the PECs, as well as on the conductivity of the medium between the two PECs. To obtain resistance, we need the potential in the region bounded by the two PECs. Assuming steady current, we have

$$\nabla \cdot \mathbf{J} = 0. \tag{4.28}$$

Since $\mathbf{J} = \sigma \mathbf{E} = -\sigma \nabla V$, Eq. (4.28) becomes

$$\nabla^2 V = 0. \tag{4.29}$$

Equation (4.29) has to be solved subject to the prescribed values of the potential on the two PECs [2–8]. Such a set of equations is called a boundary-value problem. Once V is found, resistance can be found from Eq. (4.27).

Another approach for finding resistance is as follows. Because of collisions, the work done by the source that derives the current inside the conductor is converted into heat. To find this amount of heat (i.e., power) we note that due to each electron that moves under the influence of the force, the power (in W) is given by

$$P = \mathbf{F} \cdot \mathbf{v}. \tag{4.30}$$

For a group of electrons under the influence of an electric force, the power becomes

$$P = \int_{\mathcal{V}} \rho_v \mathbf{E} \cdot \mathbf{v} \, d\mathcal{V}, \tag{4.31}$$

where $\mathcal{V}$ is a region that contains the electrons' density ρ_v. But since $\mathbf{J} = \rho_v \mathbf{v}$, one gets

$$P = \int_{\mathcal{V}} \mathbf{J} \cdot \mathbf{E} \, d\mathcal{V}, \tag{4.32}$$

which becomes

$$P = \int_{\mathcal{V}} \mathbf{J} \cdot \mathbf{E} \, d\mathcal{V} = \int_S \mathbf{J} \cdot d\mathbf{S} \int_{\mathcal{L}} \mathbf{E} \cdot d\mathbf{l} = I V_0 = I^2 R = \frac{V_0^2}{R}. \tag{4.33}$$

Hence, as a second approach, the resistance can be computed from

$$R = \frac{V_0^2}{P} = \frac{V_0^2}{\int_{\mathcal{V}} \mathbf{J} \cdot \mathbf{E}\, d\mathcal{V}} = \frac{V_0^2}{\int_{\mathcal{V}} \sigma |\nabla V|^2\, d\mathcal{V}}. \tag{4.34}$$

Whether resistance is found using Eqs. (4.27) or (4.34), the potential in the region between the two PECs must be found. In the following examples, we provide the potential function.

Example 4.3 Consider the plane $z = 0$ and the plane $z = l$. Plane $z = 0$ is held at $V = 0$, and plane $z = l$ is held at $V = V_0$. The region $0 < z < l$ has conductivity σ, and the potential function is given by

$$V(z) = \frac{V_0}{l} z.$$

Find the resistance for an area A of this configuration.

Solution. We first find $|\nabla V|^2$ as

$$\nabla V = \frac{V_0}{l} \hat{\mathbf{z}} \Rightarrow |\nabla V|^2 = \left(\frac{V_0}{l}\right)^2.$$

Then, the power is found as

$$P = \int_{\mathcal{V}} \sigma |\nabla V|^2\, d\mathcal{V} = A \int_0^l \sigma \left(\frac{V_0}{l}\right)^2 dz = \frac{\sigma A}{l} V_0^2.$$

Finally, the resistance is

$$R = \frac{V_0^2}{P} = \frac{l}{\sigma A}.$$

◁

Example 4.4 Consider the cylinder $\rho = a$ and the cylinder $\rho = b$ ($b > a$ and both are coaxial). Cylinder $\rho = a$ is held at $V = 0$, and cylinder $\rho = b$ is held at $V = V_0$. The region $a < \rho < b$ has a conductivity σ, and the potential function is given by

$$V(\rho) = \frac{V_0}{\ln(b/a)} \ln(\rho/a).$$

Find the resistance for a length l of this configuration.

Solution. We first find $|\nabla V|^2$ as

$$\nabla V = \frac{V_0}{\ln(b/a)} \frac{1}{\rho} \hat{\boldsymbol{\rho}} \Rightarrow |\nabla V|^2 = \left(\frac{V_0}{\ln(b/a)} \frac{1}{\rho}\right)^2.$$

Then, the power is found as

$$P = \int_{\mathcal{V}} \sigma |\nabla V|^2 \, d\mathcal{V} = 2\pi l \int_a^b \sigma \left(\frac{V_0}{\ln(b/a)} \frac{1}{\rho} \right)^2 \rho \, d\rho = \frac{2\pi \sigma l}{\ln(b/a)} V_0^2.$$

Finally, the resistance is

$$R = \frac{V_0^2}{P} = \frac{\ln(b/a)}{2\pi \sigma l}.$$

◁

Example 4.5 Consider the sphere $r = a$ and the sphere $r = b$ ($b > a$ and both are concentric). Sphere $r = a$ is held at $V = 0$, and sphere $r = b$ is held at $V = V_0$. The region $a < r < b$ has a conductivity σ, and the potential function is given by

$$V(r) = \frac{V_0}{\frac{1}{a} - \frac{1}{b}} \left(\frac{1}{a} - \frac{1}{r} \right).$$

Find the resistance of this configuration.

Solution. We first find $|\nabla V|^2$ as

$$\nabla V = \frac{V_0}{\frac{1}{a} - \frac{1}{b}} \frac{1}{r^2} \hat{\mathbf{r}} \Rightarrow |\nabla V|^2 = \left(\frac{V_0}{\frac{1}{a} - \frac{1}{b}} \frac{1}{r^2} \right)^2.$$

Then, the power is found as

$$P = \int_{\mathcal{V}} \sigma |\nabla V|^2 \, d\mathcal{V} = 4\pi \int_a^b \sigma \left(\frac{V_0}{\frac{1}{a} - \frac{1}{b}} \frac{1}{r^2} \right)^2 r^2 \, dr = \frac{4\pi \sigma}{\frac{1}{a} - \frac{1}{b}} V_0^2.$$

Finally, the resistance is

$$R = \frac{V_0^2}{P} = \frac{\frac{1}{a} - \frac{1}{b}}{4\pi \sigma}.$$

◁

References

1. A. Zangwill, *Modern Electrodynamics* (Cambridge University Press, 2013)
2. J.D. Jackson, *Classical Electrodynamics* (Wiley, 2007)
3. P.M. Morse, H. Feshbach, *Methods of Theoretical Physics* (McGraw-Hill, 1953)

4. O'Neil, *Engineering Mathematics* (Thomson, 2003)
5. M. Debnath, *Linear Partial Differential Equations* (Birkhauser ,2007)
6. S.J. Farlow, *Partial Differential Equations for Scientists and Engineers* (Dover Publications, 1982)
7. K.F. Riley, M.P. Hobson, S.J. Bence, *Mathematical Methods for Physics and Engineering* (Cambridge, 2006)
8. H.J. Eom, *Electromagnetic Wave Theory for Boundary-Value Problems* (Springer, 2004)

Magnetostatics

5

We saw in Sect. 2.3 that Maxwell equations decouple into two sets; one set belonging to electrostatics, and the other belonging to magnetostatics. We treated electrostatics in Chap. 3. In this chapter, we study magnetostatics. Magnetostatics, in which a source **J** gives a field **H**, is governed by

$$\left.\begin{aligned} \nabla \times \mathbf{H}(\mathbf{r}) &= \mathbf{J}(\mathbf{r}) \\ \nabla \cdot \mathbf{B}(\mathbf{r}) &= 0 \end{aligned}\right\}. \tag{5.1}$$

We learn how to compute magnetostatic field in an isotropic magnetic medium in Sect. 5.1 using (i) the so-called Biot-Savart law, (ii) Ampere law, and (iii) the method of potentials. We then discuss magnetic dipoles in Sect. 5.2, which then are used in modeling magnetic mediums. We then discuss magnetostatic perturbation by a magnetic object in Sect. 5.3, boundary conditions in Sect. 5.4, magnetic energy in Sect. 5.5, magnetic flux in Sect. 5.6, and inductance and magnetic circuits in Sect. 5.7. A useful computer program is given in the appendix at the end of the chapter.

5.1 Magnetic Field Computation

5.1.1 Solution of Maxwell Equations

In an isotropic magnetic medium (i.e., $\mathbf{B} = \mu\mathbf{H}$), Maxwell equations become

$$\left.\begin{aligned} \nabla \times \mathbf{H}(\mathbf{r}) &= \mathbf{J}(\mathbf{r}) \\ \nabla \cdot \mathbf{H}(\mathbf{r}) &= 0 \end{aligned}\right\}. \tag{5.2}$$

From Eq. $(5.2)_2$, we have

$$\mathbf{H}(\mathbf{r}) = \frac{1}{\mu}\nabla \times \mathbf{A}(\mathbf{r}), \tag{5.3}$$

H. M. Alkhoori, *Concise Introduction to Electromagnetic Fields*, Synthesis Lectures on Electromagnetics, https://doi.org/10.1007/978-3-031-60331-0_5

where $\mathbf{A}$ (in Tm) is called the magnetic vector potential. On substituting Eq. (5.3) into Eq. $(5.2)_2$, one gets

$$\nabla \times (\nabla \times \mathbf{A}) = \mu \mathbf{J}, \tag{5.4}$$

which becomes

$$\nabla(\nabla \bullet \mathbf{A}) - \nabla^2 \mathbf{A} = \mu \mathbf{J}. \tag{5.5}$$

If we let $\nabla \bullet \mathbf{A} = 0$, Eq. (5.5) becomes

$$\nabla^2 \mathbf{A} = -\mu \mathbf{J}, \tag{5.6}$$

Equation (5.6) is an inhomogeneous vector Laplace equation [1, 2]. If $\mathbf{J}$ is distributed in a region $\mathcal{V}'$, solution to Eq. (5.6) in an unbounded region is

$$\mathbf{A}(\mathbf{r}) = \int_{\mathcal{V}'} G(\mathbf{r}, \mathbf{r}') \left[-\mu \mathbf{J}(\mathbf{r}')\right] d\mathcal{V}', \tag{5.7}$$

where $G(\mathbf{r}, \mathbf{r}')$ is the Green function given by

$$G(\mathbf{r}, \mathbf{r}') = -\frac{1}{4\pi R}. \tag{5.8}$$

Substituting Eqs. (5.8) into (5.7) gives

$$\mathbf{A}(\mathbf{r}) = \frac{\mu}{4\pi} \int_{\mathcal{V}'} \frac{\mathbf{J}(\mathbf{r}')}{R} d\mathcal{V}'. \tag{5.9}$$

Finally, applying $\mathbf{H} = \frac{1}{\mu} \nabla \times \mathbf{A}$ and noting that

$$\nabla \left(\frac{1}{R}\right) = -\frac{\mathbf{R}}{R^3} \tag{5.10}$$

give

$$\mathbf{H}(\mathbf{r}) = \frac{1}{4\pi} \int_{\mathcal{V}'} \frac{\mathbf{J}(\mathbf{r}') \times \mathbf{R}}{R^3} d\mathcal{V}'. \tag{5.11}$$

Equation (5.11) is known as Biot-Savart law.

Suppose that the current is distributed over a surface $\mathcal{S}'$. Then, we make $\mathbf{J} \to \mathbf{K}$ and $d\mathcal{V}' \to d\mathcal{S}'$. Therefore,

$$\mathbf{H}(\mathbf{r}) = \frac{1}{4\pi} \int_{\mathcal{S}'} \frac{\mathbf{K}(\mathbf{r}') \times \mathbf{R}}{R^3} d\mathcal{S}'. \tag{5.12}$$

Finally, suppose that the current is distributed over a line $\mathcal{L}'$. Then, we make $\mathbf{J} \to \mathbf{I}$ and $d\mathcal{V}' \to d\mathcal{L}'$. Therefore,

$$\mathbf{H}(\mathbf{r}) = \frac{1}{4\pi} \int_{\mathcal{L}'} \frac{\mathbf{I}(\mathbf{r}') \times \mathbf{R}}{R^3} d\mathcal{L}'. \tag{5.13}$$

Example 5.1 The line $\{x = y = 0, z \in (-l, l)\}$ carries a uniform current $\mathbf{I} = I_0\hat{\mathbf{z}}$. (a) Obtain $\mathbf{H}$ at any point in space. **Hint:** $\displaystyle\int \frac{dz'}{[\rho^2 + (z - z')^2]^{3/2}} = -\frac{(z - z')}{\rho^2\sqrt{\rho^2 + (z - z')^2}}$. (b) Obtain $\mathbf{H}$ when $l \to \infty$. **Hint:** $\displaystyle\int_{-\infty}^{\infty} \frac{dz'}{[\rho^2 + (z - z')^2]^{3/2}} = \frac{2}{\rho^2}$.

Solution.

(a)

$$\mathbf{H} = \frac{1}{4\pi}\int_{\mathcal{L}'} \frac{\mathbf{I}(\mathbf{r}') \times \mathbf{R}}{R^3}\, d\mathcal{L}' = \frac{\mu_0}{4\pi}\int_{\mathcal{L}'} \frac{\mathbf{I} \times \mathbf{R}}{R^3}\, d\mathcal{L}'.$$

After noting that $d\mathbf{l}' = dz'\hat{\mathbf{z}}, d\mathcal{L}' = dz', \mathbf{r} = x\hat{\mathbf{x}} + y\hat{\mathbf{y}} + z\hat{\mathbf{z}} = \rho\hat{\boldsymbol{\rho}} + z\hat{\mathbf{z}}, \mathbf{r}' = z'\hat{\mathbf{z}}, \mathbf{R} = \rho\hat{\boldsymbol{\rho}} + (z - z')\hat{\mathbf{z}}$, $R^3 = [\rho^2 + (z - z')^2]^{3/2}$, and $\mathbf{I} \times \mathbf{R} = I_0\hat{\mathbf{z}} \times [\rho\hat{\boldsymbol{\rho}} + (z - z')\hat{\mathbf{z}}] = I_0\rho\hat{\boldsymbol{\phi}}$, we get

$$\mathbf{H} = \hat{\boldsymbol{\phi}}\,\frac{I_0\rho}{4\pi}\int_{-l}^{l} \frac{dz'}{[\rho^2 + (z - z')^2]^{3/2}} = \hat{\boldsymbol{\phi}}\,\frac{I_0}{4\pi\rho}\left[\frac{(z + l)}{\sqrt{\rho^2 + (z + l)^2}} - \frac{(z - l)}{\sqrt{\rho^2 + (z - l)^2}}\right].$$

Figure 5.1 shows streamline plots of the magnetic field everywhere when $I_o > 0$. Notice that the magnetic field lines rotate around the source.

(b)

$$\mathbf{H} = \hat{\boldsymbol{\phi}}\,\frac{I_0\rho}{4\pi}\int_{-\infty}^{\infty} \frac{dz'}{[\rho^2 + (z - z')^2]^{3/2}} = \hat{\boldsymbol{\phi}}\,\frac{I_0\rho}{4\pi}\frac{2}{\rho^2} = \frac{I_0}{2\pi\rho}\hat{\boldsymbol{\phi}}.$$

◁

Example 5.2 The loop $\{\rho = a, z = 0\}$ carries a uniform current $\mathbf{I} = I_0\hat{\boldsymbol{\phi}}$. Obtain $\mathbf{H}$ at $(0, 0, z)$.

Solution.

$$\mathbf{H} = \frac{1}{4\pi}\int_{\mathcal{L}'} \frac{\mathbf{I}(\mathbf{r}') \times \mathbf{R}}{R^3}\, d\mathcal{L}' = \frac{1}{4\pi}\int_{\mathcal{L}'} \frac{\mathbf{I} \times \mathbf{R}}{R^3}\, d\mathcal{L}'.$$

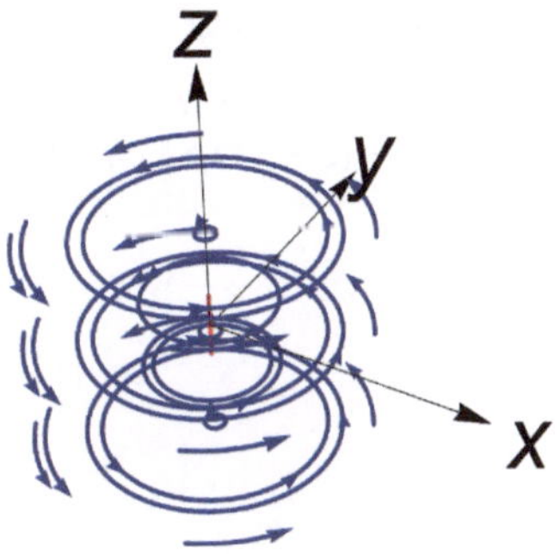

Fig. 5.1 Streamline plots of $\mathbf{H}$ due to a current $I_o > 0$. The source is represented by a red line

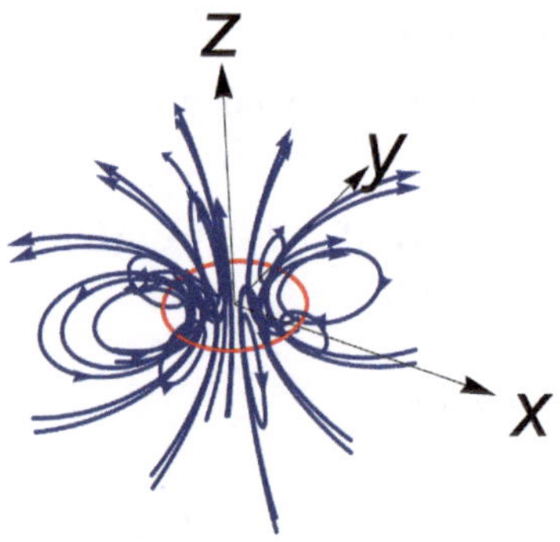

Fig. 5.2 Streamline plots of $\mathbf{H}$ due to a current $I_o > 0$. The source is represented by a red line

After noting that $d\mathbf{l}' = \rho' d\phi' \hat{\boldsymbol{\phi}}'$, $d\mathcal{L}' = \rho' d\phi'$, $\mathbf{r} = z\hat{\mathbf{z}}$, $\mathbf{r}' = \rho'\hat{\boldsymbol{\rho}}'$, $\mathbf{R} = -\rho'\hat{\boldsymbol{\rho}}' + z\hat{\mathbf{z}}$, $R^3 = (\rho'^2 + z^2)^{3/2}$, and $\mathbf{I} \times \mathbf{R} = I_0\hat{\boldsymbol{\phi}}' \times (-\rho'\hat{\boldsymbol{\rho}}' + z\hat{\mathbf{z}}) = I_0(z\hat{\boldsymbol{\rho}}' + \rho'\hat{\mathbf{z}})$, we get

$$\mathbf{H} = \frac{I_0}{4\pi} \int_0^{2\pi} \frac{z\hat{\boldsymbol{\rho}}' + \rho'\hat{\mathbf{z}}}{(\rho'^2 + z^2)^{3/2}} \rho' d\phi' \bigg|_{\rho'=a} = \frac{I_0 a}{4\pi} \int_0^{2\pi} \frac{z\hat{\boldsymbol{\rho}}' + a\hat{\mathbf{z}}}{(a^2 + z^2)^{3/2}} d\phi' = \frac{I_0 a^2}{2(a^2 + z^2)^{3/2}} \hat{\mathbf{z}}.$$

Figure 5.2 shows streamline plot of the magnetic field everywhere when $I_o > 0$. Notice that the magnetic field has a $\hat{\mathbf{z}}$ component only along the z axis, which agrees with the result we obtained. ◁

Example 5.3 The solenoid $\{\rho = a, z \in [-l, l]\}$ has N turns, and carries a uniform current $\mathbf{I} = I_0\hat{\boldsymbol{\phi}}$. (a) Obtain $\mathbf{H}$ at any point on the z axis where $|z| > l$. (b) Obtain $\mathbf{H}$ when $l \to \infty$.

Solution. (a) With the assumption that the turns are very close to each other, we can think of the problem as a finite cylinder carrying a surface current $\mathbf{K} = \frac{N}{2L} I_0\hat{\boldsymbol{\phi}}' = nI_0\hat{\boldsymbol{\phi}}'$ on the surface $\rho = a$. Let us denote the surface $z = l$ by $\mathcal{S}_1$, the surface $\rho = a$ by $\mathcal{S}_2$, and the surface $z = -l$ by $\mathcal{S}_3$. Then, assuming no current flowing on $\mathcal{S}_1$ and $\mathcal{S}_3$, we get

$$\mathbf{H} = \frac{1}{4\pi} \int_{\mathcal{S}'} \frac{\mathbf{K}(\mathbf{r}') \times \mathbf{R}}{R^3} dS' = \frac{1}{4\pi} \left(\int_{\mathcal{S}_1} 0 + \int_{\mathcal{S}_2} \frac{\mathbf{K} \times \mathbf{R}}{R^3} dS' + \int_{\mathcal{S}_3} 0 \right) = \frac{1}{4\pi} \int_{\mathcal{S}_2} \frac{\mathbf{K} \times \mathbf{R}}{R^3} dS'.$$

After noting that $d\mathbf{S}' = \rho' d\phi' dz' \hat{\boldsymbol{\rho}}'$, $dS' = \rho' d\phi' dz'$, $\mathbf{r} = z\hat{\mathbf{z}}$, $\mathbf{r}' = \rho'\hat{\boldsymbol{\rho}}' + z'\hat{\mathbf{z}}$, $\mathbf{R} = -\rho'\hat{\boldsymbol{\rho}}' + (z - z')\hat{\mathbf{z}}$, $R^3 = [\rho'^2 + (z - z')^2]^{3/2}$, and $\mathbf{K} \times \mathbf{R} = nI_0\hat{\boldsymbol{\phi}}' \times [-\rho'\hat{\boldsymbol{\rho}}' + (z - z')\hat{\mathbf{z}}] = nI_0[\rho'\hat{\mathbf{z}} + (z - z')\hat{\boldsymbol{\rho}}']$, we get

$$\mathbf{H} = \frac{nI_0}{4\pi} \int_{-l}^{l} \int_{0}^{2\pi} \frac{\rho'\hat{\mathbf{z}} + (z-z')\hat{\boldsymbol{\rho}}'}{[\rho'^2 + (z-z')^2]^{3/2}} \rho' d\phi' dz' \Big|_{\rho'=a} = \frac{nI_0 a}{4\pi} \int_{-l}^{l} \int_{0}^{2\pi} \frac{a\hat{\mathbf{z}} + (z-z')\hat{\boldsymbol{\rho}}'}{[a^2 + (z-z')^2]^{3/2}} d\phi' dz'$$

$$= \hat{\mathbf{z}} \frac{nI_0 a^2}{2} \int_{-l}^{l} \frac{dz'}{[a^2 + (z-z')^2]^{3/2}} = \hat{\mathbf{z}} \frac{nI}{2} \left[\frac{(z+l)}{\sqrt{a^2 + (z+l)^2}} - \frac{(z-l)}{\sqrt{a^2 + (z-l)^2}} \right].$$

(b) Since the expression we got is valid only along the z axis, and now the source is infinite along that axis, the field we find here must be inside the solenoid.

$$\mathbf{H} = \hat{\mathbf{z}} \frac{nI_0 a^2}{2} \int_{-\infty}^{\infty} \frac{dz'}{[a^2 + (z-z')^2]^{3/2}} = \hat{\mathbf{z}} nI_0.$$

◁

Example 5.4 The plane $z = 0$ carries a uniform surface current density $\mathbf{K} = K_0\,\hat{\mathbf{x}}$. Obtain $\mathbf{H}$ at point $(0, 0, z > 0)$. **Hint:** $\int_{-\infty}^{\infty}\int_{-\infty}^{\infty} \frac{y\,dxdy}{(x^2+y^2+z^2)^{3/2}} = 0$ and $\int_{-\infty}^{\infty}\int_{-\infty}^{\infty} \frac{dxdy}{(x^2+y^2+z^2)^{3/2}} = \frac{2\pi}{z}$.

Solution. Since the current here is a surface current,

$$\mathbf{H} = \frac{1}{4\pi} \int_{S'} \frac{\mathbf{K}(\mathbf{r}') \times \mathbf{R}}{R^3} dS' = \frac{1}{4\pi} \int_{S'} \frac{\mathbf{K} \times \mathbf{R}}{R^3} dS'.$$

Then, we have $d\mathbf{S}' = dx'dy'\hat{\mathbf{z}}$, $dS' = dx'dy'$, $\mathbf{r} = z\hat{\mathbf{z}}$, $\mathbf{r}' = x'\hat{\mathbf{x}} + y'\hat{\mathbf{y}}$, $\mathbf{R} = -x'\hat{\mathbf{x}} - y'\hat{\mathbf{y}} + z\hat{\mathbf{z}}$, $R^3 = (x'^2 + y'^2 + z^2)^{3/2}$, and $\mathbf{K} \times \mathbf{R} = K_0(-y'\hat{\mathbf{z}} - z\hat{\mathbf{y}})$. Therefore,

$$\mathbf{H} = \frac{K_0}{4\pi} \int_{-\infty}^{\infty}\int_{-\infty}^{\infty} \frac{-y'\hat{\mathbf{z}} - z\hat{\mathbf{y}}}{(x'^2 + y'^2 + z^2)^{3/2}} dx'dy' = -\frac{K_0}{2}\hat{\mathbf{y}}.$$

◁

In general, an infinite plane with a unit normal $\hat{\mathbf{n}}$ carrying a uniform current $\mathbf{K}$ will give a field

$$\mathbf{H} = \frac{1}{2}\mathbf{K} \times \hat{\mathbf{n}}. \tag{5.14}$$

5.1.2 Ampere Law

Equation $(5.2)_2$ is called Ampere law in differential form. In this section, we derive its associated integral form, which then is used for finding the magnetic field of some current distributions.

Starting from

$$\nabla \times \mathbf{H}(\mathbf{r}) = \mathbf{J}, \tag{5.15}$$

we integrate both sides of Eq. (5.15) over a closed surface $\mathcal{S}$. This gives

$$\int_{\mathcal{S}} \nabla \times \mathbf{H}(\mathbf{r}) \bullet d\mathbf{S} = \int_{\mathcal{S}} \mathbf{J} \bullet d\mathbf{S}. \tag{5.16}$$

After using Stokes theorem on the left side of Eq. (5.16), and defining

$$\int_{\mathcal{S}} \mathbf{J} \bullet d\mathbf{S} = I_{\text{enc}}, \tag{5.17}$$

we get

$$\oint_{\mathcal{L}} \mathbf{H}(\mathbf{r}) \bullet d\mathbf{l} = I_{\text{enc}}, \tag{5.18}$$

where $\mathcal{L}$ is a closed surface bounding $\mathcal{S}$. Notice that I_{enc} is the total current enclosed by $\mathcal{L}$. Equation (5.18) can be called the the integral form of Ampere law, or simply, Ampere law, and the line $\mathcal{L}$ is often referred to as amperian path.

Ampere law is true for any current distribution, but it is useful to obtain $\mathbf{H}$ only if symmetries present (just like Gauss law). In other words, if the current distribution and the shape of the object that carries the current are such that $\mathbf{H}$ is constant along $\mathcal{L}$, then it can be taken outside the integral sign, and hence, it can be evaluated.

Ampere law can be used as follows:

1. Determine the independent variables and the direction of $\mathbf{H}$.
2. From the direction of $\mathbf{H}$, choose your amperian path $\mathcal{L}$ accordingly. If the direction is $\hat{\mathbf{x}}$, $\hat{\mathbf{y}}$, or $\hat{\mathbf{z}}$, then $\mathcal{L}$ will be a rectangular path. If the direction is $\hat{\boldsymbol{\phi}}$, then $\mathcal{L}$ will be a circular path.
3. Evaluate $\oint_{\mathcal{L}} \mathbf{H}(\mathbf{r}) \bullet d\mathbf{l}$ and I_{enc}.
4. Find $\mathbf{H}$ from $\oint_{\mathcal{L}} \mathbf{H}(\mathbf{r}) \bullet d\mathbf{l} = I_{\text{enc}}$.

Example 5.5 Find $\mathbf{H}$ for an infinite line that carries a uniform current I.

Solution. By intuition, $\mathbf{H}$ must be along $\hat{\boldsymbol{\phi}}$, and thus, the amperian path here will be a circle of radius ρ. Hence, $\oint_{\mathcal{L}} \mathbf{H} \bullet d\mathbf{l} = 2\pi\rho H$, and $I_{\text{enc}} = I$. Therefore,

$$2\pi\rho H = I \Rightarrow H = \frac{I}{2\pi\rho} \Rightarrow \mathbf{H} = \frac{I}{2\pi\rho}\hat{\boldsymbol{\phi}}.$$

◁

Example 5.6 Three cylinders of radii a, b, and c are coaxial ($a < b < c$). The cylinder of radius a carries a uniform volume current density $\mathbf{J}_a = \frac{I}{\pi a^2}\hat{\mathbf{z}}$. The cylinder of radius c carries a uniform volume current density $\mathbf{J}_c = -\frac{I}{\pi(c^2-b^2)}\hat{\mathbf{z}}$. Find $\mathbf{H}$ in all regions.

Solution. By intuition, $\mathbf{H}$ must be along $\hat{\boldsymbol{\phi}}$, and thus, the amperian path here will be a circle of radius ρ. Hence, $\oint_{\mathcal{L}} \mathbf{H} \bullet d\mathbf{l} = 2\pi\rho H$. In $0 < \rho < a$, we have

$$I_{\text{enc}} = \int_S \mathbf{J}_a \bullet d\mathbf{S} = \int_0^{2\pi}\int_0^{\rho} \frac{I}{\pi a^2}\hat{\mathbf{z}} \bullet \hat{\mathbf{z}}\, \rho d\rho d\phi = I\frac{\rho^2}{a^2}.$$

Therefore,

$$2\pi\rho H = I\frac{\rho^2}{a^2} \Rightarrow H = \frac{I\rho}{2\pi a^2} \Rightarrow \mathbf{H} = \frac{I\rho}{2\pi a^2}\hat{\boldsymbol{\phi}}.$$

In $a < \rho < b$, we have

$$I_{\text{enc}} = \int_S \mathbf{J}_a \bullet d\mathbf{S} = \int_0^{2\pi}\int_0^{a} \frac{I}{\pi a^2}\hat{\mathbf{z}} \bullet \hat{\mathbf{z}}\, \rho d\rho d\phi = I.$$

Therefore,

$$2\pi\rho H = I \Rightarrow H = \frac{I}{2\pi\rho} \Rightarrow \mathbf{H} = \frac{I}{2\pi\rho}\hat{\boldsymbol{\phi}}.$$

In $b < \rho < c$, we have

$$I_{\text{enc}} = \int_S \mathbf{J}_a \bullet d\mathbf{S} + \int_S \mathbf{J}_c \bullet d\mathbf{S} = \int_0^{2\pi}\int_0^{a} \frac{I}{\pi a^2}\hat{\mathbf{z}} \bullet \hat{\mathbf{z}}\, \rho d\rho d\phi + \int_0^{2\pi}\int_b^{\rho} -\frac{I}{\pi(c^2-b^2)}\hat{\mathbf{z}} \bullet \hat{\mathbf{z}}\, \rho d\rho d\phi = I\left(1 - \frac{\rho^2-b^2}{c^2-b^2}\right).$$

Therefore,

$$2\pi\rho H = I\left(1 - \frac{\rho^2-b^2}{c^2-b^2}\right) \Rightarrow H = \frac{I}{2\pi\rho}\left(1 - \frac{\rho^2-b^2}{c^2-b^2}\right) \Rightarrow \mathbf{H} = \frac{I}{2\pi\rho}\left(1 - \frac{\rho^2-b^2}{c^2-b^2}\right)\hat{\boldsymbol{\phi}}.$$

Finally, in $c < \rho < \infty$, we have

$$I_{\text{enc}} = \int_S \mathbf{J}_a \bullet d\mathbf{S} + \int_S \mathbf{J}_c \bullet d\mathbf{S} = \int_0^{2\pi}\int_0^{a} \frac{I}{\pi a^2}\hat{\mathbf{z}} \bullet \hat{\mathbf{z}}\, \rho d\rho d\phi + \int_0^{2\pi}\int_b^{c} -\frac{I}{\pi(c^2-b^2)}\hat{\mathbf{z}} \bullet \hat{\mathbf{z}}\, \rho d\rho d\phi = 0.$$

Therefore, $\mathbf{H} = \mathbf{0}$. In summary,

$$\mathbf{H} = \begin{cases} \dfrac{I\rho}{2\pi a^2}\hat{\boldsymbol{\phi}} & 0 < \rho < a, \\ \dfrac{I}{2\pi\rho}\hat{\boldsymbol{\phi}}, & a < \rho < b, \\ \dfrac{I}{2\pi\rho}\left(1 - \dfrac{\rho^2 - b^2}{c^2 - b^2}\right)\hat{\boldsymbol{\phi}}, & b < \rho < c, \\ \mathbf{0}, & c < \rho < \infty. \end{cases}$$

◁

Example 5.7 Consider a toroid of N turns. For simplicity, we may think of a toroid as two concentric circles of radii a and b ($b > a$) on the xy plane. The toroid carries a uniform current I that points along the z axis when viewed on the xy plane. Find $\mathbf{H}$ in all regions.

Solution. By intuition, $\mathbf{H}$ must be along $\hat{\boldsymbol{\phi}}$, and thus, the amperian path here will be a circle of radius ρ. Hence, $\oint_{\mathcal{L}} \mathbf{H} \cdot d\mathbf{l} = 2\pi\rho H$. In $0 < \rho < a$, we have $I_{\text{enc}} = 0$, and thus, $\mathbf{H} = \mathbf{0}$. In $a < \rho < b$, we have $I_{\text{enc}} = NI$, and thus, $\mathbf{H} = \dfrac{NI}{2\pi\rho}\hat{\boldsymbol{\phi}}$. Finally, in $b < \rho < \infty$, we have $I_{\text{enc}} = NI - NI = 0$, and thus, $\mathbf{H} = \mathbf{0}$. So, in summary,

$$\mathbf{H} = \begin{cases} \mathbf{0}, & 0 < \rho < a, \\ \dfrac{NI}{2\pi\rho}\hat{\boldsymbol{\phi}}, & a < \rho < b, \\ \mathbf{0}, & b < \rho < \infty. \end{cases}$$

◁

Example 5.8 For the infinite solenoid of radius a problem, we found the field inside as $\mathbf{H} = nI\hat{\mathbf{z}}$. Find $\mathbf{H}$ outside.

Solution. Since $\mathbf{I} \parallel \hat{\boldsymbol{\phi}}$, then $\mathbf{H} \parallel \hat{\mathbf{z}}$. Hence, the amperian path will be a rectangle $\{\rho \in [b, c], z \in [0, l]\}$, where $c > b > a$. In general, we could assume $\mathbf{H} = H(\rho)\hat{\mathbf{z}}$. Then, $\oint \mathbf{H} \cdot d\mathbf{l} = [H(c) - H(b)]l$. But since $I_{\text{enc}} = 0$, we get $[H(c) - H(b)]l = 0 \Rightarrow H(b) = H(c)$. This indicates that $\mathbf{H}$ will not depend on ρ, and hence, is a constant. But very far away from the solenoid, $\mathbf{H}$ must be zero. So, since $H(\infty) = 0$, then $H(b) = H(c) = 0$. Therefore,

$$\mathbf{H} = \begin{cases} nI\hat{\mathbf{z}}, & 0 < \rho < a, \\ \mathbf{0}, & a < \rho < \infty. \end{cases}$$

◁

5.1.3 The Method of Potentials

In Sect. 5.1.1, we already saw that once a source $\mathbf{J}$ is specified, the magnetic vector potential $\mathbf{A}$ is obtained directly from the source, and then the magnetic field $\mathbf{H}$ can be found from the magnetic vector potential. Such an approach can be called the method of potentials, and it is summarized here. The following can be used to compute the magnetic vector potential of the various sources.

- Line current:

$$\mathbf{A}(\mathbf{r}) = \frac{\mu}{4\pi} \int_{\mathcal{L}'} \frac{\mathbf{I}(\mathbf{r}')}{R} \, d\mathcal{L}'. \tag{5.19}$$

- Surface current:

$$\mathbf{A}(\mathbf{r}) = \frac{\mu}{4\pi} \int_{\mathcal{S}'} \frac{\mathbf{K}(\mathbf{r}')}{R} \, d\mathcal{S}'. \tag{5.20}$$

- Volume current:

$$\mathbf{A}(\mathbf{r}) = \frac{\mu}{4\pi} \int_{\mathcal{V}'} \frac{\mathbf{J}(\mathbf{r}')}{R} \, d\mathcal{V}'. \tag{5.21}$$

5.2 Magnetic Dipoles

Given a source $\mathbf{J}$ occupying a region $\mathcal{V}'$, the magnetic vector potential is given by

$$\mathbf{A}(\mathbf{r}) = \frac{\mu}{4\pi} \int_{\mathcal{V}'} \frac{\mathbf{J}(\mathbf{r}')}{R} \, d\mathcal{V}'. \tag{5.22}$$

With the assumption that $\mathbf{r}$ lies along the z axis, the term $1/R$ can be approximated as

$$\frac{1}{R} = \frac{1}{r}\left(1 + \cos\theta' \frac{r'}{r} + \ldots\right) \tag{5.23}$$

If $|\mathbf{r}| >> |\mathbf{r}'|$, we may consider the first two terms and ignore the rest. Therefore,

$$\begin{aligned} \mathbf{A} &\approx \frac{\mu}{4\pi r} \int_{\mathcal{V}'} \mathbf{J}(\mathbf{r}') \, d\mathcal{V}' + \frac{\mu}{4\pi r^2} \int_{\mathcal{V}'} [r' \cos\theta' \mathbf{J}(\mathbf{r}')] \, d\mathcal{V}' \\ &= \frac{\mu}{4\pi r} \int_{\mathcal{V}'} \mathbf{J}(\mathbf{r}') \, d\mathcal{V}' + \frac{\mu}{4\pi r^2} \hat{\mathbf{r}} \cdot \int_{\mathcal{V}'} [\mathbf{r}' \mathbf{J}(\mathbf{r}')] \, d\mathcal{V}', \end{aligned} \tag{5.24}$$

where the quantity $\mathbf{J}(\mathbf{r}')\mathbf{r}'$ is a dyadic. The first integral can be regarded as a monopole term and the second integral as a dipole term. For any source $\mathbf{J}$, it happens that the first integral vanishes [3]. Hence, magnetic monopole does not exist theoretically. The second integral can be rearranged as [1, 3]

$$\mathbf{A}(\mathbf{r}) \approx -\frac{\mu}{4\pi r^2} \hat{\mathbf{r}} \times \left[\frac{1}{2} \int_{\mathcal{V}'} \mathbf{r}' \times \mathbf{J}(\mathbf{r}') \, d\mathcal{V}'\right]. \tag{5.25}$$

Upon defining $\mathbf{m}$ as

$$\mathbf{m} = \frac{1}{2} \int_{\mathcal{V}'} \mathbf{r}' \times \mathbf{J}(\mathbf{r}') \, d\mathcal{V}', \tag{5.26}$$

Eq. (5.25) can be written as

$$\mathbf{A}(\mathbf{r}) \approx \frac{\mu}{4\pi}\frac{\mathbf{m}\times\hat{\mathbf{r}}}{r^2}. \tag{5.27}$$

The quantity $\mathbf{m}$ (in Am^2) is called the magnetic dipole moment. Since field and source points do not enter into the expression of $\mathbf{m}$, we can use unprimed variables. Therefore,

$$\mathbf{m} = \frac{1}{2}\int_{\mathcal{V}} \mathbf{r}\times\mathbf{J}(\mathbf{r})\, d\mathcal{V}. \tag{5.28}$$

For a surface current $\mathbf{K}$, $\mathbf{m}$ is given by

$$\mathbf{m} = \frac{1}{2}\int_{\mathcal{S}} \mathbf{r}\times\mathbf{K}(\mathbf{r})\, d\mathcal{S}. \tag{5.29}$$

Finally, for a line current $\mathbf{I}$, $\mathbf{m}$ is given by

$$\mathbf{m} = \frac{1}{2}\int_{\mathcal{L}} \mathbf{r}\times\mathbf{I}(\mathbf{r})\, d\mathcal{L}. \tag{5.30}$$

For line currents with a closed path $\mathcal{L}$, Eq. (5.30) can be written as [3]

$$\mathbf{m} = \int_{\mathcal{S}} I(\mathbf{r})\, d\mathbf{S}, \tag{5.31}$$

where $\mathcal{S}$ is the surface bounded by $\mathcal{L}$.

Example 5.9 Find $\mathbf{m}$ for the rectangle $\{z = 0, x \in [0, a], y \in [0, b]\}$ carrying a uniform current I.

Solution.

$$\mathbf{m} = \int_{\mathcal{S}} I(\mathbf{r})\, d\mathbf{S} = \int_0^b\int_0^a I\hat{\mathbf{z}}dxdy = Iab\hat{\mathbf{z}}.$$

◁

Example 5.10 For the small loop $\{\rho = a, z = 0\}$, find $\mathbf{m}$, $\mathbf{A}$, and $\mathbf{H}$.

Solution. These are found, respectively, as

$$\mathbf{m} = \int_{\mathcal{S}} I(\mathbf{r})\, d\mathbf{S} = I\int_0^{2\pi}\int_0^a \hat{\mathbf{z}}\rho d\rho d\phi = I(\pi a^2)\hat{\mathbf{z}} = m_0\hat{\mathbf{z}},$$

$$\mathbf{A} = \frac{\mu}{4\pi}\frac{\mathbf{m}\times\hat{\mathbf{r}}}{r^2} = \frac{\mu m_0}{4\pi r^2}(\hat{\mathbf{z}}\times\hat{\mathbf{r}}) = \frac{\mu m_0}{4\pi r^2}[(\cos\theta\,\hat{\mathbf{r}} - \sin\theta\,\hat{\boldsymbol{\theta}})\times\hat{\mathbf{r}}] = \frac{\mu m_0 \sin\theta}{4\pi r^2}\hat{\boldsymbol{\phi}},$$

and

$$\mathbf{H} = \frac{1}{\mu}\nabla \times \mathbf{A} = \frac{m_0}{4\pi r^3}(2\cos\theta\,\hat{\mathbf{r}} + \sin\theta\,\hat{\boldsymbol{\theta}}).$$

◁

Since the functional form of the magnetic field of a point magnetic dipole **m** is similar to the functional form of the electric field of a point electric dipole **p**, their streamline plots must be similar.

5.3 Magnetostatic Perturbation by a Magnetic Object

Like in electrostatics, we now consider the situation where a magnetostatic field exists in free space in the presence of a magnetic object. In order to understand the interaction that takes place, we first model the magnetic object using magnetic dipoles. Then, after understanding the role of the magnetic object, we mention the overall interaction that takes place.

5.3.1 A Magnetic Object

According to a classical mechanical description of matter, an electron rotates about itself (spinning motion) and revolves around the nucleus it is attached to (orbital motion). Due to both actions, a current is generated. This current results in a magnetic dipole moment **m** [4]. Consequently, every electron has an associated magnetic dipole moment. Hence, from a magnetic point of view, an atom can be thought of as a magnetic dipole.

A magnetic object can be thought of as a group of atoms. A magnetic object is not described by **m**, but rather by a magnetization vector **M**. Suppose that inside an object, there exists a region ΔV with a total dipole moment $\Delta\mathbf{m}$. Then, the magnetization vector is defined as

$$\mathbf{M} = \lim_{\Delta\to 0}\frac{\Delta\mathbf{m}}{\Delta V} = \frac{d\mathbf{m}}{d\mathcal{V}}. \tag{5.32}$$

It plays a role similar to the polarization vector **P** in electrostatics.

Consider a magnetic object occupying a region $\mathcal{V}'$ bounded by a surface $\mathcal{S}'$ and is characterized by a magnetization vector **M**. Since **m** gives **H**, then so does **M**. Since a relation between **M** and **H** can be found using the method of potentials, let us only derive a relation for **A**. The differential perturbation potential due to one dipole moment in $\mathcal{V}'$ is given by

$$d\mathbf{A} = \frac{\mu_0}{4\pi}\frac{d\mathbf{m}\times\mathbf{R}}{R^3}. \tag{5.33}$$

Since $\mathbf{m} = \mathbf{M}\,d\mathcal{V}$, we get the perturbation potential as

$$\mathbf{A} = \frac{\mu_0}{4\pi}\int_{\mathcal{V}'}\frac{\mathbf{M}(\mathbf{r}')\times\mathbf{R}}{R^3}\,d\mathcal{V}'. \tag{5.34}$$

Internal Currents

As another point of view, using the same procedure adopted in Sect. 3.4.1, Eq. (5.34) becomes

$$\mathbf{A} = \frac{\mu_0}{4\pi}\int_{\mathcal{V}'} \frac{\nabla' \times \mathbf{M}(\mathbf{r}')}{R}\, d\mathcal{V}' + \frac{\mu_0}{4\pi}\oint_{\mathcal{S}'} \frac{\mathbf{M}(\mathbf{r}') \times \hat{\mathbf{n}}'}{R}\, d\mathcal{S}'. \tag{5.35}$$

On defining $\mathbf{J}_{\text{int}}$ as

$$\mathbf{J}_{\text{int}} = \nabla \times \mathbf{M}, \tag{5.36}$$

and $\mathbf{K}_{\text{int}}$ as

$$\mathbf{K}_{\text{int}} = \mathbf{M} \times \hat{\mathbf{n}}, \tag{5.37}$$

Equation (5.35) becomes

$$\mathbf{A} = \frac{\mu_0}{4\pi}\int_{\mathcal{V}'} \frac{\mathbf{J}_{\text{int}}(\mathbf{r}')}{R}\, d\mathcal{V}' + \frac{\mu_0}{4\pi}\int_{\mathcal{S}'} \frac{\mathbf{K}_{\text{int}}(\mathbf{r}')}{R}\, d\mathcal{S}'. \tag{5.38}$$

Equation (5.38) indicates that the magnetic field due to a magnetized object is same as that which would be produced by a volume current density $\mathbf{J}_{\text{int}}$ and a surface current density $\mathbf{K}_{\text{int}}$. These are called internal current densities.

In a situation where an external source is immersed in a magnetic medium, the internal currents can be separated from the external current once a relation between **M** and **H**, where **H** is the total field due to both external and internal currents, is given. In an isotropic magnetic medium, such a relation is given by

$$\mathbf{M} = \frac{1}{\mu_0}\mathbf{B} - \mathbf{H} = (\mu_r - 1)\mathbf{H} = \chi_m \mathbf{H}, \tag{5.39}$$

where $\chi_m = (\mu_r - 1)$ is the magnetic susceptibility scalar. As to be shown in Example 5.11, when $\chi_m > 0$, then **B** in the medium increases compared to when in free space, and vice versa. When $\chi_m > 0$, the medium is classified as paramagnetic, while when $\chi_m < 0$, the medium is classified as diamagnetic

Example 5.11 An infinite solenoid of radius a is composed of an isotropic magnetic medium of χ_m. The current passing around the solenoid is I. (a) Find **B** inside and outside the solenoid. (b) Obtain the internal currents.

Solution. (a) We first find **H** as

$$\mathbf{H} = \begin{cases} nI\hat{\mathbf{z}}, & 0 < \rho < a, \\ \mathbf{0}, & a < \rho < \infty. \end{cases}$$

Then, **B** if found as

$$\mathbf{B} = \begin{cases} \mu_0 n(1+\chi_m)I\hat{\mathbf{z}}, & 0 < \rho < a, \\ \mathbf{0}, & a < \rho < \infty. \end{cases}$$

(b) Using **B** and **H**, we find $\mathbf{M} = \chi_m n I\hat{\mathbf{z}}$. Then, $\mathbf{J}_{\text{int}} = \nabla \times \left(\chi_m n I\hat{\mathbf{z}}\right) = \mathbf{0}$, and $\mathbf{K}_{\text{int}} = \mathbf{M} \times \hat{a}_n = \mathbf{M} \times \hat{\rho} = \chi_m n I\hat{\boldsymbol{\phi}}$. When the medium is paramagnetic (i.e., $\chi_m > 0$), $\mathbf{K}_{\text{int}}$, which results in I_{int} will be parallel to I, which results in increasing **B**. When the medium is diamagnetic (i.e., $\chi_m < 0$), $\mathbf{K}_{\text{int}}$, which results in I_{int} will be antiparallel to I, which results in decreasing **B**. ◁

5.3.2 Interaction Between Magnetostatic Field and a Magnetic Object

Consider a magnetized object suspended in free space. Suppose that initially, $\mathbf{M} = \mathbf{0}$. When a source magnetic field $\mathbf{H}_{\text{source}}$ is applied, the dipoles inside the object align themselves in a certain configuration with respect to $\mathbf{H}_{\text{source}}$. These dipole are said to be magnetized. Due to their new arrangement, there will be a net **M**, which gives rise to a perturbation magnetic field $\mathbf{H}_{\text{pert}}$ outside the object, and an internal magnetic field $\mathbf{H}_{\text{int}}$ inside the object. Those fields depend on $\mathbf{H}_{\text{source}}$, as well as on the composition of the medium. Computation of the perturbation and internal magnetic fields is possible once a relation between **M** and its cause $\mathbf{H}_{\text{source}}$ is found. Perturbation of the magnetic field by an isotropic magnetic medium is similar to the perturbation of the electric field by an isotropic dielectric medium discussed in Sect. 3.4.2.

5.3.3 A Nonlinear Magnetic Medium

In linear magnetic mediums (e.g., paramagnetic and diamagnetic), magnetization is linearly proportional to **H**. Once **H** is removed, the medium is said to be demagnetized (i.e., **M** becomes zero). However, in a certain nonlinear mediums (called ferromagnetic mediums), the magnetization does not become zero, even when **H** is. The reason is that each dipole in a ferromagnetic medium tends to align in a direction similar to its neighbors.

Consider a situation where we connect an object composed of a ferromagnetic medium to a current source I. This current source produces **H**, which derives a magnetization by virtue of $\mathbf{M} = \chi_m(\mathbf{H})$. Note that $\chi_m(\mathbf{H})$ is a function of **H**, which indicates that the medium is nonlinear. We don't have an explicit expression for $\chi_m(\mathbf{H})$, but we instead can plot the magnitude of the magnetization $|\mathbf{M}| = M$ (the output) versus the current I (the input). Such a plot is often obtained experimentally (see Fig. 5.3). The process shown in Fig. 5.3 is discussed next.

- Before we turn on I, M was zero (point a).
- Once we turn on and increase I, M starts to increase (path ab).

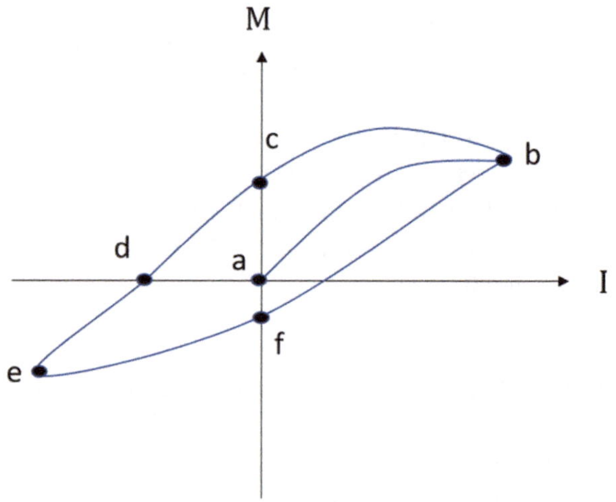

Fig. 5.3 M versus I

- At a certain point (called the saturation point), a further increase on I does not change M because all dipoles have become aligned (point b).
- If we decrease I, M will decrease, but it will not vanish even when $I = 0$ (point c).
- If we change the direction of I and increase it, M will keep decreasing until it vanishes (point d).
- A further increase on I makes up the dipoles in an opposite direction and M increases until it reaches the saturation point (point e).
- The process will be repeated again if we decrease I (path ef) and then change its direction (path fg), however, it will follow another path (path gb).

In many situations, the magnitude of the field $\mathbf{B} = \mu_0(\mathbf{H} + \mathbf{M})$ is plotted versus the magnitude of the field $\mathbf{H}$. This is done usually because in ferromagnetic mediums, $|\mathbf{M}| >> \mu_0|\mathbf{H}|$, so $|\mathbf{B}| \approx \mu_0|\mathbf{M}|$.

5.4 Boundary Conditions

In electrostatics, Maxwell equations read

$$\left.\begin{aligned} \nabla \times \mathbf{E} = \mathbf{0} \\ \nabla \cdot \mathbf{D} = \rho_v \end{aligned}\right\}. \tag{5.40}$$

The corresponding boundary conditions are

$$\left.\begin{array}{ll}\hat{\mathbf{n}} \times (\mathbf{E}_1 - \mathbf{E}_2) = \mathbf{0}, & \mathbf{r} \in \mathcal{S} \\ \hat{\mathbf{n}} \bullet (\mathbf{D}_1 - \mathbf{D}_2) = \rho_s, & \mathbf{r} \in \mathcal{S}\end{array}\right\}, \tag{5.41}$$

where $\hat{\mathbf{n}}$ is a unit normal pointing into medium 1 (i.e., from medium 2 to medium 1).

In magnetostatics, Maxwell equations are

$$\left.\begin{array}{r}\nabla \times \mathbf{H} = \mathbf{J} \\ \nabla \bullet \mathbf{B} = 0\end{array}\right\}. \tag{5.42}$$

After noting how the curl and the divergence transformed in electrostatics, the corresponding boundary conditions in magnetostatics become

$$\left.\begin{array}{ll}\hat{\mathbf{n}} \times (\mathbf{H}_1 - \mathbf{H}_2) = \mathbf{K} \Rightarrow \mathbf{H}_1^{\parallel} - \mathbf{H}_2^{\parallel} = \mathbf{K}, & \mathbf{r} \in \mathcal{S} \\ \hat{\mathbf{n}} \bullet (\mathbf{B}_1 - \mathbf{B}_2) = 0 \Rightarrow \mathbf{B}_1^{\perp} = \mathbf{B}_2^{\perp}, & \mathbf{r} \in \mathcal{S}\end{array}\right\}. \tag{5.43}$$

So, assuming both $\mathbf{H}_1$ and $\mathbf{H}_2$ are constant vectors, if $\mathbf{H}_1$ is given, then $\mathbf{H}_2$ is found simply from

$$\mathbf{H}_2 = (\mathbf{H}_1^{\parallel} - \mathbf{K}) + \frac{\mu_1}{\mu_2}\mathbf{H}_1^{\perp}. \tag{5.44}$$

Example 5.12 Two mediums are separated by the plane $y - x - 2 = 0$. Medium 1, occupied in $y - x - 2 \leq 0$, has $\mathbf{H}_1 = -2\hat{\mathbf{x}} + 6\hat{\mathbf{y}} + 4\hat{\mathbf{z}}$ and $\mu_1 = 5\mu_0$. Medium 2 has $\mu_2 = 2\mu_0$. Obtain $\mathbf{H}_2$.

Solution. After identifying the surface $f(x, y) = y - x - 2 = 0$, we get the unit normal vector as

$$\hat{\mathbf{n}} = \frac{\nabla f}{|\nabla f|} = \frac{1}{\sqrt{2}}(-\hat{\mathbf{x}} + \hat{\mathbf{y}}).$$

Then,

$$\mathbf{H}_1^{\perp} = (\mathbf{H}_1 \bullet \hat{\mathbf{n}})\hat{\mathbf{n}} = -4\hat{\mathbf{x}} + 4\hat{\mathbf{y}} \Rightarrow \mathbf{H}_2^{\perp} = \frac{\mu_1}{\mu_2}\mathbf{H}_1^{\perp} = -10\hat{\mathbf{x}} + 10\hat{\mathbf{y}},$$

and

$$\mathbf{H}_1^{\parallel} = \mathbf{H}_1 - \mathbf{H}_1^{\perp} = 2\hat{\mathbf{x}} + 2\hat{\mathbf{y}} + 4\hat{\mathbf{z}} \Rightarrow \mathbf{H}_2^{\parallel} = \mathbf{H}_1^{\parallel} = 2\hat{\mathbf{x}} + 2\hat{\mathbf{y}} + 4\hat{\mathbf{z}}.$$

Therefore,

$$\mathbf{H}_2 = -8\hat{\mathbf{x}} + 12\hat{\mathbf{y}} + 4\hat{\mathbf{z}}.$$

◁

5.5 Magnetic Flux

The magnetic flux ψ (in Wb) through a surface $\mathcal{S}$ is defined as

$$\psi = \int_{\mathcal{S}} \mathbf{B} \cdot d\mathbf{S}. \tag{5.45}$$

Another expression can be obtained after noting that $\mathbf{B} = \nabla \times \mathbf{A}$ and using Stokes theorem, which is

$$\psi = \oint_{\mathcal{L}} \mathbf{A} \cdot d\mathbf{l}, \tag{5.46}$$

where $\mathcal{L}$ is a closed path bounding $\mathcal{S}$.

Example 5.13 Let $\mathbf{A} = y\,\hat{\mathbf{x}}$ and $\mathcal{S}$ be the plane $\{x \in [-1, 1], y \in [-1, 1], z = 0\}$. Also, we define the path $\mathcal{L}$ as the boundary of $\mathcal{S}$ in counter clockwise orientation. Obtain the flux (a) using $\mathbf{A}$, and (b) using $\mathbf{B}$.

Solution.

(a)

$$\psi = \oint_{\mathcal{L}} \mathbf{A} \cdot d\mathbf{l} = \int_{1}^{-1} y\,dx\Big|_{y=1} + \int_{-1}^{1} y\,dx\Big|_{y=-1} = -4.$$

(b)

$$\mathbf{B} = \nabla \times \mathbf{A} = -\hat{\mathbf{z}} \Rightarrow \psi = \int_{\mathcal{S}} \mathbf{B} \cdot d\mathbf{S} = -\int_{-1}^{1}\int_{-1}^{1} \hat{\mathbf{z}} \cdot \hat{\mathbf{z}} dx dy = -4.$$

◁

5.6 Magnetic Energy

In electrostatics, we derived an expression for the electric energy as

$$W = \frac{1}{2}\int_{\mathcal{V}} \rho_v V \, d\mathcal{V}. \tag{5.47}$$

Using the correspondences $\rho_v \to \mathbf{J}$ and $V \to \mathbf{A}$, in magnetostatics, the magnetic energy is

$$W = \frac{1}{2}\int_{\mathcal{V}} \mathbf{J} \cdot \mathbf{A} \, d\mathcal{V}. \tag{5.48}$$

For a surface current density, W is given by

$$W = \frac{1}{2}\int_{\mathcal{S}} \mathbf{K} \cdot \mathbf{A} \, dS, \tag{5.49}$$

and for a line current by

$$W = \frac{1}{2}\int_{\mathcal{L}} \mathbf{I} \cdot \mathbf{A}\, d\mathcal{L}. \tag{5.50}$$

On expressing $\mathbf{B} = \nabla \times \mathbf{A}$ and following a procedure similar to that described in Sect. 3.1.3, Eq. (5.48) can be written as

$$W = \frac{1}{2}\int_{\mathcal{V}_\infty} \mathbf{B} \cdot \mathbf{H}\, d\mathcal{V}, \tag{5.51}$$

where $\mathcal{V}_\infty$ is the entire region.

5.7 Inductance and Magnetic Circuits

In this section, we discuss inductance as a link between current and flux. This includes mutual inductance when the current exists in an object and the flux is measured in another object, and self inductance when the current exists in an object and the flux is measured in the same object. Finally, using inductance, reluctance and magnetic circuits are introduced.

5.7.1 Mutual Inductance

Suppose that a loop $\mathcal{L}_1$ carries a current I_1 suspended in free space. This current produces $\mathbf{B}_1$. Now, let another loop $\mathcal{L}_2$ exist in space. When $\mathbf{B}_1$ passes through loop 2, the number of lines of $\mathbf{B}_1$ that passes through that loop (the flux) is given by

$$\psi_2 = \int_{S_2} \mathbf{B}_1 \cdot d\mathbf{S}_2, \tag{5.52}$$

where S_2 is a surface bounded by $\mathcal{L}_2$. Because $\mathbf{B}_1$ is produced from I_1, it has to depend on I_1 by virtue of Biot-Savart law. Consequently, ψ_2 also depends on I_1. That is,

$$\psi_2 = M_{21} I_1, \tag{5.53}$$

where the constant of proportionality, M_{21}, is called the mutual inductance (in H). Hence, the mutual inductance is the flux through loop 2 due to the current in loop 1 (i.e., the current that produces the magnetic field or flux) divided by that current.

Another way to obtain mutual inductance is as follows. The energy in loop 2 due to field 1 ($\mathbf{B}_1$ or $\mathbf{A}_1$) is given from Eq. (5.50) as

$$W_2 = \frac{I_1}{2}\oint_{\mathcal{L}_2} \mathbf{A}_1 \cdot d\mathbf{l}_2, \tag{5.54}$$

since $\mathcal{L}_2$ is a loop (i.e., a closed path). But $\oint_{\mathcal{L}_2} \mathbf{A}_1 \cdot d\mathbf{l}_2 = \psi_2$. Therefore, Eq. (5.54) becomes

$$W_2 = \frac{1}{2} I_1 \psi_2. \tag{5.55}$$

Substituting Eqs. (5.53) into (5.55) gives

$$W_2 = \frac{1}{2} M_{21} I_1^2. \tag{5.56}$$

As a summary, in general, the mutual inductance of loop i due to a current in loop j can be calculated either from the flux as

$$M_{ij} = \frac{\psi_i}{I_j}, \tag{5.57}$$

or from the energy as

$$M_{ij} = \frac{2W_i}{I_j^2}. \tag{5.58}$$

An explicit formula for M_{ij} that can be derived from the flux relation $\psi_i = M_{ij} I_j$ is found as [5]

$$M_{ij} = \frac{\mu}{4\pi} \oint_{\mathcal{L}_i} \oint_{\mathcal{L}_j} \frac{d\mathbf{l}_i \cdot d\mathbf{l}_j}{|\mathbf{r}_i - \mathbf{r}_j|}. \tag{5.59}$$

This formula shows that the mutual inductance depends on the shape of the loops, their relative distance, and the medium surrounding these loops. It also shows that $M_{ij} = M_{ji}$ (i.e., the flux through loop j due to current i is the same as the flux through loop i due to current j, regardless of the shapes of both loops).

5.7.2 Self Inductance

The self inductance (or simply the inductance) of a loop i is the flux of $\mathbf{B}_i$ divided by the current I_i. Because we don't have another loop j, we simply drop the subscript i. Hence,

$$L = \frac{\psi}{I}, \tag{5.60}$$

or

$$L = \frac{2\,W}{I^2}. \tag{5.61}$$

Notice that when fields are functions of the radial variable ρ only, as will be the case in the following examples, inductance can be found directly as

$$L = \frac{2\pi l}{I^2} \int_0^\infty \mathbf{B} \cdot \mathbf{H} \, \rho \, d\rho, \tag{5.62}$$

where l is a dimension along the z axis.

Example 5.14 A toroid with an inner radius a and an outer radius b, carries a current I with N turns. The medium inside has a permeability μ. Find the inductance for a height l of the toroid.

Solution. After finding

$$\mathbf{H} = \begin{cases} \mathbf{0}, & 0 < \rho < a, \\ \dfrac{NI}{2\pi\rho}\hat{\boldsymbol{\phi}}, & a < \rho < b, \\ \mathbf{0}, & b < \rho < \infty, \end{cases}$$

we find the inductance as

$$L = \frac{2\pi l}{I^2}\int_0^{\infty} \mathbf{B} \cdot \mathbf{H}\, \rho\, d\rho = \frac{2\pi \mu l}{I^2}\int_a^b \left(\frac{NI}{2\pi\rho}\right)^2 \rho d\rho = \frac{\mu N^2 l}{2\pi}\ln(b/a).$$

◁

Example 5.15 Two coaxial cylinders of radii a and b ($b > a$). The inner cylinder carries $\mathbf{K} = \dfrac{I}{2\pi a}\hat{\mathbf{z}}$ and the outer cylinder carries $\mathbf{K} = -\dfrac{I}{2\pi b}\hat{\mathbf{z}}$. The medium between the two cylinders has a permeability μ. Find the inductance for a length l of the cylinder.

Solution. After finding

$$\mathbf{H} = \begin{cases} \mathbf{0}, & 0 < \rho < a, \\ \dfrac{I}{2\pi\rho}\hat{\boldsymbol{\phi}}, & a < \rho < b, \\ \mathbf{0}, & b < \rho < \infty. \end{cases}$$

we find the inductance as

$$L = \frac{2\pi l}{I^2}\int_0^{\infty} \mathbf{B} \cdot \mathbf{H}\, \rho\, d\rho = \frac{2\pi \mu l}{I^2}\int_a^b \left(\frac{I}{2\pi\rho}\right)^2 \rho d\rho = \frac{\mu l}{2\pi}\ln(b/a).$$

◁

Example 5.16 In Example 5.15, let the inner cylinder carry a volume current $\mathbf{J} = \dfrac{I}{\pi a^2}\hat{\mathbf{z}}$ instead of a surface current. Also, let the medium in $\rho > b$ be free space. Find the inductance for a length l of the cylinder.

Solution. After finding

$$\mathbf{H} = \begin{cases} \dfrac{I\rho}{2\pi a^2}\hat{\boldsymbol{\phi}}, & 0 < \rho < a, \\ \dfrac{I}{2\pi\rho}\hat{\boldsymbol{\phi}}, & a < \rho < b, \\ \mathbf{0}, & b < \rho < \infty. \end{cases}$$

we find the inductance as

$$L = \frac{2\pi l}{I^2} \int_0^\infty \mathbf{B} \cdot \mathbf{H} \rho \, d\rho = \frac{2\pi l}{I^2} \left[\mu_0 \int_0^a \left(\frac{I\rho}{2\pi a^2} \right)^2 \rho d\rho + \mu \int_a^b \left(\frac{I}{2\pi \rho} \right)^2 \rho d\rho \right] = \frac{\mu_0 l}{8\pi} + \frac{\mu l}{2\pi} \ln(b/a).$$

The first part is sometimes called the internal inductance, while the second part the external inductance. ◁

5.7.3 Magnetic Circuits

For some simple configurations, the concept of magnetic circuits can be used to obtain some desired quantities. This approach is used in the analysis of electric machines. When a current is applied to an object, there will be a magnetic field generated, which will give rise to a flux. This can be described as

$$L = \frac{\psi}{I} = \frac{\mu \int_S \mathbf{B} \cdot d\mathbf{S}}{\int_{\mathcal{L}} \mathbf{B} \cdot d\mathbf{l}}, \tag{5.63}$$

If $\mathbf{B}$ is uniform, then L becomes

$$L = \frac{\mu B A}{B l} = \frac{\mu A}{l}, \tag{5.64}$$

where A is the total cross-sectional area of the object, and l is the total length (circumference) of the object. If we let

$$\mathcal{R} = \frac{1}{L} = \frac{l}{\mu A}, \tag{5.65}$$

then we get

$$\psi = \frac{1}{\mathcal{R}} I. \tag{5.66}$$

This equation tells us that, if we apply a current to an object, there will be a flux that runs through that object. This flux (output) will be proportional to the current (input) through the reluctance $\mathcal{R}$ (in H^{-1}).

Recall that when a potential V_0 is applied to an electric circuit, there will be a current I that flows inside the circuit given by $I = \frac{1}{R} V_0$, where R is the resistance. By direct analogy, when we apply a current to an object to get a flux, we can think of that object as a magnetic circuit, the current in the magnetic circuit as the potential in an electric circuit, the flux in the magnetic circuit as the current in an electric circuit, and the reluctance in the magnetic circuit as the resistance in an electric circuit. Therefore, all electric circuit analysis tools can be used in magnetic circuits to find the current or the flux.

Problems

5.1 The line $\{x = y = 0, z \in [0, l]\}$ carries a nonuniform current given by

$$I(z) = I_0 z^2,$$

where I_o is a constant. Find $\mathbf{H}$ at $(x, 0, 0)$ using Biot-Savart law. **Hint:** $\int \frac{u^2 du}{(u^2 + a^2)^{3/2}} = -\frac{u}{\sqrt{u^2 + a^2}} + \ln[u + \sqrt{a^2 + u^2}]$.

5.2 The solenoid $\{\rho = a, z \in [-l, l]\}$ has N turns, and carries a surface current

$$\mathbf{K} = K_0(\cos\alpha\, \hat{\boldsymbol{\phi}} + \sin\alpha\, \hat{\mathbf{z}}),$$

where K_0 and α are constants. Find $\mathbf{H}$ at $(0, 0, z)$ using Biot-Savart law.

5.3 Plane $z = 0$ carries $\mathbf{K}_1 = \hat{\mathbf{x}}$ and plane $z = l$ carries $\mathbf{K}_2 = -\hat{\mathbf{x}}$. Find $\mathbf{H}$ in (a) $-\infty < z < 0$, (b) $0 < z < l$, and (c) $l < z < \infty$.

5.4 A cylinder of radius a carries $\mathbf{M} = M_0 \rho \hat{\boldsymbol{\phi}}$, where M_o is a constant. Determine $\mathbf{J}_{\text{int}}$ and $\mathbf{K}_{\text{int}}$. From these currents and using Ampere law, determine $\mathbf{H}$ inside and outside the cylinder.

5.5 Two cylinders of radii a and b $(b > a)$ are coaxial. The cylinder of radius a carries a volume current density $\mathbf{J} = \frac{I_1}{\pi a^2}\hat{\mathbf{z}}$ and the cylinder of radius b carries a surface current density $\mathbf{K} = \frac{I_2}{2\pi b}\hat{\mathbf{z}}$. The region between the two cylinders is isotropic magnetic with χ_m. (a) Find $\mathbf{H}$ in all regions, and (b) locate and obtain $\mathbf{J}_{\text{int}}$ and $\mathbf{K}_{\text{int}}$.

5.6 Two mediums are separated by the plane $2z - x + 2y - 2 = 0$. Medium 1, occupied by $2z - x + 2y - 2 < 0$, has $\mu_1 = 10\mu_0$. Medium 2, occupied by $2z - x + 2y - 2 > 0$, has $\mu_2 = 5\mu_0$. At the boundary, a surface current exists, given by $\mathbf{K} = \hat{\mathbf{x}} - 2\hat{\mathbf{y}}$. Given $\mathbf{H}_1 = -2\hat{\mathbf{x}} + 6\hat{\mathbf{y}} + 4\hat{\mathbf{z}}$, find $\mathbf{H}_2$.

5.7 A hemisphere $\{r = a, \theta \in [0, \pi/2]\}$ carries a nonuniform magnetization $\mathbf{M} = M_0 \sin\theta\, \hat{\mathbf{r}}$. Locate and obtain $\mathbf{J}_{\text{int}}$ and $\mathbf{K}_{\text{int}}$.

Appendix

The following computer program can be used to produce streamline plots of Ex. 5.2.

```
\[
\bUnALT{}Mu]o = 4 \[Pi] * 10^-7;
a = 1;
i [\[Phi]_] := 4 \[Pi] {-Sin[\[Phi]], Cos[\[Phi]], 0} ;
Gf[x_, y_, z_, \[Rho]p_, \[Phi]p_, zp_] := \[Mu]o/(4 \[Pi])
    1/((x - \[Rho]p Cos[\[Phi]p])^2 + (y - \[Rho]p Sin[\[Phi]p])^2 + \
(z - zp)^2)^(1/2);
Avec[x_, y_, z_] :=
  NIntegrate[
   i [\[Phi]p] Gf[x, y, z, a, \[Phi]p, 0], {\[Phi]p, 0, 2 \[Pi]}];
Hfield[x_, y_, z_] = 1/\[Mu]o Curl[Avec[x, y, z], {x, y, z}];
Hx[x_, y_, z_] := Hfield[x, y, z] . {1, 0, 0};
Hy[x_, y_, z_] := Hfield[x, y, z] . {0, 1, 0};
Hz[x_, y_, z_] := Hfield[x, y, z] . {0, 0, 1};
data = Table[{{x, y, z}, {Hx[x, y, z], Hy[x, y, z],
     Hz[x, y, z]}}, {x, -2.1, 2, 0.2}, {y, -2.1, 2, 0.2}, {z, -2.1, 2,
     0.2}];\eUnALT{}
\]aa = 2;
Fig = Show[
  Graphics3D@{Arrow[{{0, 0, 0}, {2 aa, 0, 0}}]},
  Graphics3D@{Arrow[{{0, 0, 0}, {0, 2 aa, 0}}]},
  Graphics3D@{Arrow[{{0, 0, 0}, {0, 0, 2 aa}}]},
  Graphics3D@{Text[x, {2 aa + 0.5, 0, 0}]},
  Graphics3D@{Text[y, {0, 2 aa + 0.5, 0}]},
  Graphics3D@{Text[z, {0, 0, 2 aa + 0.5}]}, ListStreamPlot3D[data,
   ImageSize -> 72*9, AspectRatio -> 1, StreamColorFunction -> None,
   StreamStyle -> {Blue}, StreamMarkers -> "Arrow",
   StreamScale -> Full, StreamPoints -> 30,
   Epilog -> {Inset[Style["", FontSize -> 60], Scaled[{0.1, 0.1}]]}],
  ParametricPlot3D[{a Cos[t], a Sin[t], 0}, {t, 0, 2 \[
\bUnALT{}Pi]},
   PlotStyle -> {FontFamily -> "Times New Roman", 50, Red}],
  Boxed -> False, BaseStyle -> FontSize -> 40]
```

References

1. J.D. Jackson, *Classical Electrodynamics* (Wiley, 2007)
2. P.M. Morse, H. Feshbach, *Methods of Theoretical Physics* (McGraw-Hill, 1953)
3. A. Zangwill, *Modern Electrodynamics* (Cambridge University Press, 2013)
4. C.A. Balanis, *Engineering Electromagnetics* (Wiley, 2012)
5. D.J. Griffiths, *Introduction to Electrodynamics* (Pearson, 2017)

Part III
Time-Dependent Electromagnetics

Now that electrostatics and magnetostatics have been discussed, this part considers time-dependent electromagnetics in 5 chapters. In order to facilitate the presentation, we consider source-free regions only in this part. Chapter 6 presents the transition from time independent to time dependent regimes. Chapters 7 and 8 discuss the propagation of electromagnetic fields in unbounded space, and in the presence of a boundary of an infinite extent, respectively. Finally, Chaps. 9 and 10 discuss propagation of electromagnetic fields in bounded structures, better known as transmission lines and waveguides.

Quasi-static and Time-Dependent Fields 6

Before turning to time-dependent electromagnetics, there exists a transitional subject between time-dependent and time-independent electromagnetics, called quasi statics. Quasi statics includes quasi electrostatics and quasi magnetostatics, the later which includes Faraday law. Due to its various applications in electric machines, we only discuss quasi magnetostatics in Sect. 6.1. We then discuss time-dependent electromagnetics in Sect. 6.2 covering the Poynting theorem, which describes the conservation of energy in electromagnetics.

6.1 Quasi Statics

In a source-free region, time-domain Maxwell equations are

$$\left.\begin{aligned} \nabla \times \mathscr{E}(\mathbf{r}, t) + \frac{\partial \mathscr{B}(\mathbf{r}, t)}{\partial t} &= \mathbf{0} \\ \nabla \times \mathscr{H}(\mathbf{r}, t) - \frac{\partial \mathscr{D}(\mathbf{r}, t)}{\partial t} &= \mathbf{0} \\ \nabla \bullet \mathscr{D}(\mathbf{r}, t) &= 0 \\ \nabla \bullet \mathscr{B}(\mathbf{r}, t) &= 0 \end{aligned}\right\}. \tag{6.1}$$

When fields are time independent, then both $\frac{\partial \mathscr{B}}{\partial t}$ and $\frac{\partial \mathscr{D}}{\partial t}$ are zero, which brings us to the static regime discussed in Part II. If it happens that fields are weakly time dependent in a way that $\frac{\partial \mathscr{B}}{\partial t}$ is more significant than $\frac{\partial \mathscr{D}}{\partial t}$, then we can ignore $\frac{\partial \mathscr{D}}{\partial t}$. Such a regime is called quasi magnetostatics. Similarly, if it happens that fields are weakly time dependent in a way that $\frac{\partial \mathscr{D}}{\partial t}$ is more significant than $\frac{\partial \mathscr{B}}{\partial t}$, then we can ignore $\frac{\partial \mathscr{B}}{\partial t}$. Such a regime is called quasi

H. M. Alkhoori, *Concise Introduction to Electromagnetic Fields*, Synthesis Lectures on Electromagnetics, https://doi.org/10.1007/978-3-031-60331-0_6

electrostatics. Here, we only cover quasi magnetostatics. Quasi electrostatics can be found elsewhere in the literature [1].

6.1.1 Quasi Magnetostatics

In quasi magnetostatics, Maxwell equations become

$$\left.\begin{aligned} \nabla \times \boldsymbol{\mathscr{E}}(\mathbf{r},t) + \frac{\partial \boldsymbol{\mathscr{B}}(\mathbf{r},t)}{\partial t} &= \mathbf{0} \\ \nabla \times \boldsymbol{\mathscr{H}}(\mathbf{r},t) &= \mathbf{0} \\ \nabla \bullet \boldsymbol{\mathscr{D}}(\mathbf{r},t) &= 0 \\ \nabla \bullet \boldsymbol{\mathscr{B}}(\mathbf{r},t) &= 0 \end{aligned}\right\}. \tag{6.2}$$

Equation (6.2)$_1$ is known as Faraday law in differential form. Here, we obtain the integral form as follows. On integrating both sides of Eq. (6.1)$_1$ over a surface $\mathcal{S}$, one gets

$$\int_{\mathcal{S}} \nabla \times \boldsymbol{\mathscr{E}}(\mathbf{r},t) \bullet d\mathbf{S} = -\int_{\mathcal{S}} \frac{\partial \boldsymbol{\mathscr{B}}(\mathbf{r},t)}{\partial t} \bullet d\mathbf{S}. \tag{6.3}$$

On making use of Stokes theorem on the left side and assuming that $\mathcal{S}$ is stationary, one gets

$$\oint_{\mathcal{L}} \boldsymbol{\mathscr{E}}(\mathbf{r},t) \bullet d\mathbf{l} = -\frac{d}{dt}\int_{\mathcal{S}} \boldsymbol{\mathscr{B}}(\mathbf{r},t) \bullet d\mathbf{S}. \tag{6.4}$$

This is the integral form of Faraday law. If the surface $\mathcal{S}$ is moving with a velocity $\mathbf{v}$, then it can be shown that Eq. (6.4) becomes

$$\oint_{\mathcal{L}} \boldsymbol{\mathscr{E}}(\mathbf{r},t) \bullet d\mathbf{l} = -\int_{\mathcal{S}} \left[\frac{\partial \boldsymbol{\mathscr{B}}(\mathbf{r},t)}{\partial t} + \nabla \times (\boldsymbol{\mathscr{B}}(\mathbf{r},t) \times \mathbf{v})\right] \bullet d\mathbf{S}. \tag{6.5}$$

It should be emphasized that by properly switching the coordinates (Galilean transformation for nonrelativistic motion, or Lorentz transformation for relativistic motion), the motion of the surface can be treated without the term $\boldsymbol{\mathscr{B}}(\mathbf{r},t) \times \mathbf{v}$ [1–3].

A circuit version of Faraday law can be obtained as follows. On making

$$\mathcal{E}(t) = \oint_{\mathcal{L}} \boldsymbol{\mathscr{E}}(\mathbf{r},t) \bullet d\mathbf{l}, \tag{6.6}$$

and

$$\psi(t) = \int_{\mathcal{S}} \boldsymbol{\mathscr{B}}(\mathbf{r},t) \bullet d\mathbf{S}, \tag{6.7}$$

Equation (6.4) becomes

$$\mathcal{E}(t) = -\frac{d\psi(t)}{dt}. \tag{6.8}$$

Fig. 6.1 For Example 6.1

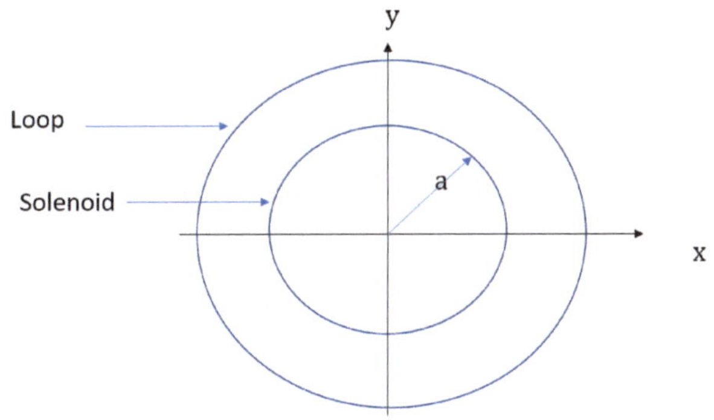

The term $\mathcal{E}(t)$ is called the electromotive force. Faraday law shows that if a field $\mathscr{B}$ passes through a loop $\mathcal{L}$, then an electromotive force $\mathcal{E}$ will be induced around the loop if $\mathscr{B}$ is time dependent, or if the loop moves with a velocity $\mathbf{v}$. Notice that either actions gives a time-dependent flux $\psi(t)$ responsible for inducing $\mathcal{E}(t)$ around the loop. Since there is an electromotive force, then, there will be an induced current flowing around the loop. This induced current will flow in a direction such that it gives another $\mathscr{B}$ field that opposes the original $\mathscr{B}$ field that caused the induced current to follow. This is the reason for the minus sign in Eq. (6.8), and it is called Len'z law.

Example 6.1 An infinite solenoid of radius a carries a slowly time-varying current $I(t)$. A loop coaxial with the solenoid and of a radius larger than a is placed (see Fig. 6.1). Determine $\mathcal{E}$ around the loop.

Solution After noting that **H** still satisfies the static version of Maxwell equations, we get

$$\mathscr{B}(t) = \begin{cases} \mu_0 n I(t)\hat{\mathbf{z}}, & 0 < \rho < a, \\ \mathbf{0}, & a < \rho < \infty, \end{cases}$$

we find $\mathcal{E}$ as

$$\mathcal{E} = -\int_S \left[\frac{\partial \mathscr{B}}{\partial t} + \nabla \times (\mathscr{B} \times \mathbf{v})\right] \cdot d\mathbf{S} = -\int_S \frac{\partial \mathscr{B}}{\partial t} \cdot d\mathbf{S}$$

$$= -\int_0^{2\pi}\int_0^{a} \mu_0 n \frac{dI(t)}{dt} \rho d\rho d\phi - \int_0^{2\pi}\int_a^{\rho} 0 \rho d\rho d\phi = -\mu_0 n \pi a^2 \frac{dI(t)}{dt}$$

◁

Example 6.2 A rectangular loop $\{x \in [0, a], y \in [0, b]\}$ is exposed to a field $\mathscr{B} = B_0 x \hat{\mathbf{z}}$, where B_0 is a constant. The loop moves with a velocity $\mathbf{v} = v_0 \hat{\mathbf{x}}$. Determine $\mathcal{E}$.

Solution

$$\mathcal{E} = -\int_S \left[\frac{\partial \mathcal{B}}{\partial t} + \nabla \times (\mathcal{B} \times \mathbf{v})\right] \bullet d\mathbf{S} = -\int_S \nabla \times (\mathcal{B} \times \mathbf{v}) \bullet d\mathbf{S}.$$

Since $\mathcal{B} \times \mathbf{v} = B_0 v_0 x \hat{\mathbf{y}}$, we get $\nabla \times (\mathcal{B} \times \mathbf{v}) = B_0 v_0 \hat{\mathbf{z}}$. Then,

$$\mathcal{E} = B_0 v_0 \int_0^a \int_0^b \hat{\mathbf{z}} \bullet \hat{\mathbf{z}} dx dy = -B_0 v_0 ab$$

◁

6.2 Electromagnetics and the Poynting Theorem

As stated in Chap. 2, the ultimate sources of electromagnetic fields are the electric and magnetic currents. In this section, we discuss conservation of energy theorem in electromagnetics, better known as the Poynting theorem, whereby energy transfer from sources to fields is illustrated. To facilitate the presentation, let us ignore the magnetic currents in our derivation. Also, let us suppress the spatial/temporal dependence from the argument of the fields and source for the sake of brevity.

6.2.1 Poynting Theorem

Let a current source $\mathcal{J}$ occupy a region $\mathcal{V}$ bounded by a closed surface $\mathcal{S}$. The identity of the medium is irrelevant, but it must be non dispersive. If a medium is dissipative, then there will exist a current $\mathcal{J}_d$ that accounts for the dissipation. Maxwell curl equations then become

$$\left.\begin{aligned} \nabla \times \mathcal{E} + \frac{\partial \mathcal{B}}{\partial t} &= \mathbf{0} \\ \nabla \times \mathcal{H} - \frac{\partial \mathcal{D}}{\partial t} - \mathcal{J}_d &= \mathcal{J} \end{aligned}\right\}. \tag{6.9}$$

If we dot multiply Eq. $(6.9)_1$ by $\mathcal{H}$, and Eq. $(6.9)_2$ by $\mathcal{E}$, we get

$$\left.\begin{aligned} \mathcal{H} \bullet (\nabla \times \mathcal{E}) + \mathcal{H} \bullet \frac{\partial \mathcal{B}}{\partial t} &= \mathbf{0} \\ \mathcal{E} \bullet (\nabla \times \mathcal{H}) - \mathcal{E} \bullet \frac{\partial \mathcal{D}}{\partial t} - \mathcal{E} \bullet \mathcal{J}_d &= \mathcal{E} \bullet \mathcal{J} \end{aligned}\right\}. \tag{6.10}$$

Subtracting both equations, we get

$$\mathcal{H} \bullet (\nabla \times \mathcal{E}) - \mathcal{E} \bullet (\nabla \times \mathcal{H}) + \mathcal{E} \bullet \frac{\partial \mathcal{D}}{\partial t} + \mathcal{H} \bullet \frac{\partial \mathcal{B}}{\partial t} + \mathcal{E} \bullet \mathcal{J}_d = -\mathcal{E} \bullet \mathcal{J}. \tag{6.11}$$

After noting that

$$\mathscr{H} \cdot (\nabla \times \mathscr{E}) - \mathscr{E} \cdot (\nabla \times \mathscr{H}) = \nabla \cdot (\mathscr{E} \times \mathscr{H}), \tag{6.12}$$

and after defining the vector

$$\mathscr{W} = \mathscr{E} \times \mathscr{H} \tag{6.13}$$

as the Poynting vector (in W m^{-2}), one gets

$$\nabla \cdot \mathscr{W} + \mathscr{E} \cdot \frac{\partial \mathscr{D}}{\partial t} + \mathscr{H} \cdot \frac{\partial \mathscr{B}}{\partial t} + \mathscr{E} \cdot \mathscr{J}_d = -\mathscr{E} \cdot \mathscr{J} \tag{6.14}$$

Equation (6.14) is the Poynting theorem in differential form.

To derive the corresponding integral form, let us integrate Eq. (6.14) over the region $\mathcal{V}$. Then,

$$\int_{\mathcal{V}} \nabla \cdot \mathscr{W} \, d\mathcal{V} + \int_{\mathcal{V}} \left(\mathscr{E} \cdot \frac{\partial \mathscr{D}}{\partial t} + \mathscr{H} \cdot \frac{\partial \mathscr{B}}{\partial t} \right) d\mathcal{V} + \int_{\mathcal{V}} \mathscr{E} \cdot \mathscr{J}_d \, d\mathcal{V} = - \int_{\mathcal{V}} \mathscr{E} \cdot \mathscr{J} \, d\mathcal{V}. \tag{6.15}$$

On making use of the divergence theorem on the first term on the left side, one gets

$$\oint_S \mathscr{W} \cdot d\mathbf{S} + \int_{\mathcal{V}} \left(\mathscr{E} \cdot \frac{\partial \mathscr{D}}{\partial t} + \mathscr{H} \cdot \frac{\partial \mathscr{B}}{\partial t} \right) d\mathcal{V} + \int_{\mathcal{V}} \mathscr{E} \cdot \mathscr{J}_d \, d\mathcal{V} = - \int_{\mathcal{V}} \mathscr{E} \cdot \mathscr{J} \, d\mathcal{V}, \tag{6.16}$$

where S is the surface enclosing $\mathcal{V}$. Equation (6.16) is the Poynting theorem in differential form.

In order to understand the Poynting theorem, let us denote

$$P_{EM} = \oint_S \mathscr{W} \cdot d\mathbf{S} \tag{6.17}$$

as the exiting power,

$$P_d = \int_{\mathcal{V}} \mathscr{E} \cdot \mathscr{J}_d \, d\mathcal{V} \tag{6.18}$$

as the dissipated power, and

$$P_s = \int_{\mathcal{V}} \mathscr{E} \cdot \mathscr{J} \, d\mathcal{V} \tag{6.19}$$

as the supplied power. The terms $\mathscr{E} \cdot \frac{\partial \mathscr{D}}{\partial t}$ and $\mathscr{H} \cdot \frac{\partial \mathscr{B}}{\partial t}$ can be interpreted as energy densities as follows. In free space, for instance, we have

$$\mathscr{E} \cdot \frac{\partial \mathscr{D}}{\partial t} = \varepsilon_0 \mathscr{E} \cdot \frac{\partial \mathscr{E}}{\partial t} = \frac{1}{2} \varepsilon_0 \frac{\partial \mathscr{E}^2}{\partial t} = \frac{\partial}{\partial t} \left(\frac{1}{2} \varepsilon_0 \mathscr{E}^2 \right) \tag{6.20}$$

and

$$\mathscr{H} \cdot \frac{\partial \mathscr{B}}{\partial t} = \mu_0 \mathscr{H} \cdot \frac{\partial \mathscr{H}}{\partial t} = \frac{1}{2} \mu_0 \frac{\partial \mathscr{H}^2}{\partial t} = \frac{\partial}{\partial t} \left(\frac{1}{2} \mu_0 \mathscr{H}^2 \right). \tag{6.21}$$

Therefore,

$$\int_{\mathcal{V}} \left(\boldsymbol{\mathscr{E}} \cdot \frac{\partial \boldsymbol{\mathscr{D}}}{\partial t} + \boldsymbol{\mathscr{H}} \cdot \frac{\partial \boldsymbol{\mathscr{B}}}{\partial t} \right) d\mathcal{V} = \frac{d}{dt} \int_{\mathcal{V}} \left(\frac{1}{2}\varepsilon_0 \mathscr{E}^2 + \frac{1}{2}\mu_0 \mathscr{H}^2 \right) d\mathcal{V} = \frac{d}{dt}(W_e + W_m), \tag{6.22}$$

where W_e is the electric energy and W_m is the magnetic energy. Although that was shown in free space, it holds in any non-dispersive medium. Therefore, Eq. (6.16) becomes

$$P_{EM} + \frac{d}{dt}(W_e + W_m) + P_d = -P_s. \tag{6.23}$$

Equation (6.23) states that the supplied power P_s to a region $\mathcal{V}$ is equal to (i) the power exiting the region P_{EM}, (ii) the rate of change of the electromagnetic energy $\frac{d}{dt}(W_e + W_m)$, and (iii) the power being dissipated P_d throughout the region. Hence, the Poynting theorem is nothing but conservation of power law.

Example 6.3 Given that

$$\boldsymbol{\mathscr{E}} = \mathscr{E}_o \cos(\omega t - kz)\hat{\mathbf{x}},$$

and

$$\boldsymbol{\mathscr{H}} = \mathscr{H}_o \cos(\omega t - kz)\hat{\mathbf{y}},$$

obtain $\boldsymbol{\mathscr{W}}(\mathbf{r}, t)$.

Solution

$$\boldsymbol{\mathscr{W}} = \boldsymbol{\mathscr{E}} \times \boldsymbol{\mathscr{H}} = \mathscr{E}_o \mathscr{H}_o \cos^2(\omega t - kz)\hat{\mathbf{z}}.$$

◁

6.2.2 Average and Complex Poynting Vectors

For two scalars

$$v(t) = \text{Re}\{V e^{j\omega t}\},$$

and

$$i(t) = \text{Re}\{I e^{j\omega t}\}, \tag{6.24}$$

we define the instantaneous power $p(t)$ as

$$p(t) = v(t)i(t). \tag{6.25}$$

Then, if we integrate $p(t)$ over an entire period $T = 2\pi/\omega$, we obtain the average power P as

$$P = \frac{1}{2}\text{Re}\{V I^*\} \tag{6.26}$$

where $*$ denotes the conjugate operator. Note that the quantity $\frac{1}{2} V I^*$ is called the complex power.

Similarly, for two vectors

$$\mathscr{E}(\mathbf{r},t) = \text{Re}\left\{\mathbf{E}(\mathbf{r})e^{j\omega t}\right\},$$

and

$$\mathscr{H}(\mathbf{r},t) = \text{Re}\left\{\mathbf{H}(\mathbf{r})e^{j\omega t}\right\}, \tag{6.27}$$

we already have defined the instantaneous Poynting vector $\mathscr{W}(\mathbf{r},t)$ as

$$\mathscr{W}(\mathbf{r},t) = \mathscr{E}(\mathbf{r},t) \times \mathscr{H}(\mathbf{r},t) \tag{6.28}$$

Now, if we integrate $\mathscr{W}(\mathbf{r},t)$ over an entire period $T = 2\pi/\omega$, we obtain the average Poynting vector $\mathbf{W}_{\text{ave}}(\mathbf{r})$ as

$$\mathbf{W}_{\text{ave}}(\mathbf{r}) = \frac{1}{2}\text{Re}\{\mathbf{E}(\mathbf{r}) \times \mathbf{H}^*(\mathbf{r})\}. \tag{6.29}$$

This leads us defining the complex Poynting vector $\mathbf{W}(\mathbf{r})$ as

$$\mathbf{W}(\mathbf{r}) = \frac{1}{2}\mathbf{E}(\mathbf{r}) \times \mathbf{H}^*(\mathbf{r}). \tag{6.30}$$

Given a surface $\mathcal{S}$, the total average power P_{ave} (the frequency-domain counterpart of P_{EM}) crossing that surface is obtained as

$$P_{\text{ave}} = \int_{\mathcal{S}} \mathbf{W}_{\text{ave}} \bullet d\mathbf{S}. \tag{6.31}$$

Example 6.4 Given

$$\mathbf{E} = 0.1e^{-j2z}\hat{\mathbf{x}},$$

and

$$\mathbf{H} = 0.02e^{-j2z}\hat{\mathbf{y}},$$

obtain (a) $\mathbf{W}_{\text{ave}}$, (b) P_{ave} when $\mathcal{S}$ is the plane $\{z = 0, x \in [0, 2], y \in [0, 3]\}$, and (c) P_{ave} when $\mathcal{S}$ is the plane $\{x = 0, y \in [0, 2], z \in [0, 3]\}$.

Solution

(a)

$$\mathbf{W}_{\text{ave}} = \frac{1}{2}\text{Re}\{\mathbf{E} \times \mathbf{H}^*\} = \frac{1}{2}\text{Re}\{0.1e^{-j2z}\hat{\mathbf{x}} \times 0.02e^{j2z}\hat{\mathbf{y}}\} = 10^{-3}\,\hat{\mathbf{z}}.$$

(b)

$$P_{\text{ave}} = \int_{\mathcal{S}} \mathbf{W}_{\text{ave}} \bullet d\mathbf{S} = \int_0^2 \int_0^3 10^{-3}\hat{\mathbf{z}} \bullet \hat{\mathbf{z}} dx\, dy = 6 \times 10^{-3}.$$

(c)

$$P_{\text{ave}} = \int_0^2 \int_0^3 10^{-3} \hat{\mathbf{z}} \bullet \hat{\mathbf{x}} dy\, dz = 0.$$

◁

Problems

6.1 A rectangular loop $\{x \in [0, a], y \in [0, b], z = 0\}$ is exposed to a field $\mathscr{B} = B_0 y \hat{\mathbf{z}}$, where B_0 is a constant. The loop moves with a velocity $\mathbf{v} = v_0 \hat{\mathbf{y}}$. Determine $\mathcal{E}$.

6.2 Two coaxial cylinders of radii a and b ($b > a$) have the following field expression

$$\mathscr{B}(t) = \begin{cases} \mathbf{0}, & 0 < \rho < a, \\ \mu_o I(t) \rho\, \hat{\mathbf{z}}, & a < \rho < b, \\ \dfrac{\mu_o I(t)}{\rho} \hat{\mathbf{z}}, & b < \rho < \infty. \end{cases}$$

Obtain $\mathcal{E}$ around a coaxial loop of radius ρ placed (i) inside the inner cylinder $0 < \rho < a$, (ii) between the two cylinders $a < \rho < b$, and (iii) outside the outer cylinder $b < \rho < \infty$.

6.3 Given

$$\mathbf{E} = E_o \sin x e^{-jz} \hat{\mathbf{y}},$$

and

$$\mathbf{H} = H_o \sin x e^{-jz} \hat{\mathbf{x}},$$

obtain (a) $\mathbf{W}_{\text{ave}}$, and (b) P_{ave} when $\mathcal{S}$ is the plane $\{x \in [0, \pi], y \in [0, \pi], z = 0\}$.

References

1. A. Zangwill, *Modern Electrodynamics* (Cambridge University Press, 2013)
2. J.D. Jackson, *Classical Electrodynamics* (Wiley, 2007)
3. E.J. Rothwell, M.J. Cloud, *Electromagnetics*, 3rd edn. (CRC Press, 2018)

7 Plane-Wave Propagation

Electromagnetic phenomenon is described by Maxwell equations, and solutions of **E** and **H** from their sources can be obtained on solving these equations; a procedure that will be discussed in Part IV. In this chapter, we limit our discussion to source-free regions. Also, let us initially assume that the region under consideration is free space. Extension to isotropic dielectric-magnetic medium will be easily made in frequency domain. Under these assumptions, Maxwell equations become

$$\left.\begin{aligned} \nabla \times \mathscr{E}(\mathbf{r}, t) &= -\mu_0 \frac{\partial \mathscr{H}(\mathbf{r}, t)}{\partial t} \\ \nabla \times \mathscr{H}(\mathbf{r}, t) &= \varepsilon_0 \frac{\partial \mathscr{E}(\mathbf{r}, t)}{\partial t} \\ \nabla \bullet \mathscr{E}(\mathbf{r}, t) &= 0 \\ \nabla \bullet \mathscr{H}(\mathbf{r}, t) &= 0 \end{aligned}\right\}. \tag{7.1}$$

Source-free solutions to Eq. (7.1) are called plane waves. Notice that we mentioned that Maxwell equations are solved in a region free of sources, implying that the sources must exist somewhere outside our region of interest. In Sect. 7.1, we derive the homogeneous vector and scalar wave equations from source-free Maxwell equations. After solving the wave equation in an unbounded region, solution of the source-free Maxwell equations (i.e., plane waves) in an isotropic dielectric-magnetic medium is discussed in Sect. 7.2, with a restriction to frequency domain. Dissipation effect is considered in Sect. 7.3, and finally, polarization state of plane waves is discussed in Sect. 7.4.

H. M. Alkhoori, *Concise Introduction to Electromagnetic Fields*, Synthesis Lectures on Electromagnetics, https://doi.org/10.1007/978-3-031-60331-0_7

7.1 Source-Free Maxwell Equations

7.1.1 The Homogeneous Wave Equation

The four equations in Eq. (7.1) constitute a set of coupled, first-order, vector partial differential equations. To find $\boldsymbol{\mathscr{E}}(\mathbf{r}, t)$ and $\boldsymbol{\mathscr{H}}(\mathbf{r}, t)$, we need to decouple these equations. This is implemented next to obtain an equation involving $\boldsymbol{\mathscr{E}}(\mathbf{r}, t)$ only. A similar procedure can be adopted to obtain an equation involving $\boldsymbol{\mathscr{H}}(\mathbf{r}, t)$ only.

Let us take the curl of Eq. $(7.1)_1$. This gives

$$\nabla \times [\nabla \times \boldsymbol{\mathscr{E}}(\mathbf{r}, t)] = -\mu_0 \nabla \times \frac{\partial \boldsymbol{\mathscr{H}}(\mathbf{r}, t)}{\partial t}. \tag{7.2}$$

Then, from Eq. $(7.1)_2$, we get

$$\nabla \times [\nabla \times \boldsymbol{\mathscr{E}}(\mathbf{r}, t)] = -\mu_0 \varepsilon_0 \frac{\partial^2 \boldsymbol{\mathscr{E}}(\mathbf{r}, t)}{\partial t^2}. \tag{7.3}$$

To simplify further, using

$$\nabla \times [\nabla \times \boldsymbol{\mathscr{E}}(\mathbf{r}, t)] = \nabla[\nabla \bullet \boldsymbol{\mathscr{E}}(\mathbf{r}, t)] - \nabla^2 \boldsymbol{\mathscr{E}}(\mathbf{r}, t), \tag{7.4}$$

together with Eq. $(7.1)_3$, we get

$$-\nabla^2 \boldsymbol{\mathscr{E}}(\mathbf{r}, t) = -\mu_0 \varepsilon_0 \frac{\partial^2 \boldsymbol{\mathscr{E}}(\mathbf{r}, t)}{\partial t^2}, \tag{7.5}$$

which becomes

$$\nabla^2 \boldsymbol{\mathscr{E}}(\mathbf{r}, t) - \mu_0 \varepsilon_0 \frac{\partial^2 \boldsymbol{\mathscr{E}}(\mathbf{r}, t)}{\partial t^2} = \mathbf{0}. \tag{7.6}$$

This second-order partial differential equation is called the homogeneous vector wave equation. Since

$$\boldsymbol{\mathscr{E}}(\mathbf{r}, t) = \mathscr{E}_x(\mathbf{r}, t)\, \hat{\mathbf{x}} + \mathscr{E}_y(\mathbf{r}, t)\, \hat{\mathbf{y}} + \mathscr{E}_z(\mathbf{r}, t)\, \hat{\mathbf{z}}, \tag{7.7}$$

solving this vector wave equation is equivalent to solving the following three scalar wave equations,

$$\nabla^2 \mathscr{E}_x(\mathbf{r}, t) - \mu_0 \varepsilon_0 \frac{\partial^2 \mathscr{E}_x(\mathbf{r}, t)}{\partial t^2} = 0, \tag{7.8}$$

$$\nabla^2 \mathscr{E}_y(\mathbf{r}, t) - \mu_0 \varepsilon_0 \frac{\partial^2 \mathscr{E}_y(\mathbf{r}, t)}{\partial t^2} = 0, \tag{7.9}$$

and

$$\nabla^2 \mathscr{E}_z(\mathbf{r}, t) - \mu_0 \varepsilon_0 \frac{\partial^2 \mathscr{E}_z(\mathbf{r}, t)}{\partial t^2} = 0. \tag{7.10}$$

By adopting a similar procedure, one can show that

$$\nabla^2 \boldsymbol{\mathscr{H}}(\mathbf{r},t) - \mu_0 \varepsilon_0 \frac{\partial^2 \boldsymbol{\mathscr{H}}(\mathbf{r},t)}{\partial t^2} = \mathbf{0}, \tag{7.11}$$

which is equivalent to solving

$$\nabla^2 \mathscr{H}_x(\mathbf{r},t) - \mu_0 \varepsilon_0 \frac{\partial^2 \mathscr{H}_x(\mathbf{r},t)}{\partial t^2} = 0, \tag{7.12}$$

$$\nabla^2 \mathscr{H}_y(\mathbf{r},t) - \mu_0 \varepsilon_0 \frac{\partial^2 \mathscr{H}_y(\mathbf{r},t)}{\partial t^2} = 0, \tag{7.13}$$

and

$$\nabla^2 \mathscr{H}_z(\mathbf{r},t) - \mu_0 \varepsilon_0 \frac{\partial^2 \mathscr{H}_z(\mathbf{r},t)}{\partial t^2} = 0. \tag{7.14}$$

7.1.2 Solution of the Scalar Wave Equation

Since each Cartesian component of $\boldsymbol{\mathscr{E}}(\mathbf{r},t)$ satisfies

$$\nabla^2 \mathscr{E}(\mathbf{r},t) - \mu_0 \varepsilon_0 \frac{\partial^2 \mathscr{E}(\mathbf{r},t)}{\partial t^2} = 0, \tag{7.15}$$

solving for one component only is sufficient. For simplicity, let us assume $\mathscr{E}(\mathbf{r},t) \equiv \mathscr{E}(z,t)$; we will relax this assumption later. Then, we solve

$$\frac{\partial^2 \mathscr{E}(z,t)}{\partial z^2} - \mu_0 \varepsilon_0 \frac{\partial^2 \mathscr{E}(z,t)}{\partial t^2} = 0. \tag{7.16}$$

Assuming time-harmonic variation

$$\mathscr{E}(z,t) = \mathrm{Re}\left\{E(z)e^{j\omega t}\right\}, \tag{7.17}$$

we get

$$\frac{d^2 E(z)}{dz^2} + k_0^2 E(z) = 0 \tag{7.18}$$

where $k_0 = \omega\sqrt{\mu_0 \varepsilon_0}$ is the free-space wave number (in rad m^{-1}). Therefore, we get a second-order ordinary differential equation with solution

$$E(z) = E_o^+ e^{-jk_0 z} + E_o^- e^{jk_0 z}, \tag{7.19}$$

where E_o^+ and E_o^- are arbitrary constants. The reason for the plus and minus sign superscripts shall become evident soon. In time domain, solution becomes

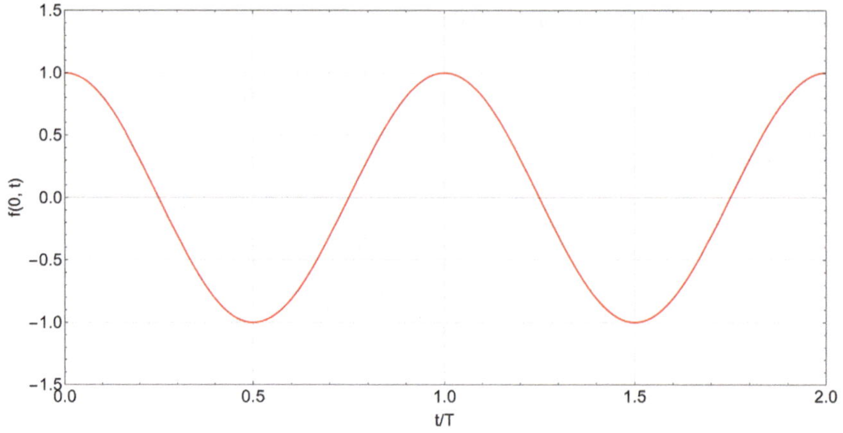

Fig. 7.1 $f(0, t)$ versus $t \in [0, 2]T$

$$\mathscr{E}(z, t) = E_o^+ \cos(\omega t - k_0 z) + E_o^- \cos(\omega t + k_0 z) . \tag{7.20}$$

Since this is the solution of the wave equation, each term in the solution can be called a wave function.

We next investigate properties of the wave function. Let $f(z, t) = \cos(\omega t - k_0 z)$, with $k_0 = 2\pi/\lambda_o$, where λ_o (in m) is called the free-space wavelength, and $\omega = 2\pi/T$, where T (in s) is called the period. Figure 7.1 shows a plot of $f(0, t)$ for $t \in [0, 2]T$. We see that the wave function repeats itself every T seconds. This leads to defining the linear frequency $f = 1/T$. Now, the plot of $f(z, 0)$ for $z \in [0, 2]\lambda_o$ is exactly similar to the plot in Fig. 7.1, except that the horizontal scale is changed to z/λ_o. In such a case, we see that the wave function repeats itself every λ_o meters. The distance λ_o can be thought of as the spatial period. Equivalently, we define the spatial angular frequency, or simply the wave number, as $k_0 = 2\pi/\lambda_o$. We next examine the variation against both space and time. Figure 7.2 shows plots of $f(z, t)$ versus $z \in [0, 2]\lambda_o$ when $t \in \{0, 1/8, 1/4\}T$. By examining the three curves, we see that as t increases, the wave function moves to the right (i.e., along the positive z axis) with a phase velocity $v_p = \lambda/T = \omega/k_0$ (in ms^{-1}). This motion is called propagation of the wave. So, we say $f(z, t)$ propagates along the positive z axis with a phase velocity v_p.

If we examine $f(z, t) = \cos(\omega t + k_0 z)$, we will find that it represents a wave propagating along the negative z axis with a phase velocity v_p. Therefore, the solution

$$\mathscr{E}(z, t) = E_o^+ \cos(\omega t - k_0 z) + E_o^- \cos(\omega t + k_0 z) . \tag{7.21}$$

is composed of two waves; one with an amplitude E_o^+ propagating along the positive z axis, and the other with an amplitude E_o^- propagating along the negative z axis. Both waves are propagating with the same phase velocity v_p.

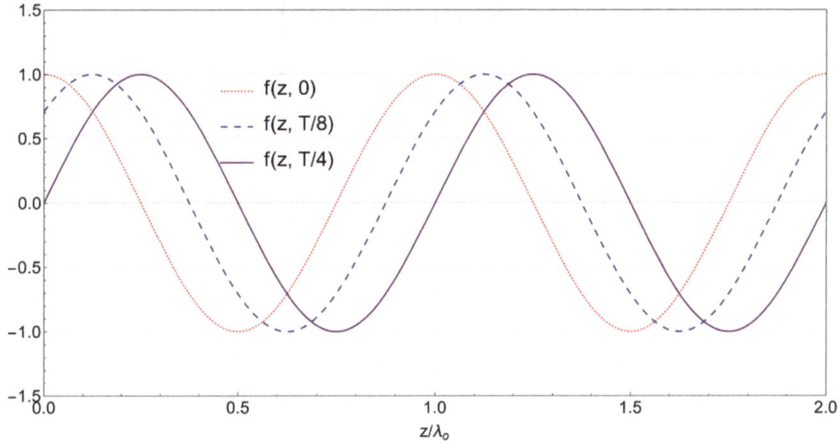

Fig. 7.2 $f(z,t)$ versus $z \in [0,2]\lambda_o$ when $t \in \{0, 1/8, 1/4\}T$

Example 7.1 Given that the wave

$$\mathscr{E}(z,t) = \cos(3 \times 10^8 t - k_0 z)$$

is propagating in free space with a phase velocity $v_p = 3 \times 10^8\ \mathrm{m\,s^{-1}}$, obtain (a) the wave amplitude, (b) the propagation direction, (c) the wave number k_0, and (d) the wavelength λ_o.

Solution (a) The amplitude is unity. (b) The wave is propagating along the positive z axis. (c) $k_0 = \omega/v_p = 1$. (d) $\lambda_o = 2\pi/k = 2\pi$. ◁

7.2 Solution of the Frequency-Domain, Source-Free Maxwell Equations

We have seen that Maxwell equations in free space lead to the vector wave equation. Then, in Cartesian coordinates, vector wave equation is equivalent to three scalar wave equations. We have solved the scalar wave equation in an unbounded domain and for time-harmonic temporal variation. However, we have not completely solved Maxwell equations yet. This is addressed in this section. In order to consider isotropic dielectric-magnetic medium, let us limit our discussion to frequency domain.

Frequency-domain, source-free Maxwell equations in an isotropic dielectric-magnetic medium are

$$\left.\begin{aligned}\nabla\times\mathbf{E}(\mathbf{r})&=-j\omega\mu\mathbf{H}(\mathbf{r})\\\nabla\times\mathbf{H}(\mathbf{r})&=j\omega\varepsilon\mathbf{E}(\mathbf{r})\\\nabla\bullet\mathbf{E}(\mathbf{r})&=0\\\nabla\bullet\mathbf{H}(\mathbf{r})&=0\end{aligned}\right\}. \tag{7.22}$$

For simplicity at this time, if we assume that the wave is propagating along the z axis with only an x component in $\mathbf{E}(z)$ [i.e., $\mathbf{E}(z)=E(z)\hat{\mathbf{x}}$], then according to the previous section's discussion, $E(z)$ will satisfy

$$\frac{d^2E(z)}{dz^2}+k^2E(z)=0, \tag{7.23}$$

where $k=k_0\sqrt{\mu_r}\sqrt{\varepsilon_r}$ is the medium wave number. Notice that $\sqrt{\varepsilon_r}\sqrt{\mu_r}$ can be written as $\sqrt{\varepsilon_r\mu_r}$ only when both ε_r and μ_r are real positive numbers. Then, the electric field phasor becomes

$$\mathbf{E}(z)=\left(E_o^+e^{-jkz}+E_o^-e^{jkz}\right)\hat{\mathbf{x}}. \tag{7.24}$$

Let us consider the positively-traveling wave only. Hence,

$$\mathbf{E}(z)=E_o^+e^{-jkz}\,\hat{\mathbf{x}}. \tag{7.25}$$

In order to relax the assumption $\mathbf{E}(z)=E(z)\hat{\mathbf{x}}$ later, we will rewrite $\mathbf{E}(z)$ using vector notation as

$$\mathbf{E}(z)=\mathbf{E}_o\,e^{-j\mathbf{k}\bullet\mathbf{r}} \tag{7.26}$$

where

$$\mathbf{E}_o=E_o^+\,\hat{\mathbf{x}} \tag{7.27}$$

can be called the polarization state vector, and

$$\mathbf{k}=k\,\hat{\mathbf{k}}=k\,\hat{\mathbf{z}}, \tag{7.28}$$

is called the propagation vector, with $\hat{\mathbf{k}}$ being the direction of the wave propagation. With this form of $\mathbf{E}(z)$, it will be shown (see Example 7.2) that $\nabla\equiv-j\mathbf{k}$. Hence, Maxwell equations become

$$\left.\begin{aligned}\mathbf{k}\times\mathbf{E}(\mathbf{r})&=\omega\mu\mathbf{H}(\mathbf{r})\\\mathbf{k}\times\mathbf{H}(\mathbf{r})&=-\omega\varepsilon\mathbf{E}(\mathbf{r})\\\mathbf{k}\bullet\mathbf{E}(\mathbf{r})&=0\\\mathbf{k}\bullet\mathbf{H}(\mathbf{r})&=0\end{aligned}\right\}. \tag{7.29}$$

Now, the magnetic field phasor $\mathbf{H}(\mathbf{r})$ can be found easily from Eq. (7.29) as

$$\mathbf{H}(\mathbf{r})=\frac{1}{\omega\mu}\mathbf{k}\times\mathbf{E}(\mathbf{r})=\frac{k}{\omega\mu}E_o^+e^{-jkz}\,\hat{\mathbf{y}}=\frac{1}{\sqrt{\mu/\varepsilon}}E_o^+e^{-jkz}\,\hat{\mathbf{y}}=H_o^+e^{-jkz}\,\hat{\mathbf{y}}, \tag{7.30}$$

where

$$H_o^+ = \frac{1}{\sqrt{\mu/\varepsilon}} E_o^+, \tag{7.31}$$

Notice that since E is in V m^{-1} and H is in A m^{-1}, the ratio between them is in Ω. Thus, we define the intrinsic impedance (or wave impedance) η as

$$\eta = \frac{E_o^+}{H_o^+} = \sqrt{\frac{\mu}{\varepsilon}} = \sqrt{\frac{\mu_r}{\varepsilon_r}}\eta_0, \tag{7.32}$$

where $\eta_0 = \sqrt{\mu_0/\varepsilon_0} = 120\pi$ is the free-space intrinsic impedance (in Ω). So, solutions of Maxwell equations can be written as

$$\left.\begin{aligned} \mathbf{E}(\mathbf{r}) &= \mathbf{E}_o e^{-j\mathbf{k}\bullet\mathbf{r}} \\ \mathbf{H}(\mathbf{r}) &= \frac{1}{\eta}\left(\hat{\mathbf{k}} \times \mathbf{E}_o\right) e^{-j\mathbf{k}\bullet\mathbf{r}} \end{aligned}\right\}. \tag{7.33}$$

The average Poynting vector $\mathbf{W}_{\text{ave}}$ for the wave can be obtained as

$$\mathbf{W}_{\text{ave}} = \frac{1}{2}\text{Re}\{\mathbf{E}(\mathbf{r}) \times \mathbf{H}(\mathbf{r})^*\} = \frac{1}{2\eta}\text{Re}\left\{\mathbf{E}_o \times \left(\hat{\mathbf{k}} \times \mathbf{E}_o^*\right)\right\}. \tag{7.34}$$

Using $\mathbf{A} \times (\mathbf{B} \times \mathbf{C}) = \mathbf{B}(\mathbf{C} \bullet \mathbf{A}) - \mathbf{C}(\mathbf{A} \bullet \mathbf{B})$ and after noting that $\hat{\mathbf{k}} \bullet \mathbf{E} = 0$, we get

$$\mathbf{W}_{\text{ave}} = \frac{|E_o|^2}{2\eta}\hat{\mathbf{k}}, \tag{7.35}$$

where

$$|E_o| = \sqrt{\mathbf{E}_o \bullet \mathbf{E}_o^*} = \sqrt{|E_{ox}|^2 + |E_{oy}|^2 + |E_{oz}|^2}. \tag{7.36}$$

Notice that Poynting vector is parallel to the propagation direction.

Example 7.2 Using the identity

$$\nabla \times (V\mathbf{A}) = V\nabla \times \mathbf{A} - \mathbf{A} \times \nabla V,$$

show that if

$$\mathbf{E}(\mathbf{r}) = \mathbf{E}_o e^{-j\mathbf{k}\bullet\mathbf{r}},$$

where $\mathbf{E}_o$ is a constant vector, then

$$\nabla \times \mathbf{E}(\mathbf{r}) = -j\mathbf{k} \times \mathbf{E}(\mathbf{r}).$$

Solution

$$\nabla \times \mathbf{E}(\mathbf{r}) = \nabla \times \left(\mathbf{E}_o e^{-j\mathbf{k}\bullet\mathbf{r}}\right) = e^{-j\mathbf{k}\bullet\mathbf{r}}\nabla \times \mathbf{E}_o - \mathbf{E}_o \times \nabla e^{-j\mathbf{k}\bullet\mathbf{r}}.$$

Since $\mathbf{E}_o$ is constant, $\nabla \times \mathbf{E}_o = \mathbf{0}$. Also, without loss of generality, if we let $\mathbf{k} = k\,\hat{\mathbf{x}}$, we have

$$\nabla e^{-j\mathbf{k}\bullet\mathbf{r}} = \nabla e^{-jkx} = \hat{\mathbf{x}}\frac{\partial}{\partial x}e^{-jkx} = -jke^{-jkx}\,\hat{\mathbf{x}} = -j\mathbf{k}e^{-jkx} = -j\mathbf{k}e^{-j\mathbf{k}\bullet\mathbf{r}}.$$

Note that this holds for any $\mathbf{k}$. Therefore,

$$\nabla \times \left(\mathbf{E}_o e^{-j\mathbf{k}\bullet\mathbf{r}}\right) = -j\mathbf{k} \times \mathbf{E}_o e^{-j\mathbf{k}\bullet\mathbf{r}} = -j\mathbf{k} \times \mathbf{E}(\mathbf{r}).$$

◁

So, we conclude that, if the spatial variation of a field is $e^{-j\mathbf{k}\bullet\mathbf{r}}$, the ∇ operator becomes the constant vector $-j\mathbf{k}$. That is,

$$\nabla \Rightarrow -j\mathbf{k}. \tag{7.37}$$

Remarks

With

$$\left.\begin{aligned} \mathbf{k} \times \mathbf{E}(\mathbf{r}) &= \omega\mu\mathbf{H}(\mathbf{r}) \\ \mathbf{k} \times \mathbf{H}(\mathbf{r}) &= -\omega\varepsilon\mathbf{E}(\mathbf{r}) \\ \mathbf{k}\bullet\mathbf{E}(\mathbf{r}) &= 0 \\ \mathbf{k}\bullet\mathbf{H}(\mathbf{r}) &= 0 \end{aligned}\right\} \tag{7.38}$$

in mind, the following holds true.

- An electromagnetic field in an isotropic dielectric-magnetic medium is a wave composed of an electric field $\mathbf{E}(\mathbf{r})$ and a magnetic field $\mathbf{H}(\mathbf{r})$.
- Both of $\mathbf{E}(\mathbf{r})$ and $\mathbf{H}(\mathbf{r})$ are orthogonal to each other, and their magnitudes and phases are related by the intrinsic impedance. The orthogonality between $\mathbf{E}(\mathbf{r})$ and $\mathbf{H}(\mathbf{r})$ can be derived directly from
$$\mathbf{k} \times \mathbf{E}(\mathbf{r}) = \omega\mu\mathbf{H}(\mathbf{r}) \tag{7.39}$$
by taking $\mathbf{E}(\mathbf{r})\,\bullet$ of both sides, which after using the identity $\mathbf{A}\bullet(\mathbf{B}\times\mathbf{C}) = \mathbf{B}\bullet(\mathbf{C}\times\mathbf{A})$ yields $\mathbf{E}(\mathbf{r})\bullet\mathbf{H}(\mathbf{r}) = 0$.
- Since an electromagnetic field is a wave, and a wave propagates, the field propagates along the direction of $\mathbf{k}$. The velocity of propagation v_p can be made as $v_p = c/n$, where $c = 1/\sqrt{\mu_0\varepsilon_0} = 3 \times 10^8\ \mathrm{ms}^{-1}$ is the speed of light and $n = \sqrt{\varepsilon_r}\sqrt{\mu_r}$ is the refractive index.
- From Eqs. $(7.38)_3$ and $(7.38)_4$, it can be seen that $\mathbf{k}$ is normal to each of $\mathbf{E}(\mathbf{r})$ and $\mathbf{H}(\mathbf{r})$. Thus, an electromagnetic wave can be visualized as shown in Fig. 7.3. Such a configuration is called Transverse Electromagnetic (TEM) configuration (or mode).

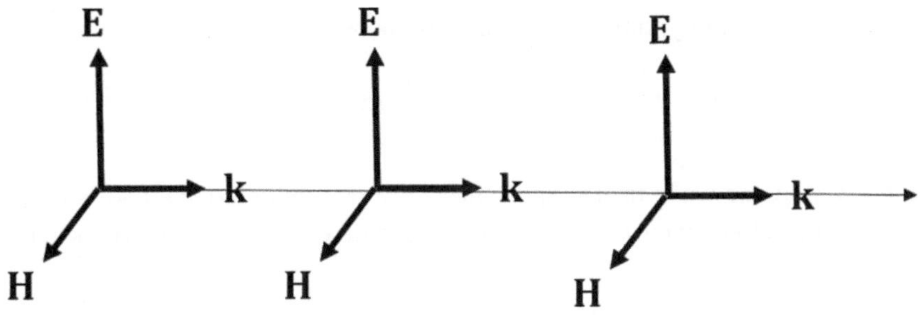

Fig. 7.3 A plane wave

- Due to the form of the electromagnetic wave derived in this section, if we plot **E**, **H**, and **k** on a plane, we will find that all these vectors are orthogonal to each other (**E** and **H** will lie on the plane, and **k** will be normal to the plane). If we move along the direction of propagation **k**, we will find that they are again orthogonal to each other, with the later plane being parallel to the former plane, as shown in Fig. 7.3. Such a wave is called a plane wave. If, in addition, the amplitudes of **E** and **H** are same on all planes, such wave is called a uniform plane wave.

Finally, remember that all these remarks hold in an isotropic dielectric-magnetic medium only.

Summary

Given a general propagation vector

$$\mathbf{k} = k_x\,\hat{\mathbf{x}} + k_y\,\hat{\mathbf{y}} + k_z\,\hat{\mathbf{z}}, \tag{7.40}$$

the corresponding electromagnetic field can be determined in an isotropic dielectric-magnetic medium from

$$\left.\begin{aligned} \mathbf{k}\times\mathbf{E}(\mathbf{r}) &= \omega\mu\mathbf{H}(\mathbf{r}) \\ \mathbf{k}\times\mathbf{H}(\mathbf{r}) &= -\omega\varepsilon\mathbf{E}(\mathbf{r}) \\ \mathbf{k}\bullet\mathbf{E}(\mathbf{r}) &= 0 \\ \mathbf{k}\bullet\mathbf{H}(\mathbf{r}) &= 0 \end{aligned}\right\}, \tag{7.41}$$

which yield

$$\mathbf{E}(\mathbf{r}) = \mathbf{E}_o e^{-j\mathbf{k}\bullet\mathbf{r}}, \tag{7.42}$$

with $\mathbf{E}_o$ satisfying $\mathbf{k}\bullet\mathbf{E}_o = 0$, and

$$\mathbf{H}(\mathbf{r}) = \frac{1}{\eta}\hat{\mathbf{k}}\times\mathbf{E}(\mathbf{r}). \tag{7.43}$$

The associated average Poynting vector can be found as

$$\mathbf{W}_{\text{ave}} = \frac{|E_o|^2}{2\eta}\,\hat{\mathbf{k}}. \tag{7.44}$$

Example 7.3 A plane wave in free space has $E_o = 1$, is x polarized, and is propagating along the y axis at a frequency $f = 300$ MHz. Obtain (a) $\mathbf{k}$, (b) $\mathbf{E}(\mathbf{r})$, (c) $\mathbf{H}(\mathbf{r})$, and (d) $\mathbf{W}_{\text{ave}}$.

Solution (a) $k = \omega/c = 2\pi$. Then,

$$\mathbf{k} = 2\pi\,\hat{\mathbf{y}}.$$

(b)

$$\mathbf{E}(\mathbf{r}) = e^{-j2\pi y}\,\hat{\mathbf{x}}.$$

(c)

$$\mathbf{H}(\mathbf{r}) = \frac{1}{\eta_0}\hat{\mathbf{k}} \times \mathbf{E}(\mathbf{r}) = \frac{1}{\eta_0}\left(\hat{\mathbf{k}} \times \mathbf{E}_o\right) e^{-j\mathbf{k}\bullet\mathbf{r}} = \frac{1}{120\pi}\left(\hat{\mathbf{y}} \times \hat{\mathbf{x}}\right) e^{-j2\pi y} = -\frac{1}{120\pi} e^{-j2\pi y}\,\hat{\mathbf{z}}.$$

(d)

$$\mathbf{W}_{\text{ave}} = \frac{|E_o|^2}{2\eta_0}\,\hat{\mathbf{k}} = \frac{1}{240\pi}\hat{\mathbf{y}}.$$

◁

Example 7.4 A plane wave in free space has

$$\mathbf{E}(\mathbf{r}) = (\hat{\mathbf{x}} - j\hat{\mathbf{y}})\,e^{j2z}.$$

Obtain (a) $\mathbf{k}$, (b) $\mathbf{H}(\mathbf{r})$, and (c) $\mathbf{W}_{\text{ave}}$.

Solution

(a)

$$\mathbf{k} = -2\hat{\mathbf{z}}.$$

(b)

$$\mathbf{H}(\mathbf{r}) = \frac{1}{\eta_0}\hat{\mathbf{k}} \times \mathbf{E}(\mathbf{r}) = \frac{1}{\eta_0}\left(\hat{\mathbf{k}} \times \mathbf{E}_o\right) e^{-j\mathbf{k}\bullet\mathbf{r}} = \frac{1}{120\pi}\left[-\hat{\mathbf{z}} \times (\hat{\mathbf{x}} - j\hat{\mathbf{y}})\right] e^{j2z} = -\frac{1}{120\pi}(j\hat{\mathbf{x}} + \hat{\mathbf{y}})e^{j2z}.$$

(c)

$$|E_o| = \sqrt{\mathbf{E}_o \bullet \mathbf{E}_o^*} = \sqrt{(\hat{\mathbf{x}} - j\hat{\mathbf{y}}) \bullet (\hat{\mathbf{x}} + j\hat{\mathbf{y}})} = \sqrt{2} \Rightarrow \mathbf{W}_{\text{ave}} = \frac{|E_o|^2}{2\eta_0}\,\hat{\mathbf{k}} = -\frac{1}{120\pi}\,\hat{\mathbf{z}}.$$

◁

Example 7.5 A plane wave in free space has

$$\mathbf{E}(\mathbf{r}) = 10\, e^{j(0.866y+0.5z)}\, \hat{\mathbf{x}}.$$

Obtain (a) $\mathbf{k}$ and $\hat{\mathbf{k}}$, (b) $\mathbf{H}(\mathbf{r})$, and (c) $\mathbf{W}_{\text{ave}}$.

Solution

(a)

$$\mathbf{k} = -(0.866\, \hat{\mathbf{y}} + 0.5\, \hat{\mathbf{z}}).$$

Since $k = \sqrt{0.866^2 + 0.5^2} = 1$, then

$$\hat{\mathbf{k}} = \frac{\mathbf{k}}{k_o} = \mathbf{k} = -(0.866\, \hat{\mathbf{y}} + 0.5\, \hat{\mathbf{z}}).$$

(b)

$$\begin{aligned}\mathbf{H}(\mathbf{r}) &= \frac{1}{\eta_0}\hat{\mathbf{k}} \times \mathbf{E}(\mathbf{r}) = \frac{1}{\eta_0}\left(\hat{\mathbf{k}} \times \mathbf{E}_o\right) e^{-j\mathbf{k}\bullet\mathbf{r}} \\ &= -\frac{1}{120\pi}\left[(0.866\, \hat{\mathbf{y}} + 0.5\, \hat{\mathbf{z}}) \times 10\, \hat{\mathbf{x}}\right] e^{j(0.866y+0.5z)} \\ &= (-0.0132\, \hat{\mathbf{y}} + 0.0229\, \hat{\mathbf{z}}) e^{j(0.866y+0.5z)}.\end{aligned}$$

(c)

$$\mathbf{W}_{\text{ave}} = \frac{|E_o|^2}{2\eta_0}\, \hat{\mathbf{k}} = -\frac{10^2}{240\pi}(0.866\, \hat{\mathbf{y}} + 0.5\, \hat{\mathbf{z}}) = -(0.115\, \hat{\mathbf{y}} + 0.066\, \hat{\mathbf{z}}).$$

◁

Example 7.6 A plane wave in free space has

$$\mathbf{H}(\mathbf{r}) = 0.001\hat{\mathbf{x}}\, e^{-j4y}.$$

Obtain (a) $\mathbf{k}$, (b) $\mathbf{E}(\mathbf{r})$, and (c) $\mathbf{W}_{\text{ave}}$.

Solution

(a)

$$\mathbf{k} = 4\hat{\mathbf{y}}.$$

(b)

$$\mathbf{E}(\mathbf{r}) = -\frac{1}{\omega\varepsilon}\mathbf{k} \times \mathbf{H}(\mathbf{r}) = -\frac{k}{\omega\varepsilon}\hat{\mathbf{k}} \times \mathbf{H}(\mathbf{r}) = -\eta_0\hat{\mathbf{k}} \times \mathbf{H}(\mathbf{r}) = 0.12\pi\, \hat{\mathbf{z}}\, e^{-j4y}.$$

(c)

$$\mathbf{W}_{\text{ave}} = \frac{|E_o|^2}{2\eta_0}\, \hat{\mathbf{k}} = 1.8 \times 10^{-4}\, \hat{\mathbf{y}}.$$

◁

7.3 Propagation in a Dissipative Dielectric-Magnetic Medium

In this section, we shed light on the dissipation effect in propagation. To model dissipation, either ε, μ, or both, should be complex. For simplicity, let us keep μ real, and make

$$\varepsilon \rightarrow \varepsilon - j\frac{\sigma}{\omega}, \tag{7.45}$$

where σ is the conductivity of the medium. Therefore, frequency-domain Maxwell equations become

$$\left.\begin{aligned} \nabla \times \mathbf{E}(\mathbf{r}) &= -j\omega\mu\mathbf{H}(\mathbf{r}) \\ \nabla \times \mathbf{H}(\mathbf{r}) &= (\sigma + j\omega\varepsilon)\mathbf{E}(\mathbf{r}) \\ \nabla \bullet \mathbf{E}(\mathbf{r}) &= 0 \\ \nabla \bullet \mathbf{H}(\mathbf{r}) &= 0 \end{aligned}\right\}. \tag{7.46}$$

Then, the equation of $\mathbf{E}$ becomes

$$\nabla^2\mathbf{E}(\mathbf{r}) - j\omega\mu(\sigma + j\omega\varepsilon)\mathbf{E}(\mathbf{r}) = \mathbf{0}. \tag{7.47}$$

If we assume $\mathbf{E}(\mathbf{r}) = E(z)\,\hat{\mathbf{x}}$, then we get

$$\frac{d^2E(z)}{dz^2} - j\omega\mu(\sigma + j\omega\varepsilon)E(z) = 0. \tag{7.48}$$

Let

$$\gamma = \sqrt{j\omega\mu(\sigma + j\omega\varepsilon)} \tag{7.49}$$

be called the propagation constant (in s^{-1}). Then, we solve

$$\frac{d^2E(z)}{dz^2} - \gamma^2 E(z) = 0. \tag{7.50}$$

The solution can be written as

$$\mathbf{E}(\mathbf{r}) = \left(E_o^+ e^{-\gamma z} + E_o^- e^{\gamma z}\right)\hat{\mathbf{x}}. \tag{7.51}$$

Let us consider the positively-traveling wave only. Then,

$$\mathbf{E}(\mathbf{r}) = E_o^+ e^{-\gamma z}\,\hat{\mathbf{x}}, \tag{7.52}$$

which can be written as

$$\mathbf{E}(\mathbf{r}) = \mathbf{E}_o e^{-\boldsymbol{\gamma}\bullet\mathbf{r}}, \tag{7.53}$$

where $\boldsymbol{\gamma} = \gamma\, \hat{\mathbf{k}}$. Using $\nabla \to -\boldsymbol{\gamma}$ in Eq. (7.46), one gets

$$\left.\begin{aligned} \boldsymbol{\gamma} \times \mathbf{E}(\mathbf{r}) &= j\omega\mu\mathbf{H}(\mathbf{r}) \\ \boldsymbol{\gamma} \times \mathbf{H}(\mathbf{r}) &= -(\sigma + j\omega\varepsilon)\mathbf{E}(\mathbf{r}) \\ \boldsymbol{\gamma} \cdot \mathbf{E}(\mathbf{r}) &= 0 \\ \boldsymbol{\gamma} \cdot \mathbf{H}(\mathbf{r}) &= 0 \end{aligned}\right\}. \tag{7.54}$$

Then, the magnetic field phasor is found as

$$\mathbf{H}(\mathbf{r}) = \frac{\gamma}{j\omega\mu}\hat{\mathbf{k}} \times \mathbf{E}(\mathbf{r}). \tag{7.55}$$

Now, we define

$$\eta = \frac{j\omega\mu}{\gamma} = \sqrt{\frac{j\omega\mu}{\sigma + j\omega\varepsilon}} \tag{7.56}$$

as the intrinsic impedance of the medium. So,

$$\mathbf{H}(\mathbf{r}) = \frac{1}{\eta}\,\hat{\mathbf{k}} \times \mathbf{E}(\mathbf{r}). \tag{7.57}$$

Notice that since γ is a complex quantity, it can be written as

$$\gamma = \alpha + jk, \tag{7.58}$$

where

$$\alpha = \omega\sqrt{\frac{\mu\varepsilon}{2}\left[\sqrt{1 + \left[\frac{\sigma}{\omega\varepsilon}\right]^2} - 1\right]} \tag{7.59}$$

is called the attenuation factor (in $Np\, m^{-1}$), and

$$k = \omega\sqrt{\frac{\mu\varepsilon}{2}\left[\sqrt{1 + \left[\frac{\sigma}{\omega\varepsilon}\right]^2} + 1\right]} \tag{7.60}$$

is the wave number of the medium. Likewise, since η is a complex quantity, it can be written as

$$\eta = |\eta| e^{j\theta_\eta}, \tag{7.61}$$

where

$$|\eta| = \frac{\sqrt{\mu/\varepsilon}}{\left[1 + \left(\frac{\sigma}{\omega\varepsilon}\right)^2\right]^{1/4}}, \tag{7.62}$$

and

$$\theta_\eta = \frac{1}{2}\tan^{-1}\frac{\sigma}{\omega\varepsilon}. \tag{7.63}$$

The quantity $\dfrac{\sigma}{\omega\varepsilon}$ is often called the loss tangent. Notice that we wrote γ in rectangular form and η in polar form for later convenience. Therefore,

$$\mathbf{E}(\mathbf{r}) = \mathbf{E}_o e^{-\boldsymbol{\gamma}\bullet\mathbf{r}} = \mathbf{E}_o e^{-\alpha\hat{\mathbf{k}}\bullet\mathbf{r}} e^{-jk\hat{\mathbf{k}}\bullet\mathbf{r}}, \tag{7.64}$$

and

$$\mathbf{H}(\mathbf{r}) = \frac{1}{|\eta|}(\hat{\mathbf{k}} \times \mathbf{E}_o) e^{-\alpha\hat{\mathbf{k}}\bullet\mathbf{r}} e^{-jk\hat{\mathbf{k}}\bullet\mathbf{r}} e^{-j\theta_\eta}. \tag{7.65}$$

The average Poynting vector $\mathbf{W}_{\text{ave}}$ for the wave can be obtained as

$$\mathbf{W}_{\text{ave}} = \frac{1}{2}\text{Re}\{\mathbf{E}(\mathbf{r}) \times \mathbf{H}(\mathbf{r})^*\} = \frac{1}{2}\text{Re}\left\{\frac{1}{|\eta|}\mathbf{E}_o e^{-\alpha\hat{\mathbf{k}}\bullet\mathbf{r}} \times \left(\hat{\mathbf{k}} \times \mathbf{E}_o^* e^{-\alpha\hat{\mathbf{k}}\bullet\mathbf{r}} e^{j\theta_\eta}\right)\right\}. \tag{7.66}$$

After noting that $\hat{\mathbf{k}} \bullet \mathbf{E} = 0$, we get

$$\mathbf{W}_{\text{ave}} = \frac{|E_o|^2 \cos\theta_\eta}{2|\eta|} e^{-2\alpha\hat{\mathbf{k}}\bullet\mathbf{r}}\,\hat{\mathbf{k}}. \tag{7.67}$$

To see the effect of the medium in propagation of waves, let us for simplicity set $\hat{\mathbf{k}} = \hat{\mathbf{z}}$ and $\mathbf{E}_o = E_o\hat{\mathbf{x}}$. Then

$$\mathbf{E}(\mathbf{r}) = E_o e^{-\alpha z} e^{-jkz}\,\hat{\mathbf{x}}, \tag{7.68}$$

and

$$\mathbf{H}(\mathbf{r}) = \frac{E_o}{|\eta|} e^{-\alpha z} e^{-jkz} e^{-j\theta_\eta}\,\hat{\mathbf{y}}. \tag{7.69}$$

Therefore,

$$\mathscr{E}(z,t) = E_o e^{-\alpha z} \cos(\omega t - kz)\,\hat{\mathbf{x}}, \tag{7.70}$$

and

$$\mathscr{H}(z,t) = \frac{E_o}{|\eta|} e^{-\alpha z} \cos\left(\omega t - kz - \theta_\eta\right)\,\hat{\mathbf{y}}. \tag{7.71}$$

We see that the wave attenuates due to the attenuation factor α. The distance δ in which an electromagnetic wave penetrates a medium is called the skin depth, and is given by

$$\delta = \frac{1}{\alpha}. \tag{7.72}$$

Also, the phase velocity in the medium

$$v_p = \frac{\omega}{k} = \left\{\frac{\mu\varepsilon}{2}\left[\sqrt{1 + \left[\frac{\sigma}{\omega\varepsilon}\right]^2} + 1\right]\right\}^{-1/2} \tag{7.73}$$

is affected because of the losses. Finally, notice that the electric field and the magnetic field are out of phase by an angle θ_η.

As a special case, a medium is classified as good conductor if $\sigma >> \omega\varepsilon$. Using Taylor series, expressions for wave properties can be obtained approximately as

$$\alpha = \sqrt{\frac{\omega\mu\sigma}{2}},\, k = \sqrt{\frac{\omega\mu\sigma}{2}},\, v_p = \sqrt{\frac{2\omega}{\mu\sigma}},\, |\eta| = \sqrt{\frac{\omega\mu}{\sigma}}, \text{ and } \theta_\eta = \pi/4. \tag{7.74}$$

Notice that the skin depth δ for good conductors can be approximated as

$$\delta = \frac{1}{\alpha} = \sqrt{\frac{2}{\omega\mu\sigma}} = \frac{1}{\sqrt{\pi f \sigma \mu}}. \tag{7.75}$$

Example 7.7 A plane wave in a medium has

$$\mathscr{H}(\mathbf{r}, t) = 0.01 e^{-\alpha y} \cos(\omega t - y)\, \hat{\mathbf{x}}.$$

Given that $\eta = 100 e^{j\pi/6}$, obtain (a) α, (b) $\mathbf{E}$, and $\mathbf{W}_{\text{ave}}$.

Solution (a) Using the ratio

$$\frac{\alpha}{k} = \frac{\omega\sqrt{\frac{\mu\varepsilon}{2}\left[\sqrt{1+\left[\frac{\sigma}{\omega\varepsilon}\right]^2} - 1\right]}}{\omega\sqrt{\frac{\mu\varepsilon}{2}\left[\sqrt{1+\left[\frac{\sigma}{\omega\varepsilon}\right]^2} + 1\right]}} = \frac{\left[\sqrt{1+\left[\frac{\sigma}{\omega\varepsilon}\right]^2} - 1\right]^{1/2}}{\left[\sqrt{1+\left[\frac{\sigma}{\omega\varepsilon}\right]^2} + 1\right]^{1/2}},$$

and after noting that

$$\frac{\sigma}{\omega\varepsilon} = \tan 2\theta_\eta = \tan \pi/3 = \sqrt{3},$$

we get

$$\frac{\alpha}{k} = \frac{1}{\sqrt{3}} \Rightarrow \alpha = 0.577.$$

(b)

$$\mathbf{E}(\mathbf{r}) = -\eta\, \hat{\mathbf{k}} \times \mathbf{H}(\mathbf{r}) = -100 e^{j\pi/6} \hat{\mathbf{y}} \times \hat{\mathbf{x}}\, 0.01 e^{-0.577y} e^{-jy} = e^{-0.577y} e^{-jy} e^{j\pi/6}\, \hat{\mathbf{z}}.$$

(c)

$$\mathbf{W}_{\text{ave}} = \frac{1}{2}\text{Re}\{\mathbf{E}(\mathbf{r}) \times \mathbf{H}^*(\mathbf{r})\} = \frac{1}{2}\text{Re}\{e^{-0.577y} e^{-jy} e^{j\pi/6}\, \hat{\mathbf{z}} \times 0.01 e^{-0.577y} e^{jy} \hat{\mathbf{x}}\} = 8.66 \times 10^{-3} e^{-1.15y}\, \hat{\mathbf{y}}.$$

◁

Example 7.8 Given

$$\mathscr{H}(\mathbf{r}, t) = \cos(\omega t - z)\hat{\mathbf{x}} + \sin(\omega t - z)\hat{\mathbf{y}},$$

Obtain **E** when $\eta = 80\pi$.

Solution Conversion to phasor domain gives

$$\mathbf{H}(\mathbf{r}) = e^{-jz}\hat{\mathbf{x}} + e^{-j(z+\pi/2)}\hat{\mathbf{y}}.$$

Then,

$$\begin{aligned}\mathbf{E}(\mathbf{r}) &= -\eta\,\hat{\mathbf{k}} \times \mathbf{H}(\mathbf{r}) \\ &= -80\pi\hat{\mathbf{z}} \times \left[e^{-jz}\hat{\mathbf{x}} + e^{-j(z+\pi/2)}\hat{\mathbf{y}}\right] \\ &= 80\pi e^{-jz}\hat{\mathbf{y}} + 80\pi e^{-j(z+\pi/2)}\hat{\mathbf{x}}.\end{aligned}$$

Conversion to time domain gives

$$\mathscr{E}(\mathbf{r}, t) = 80\pi\cos(\omega t - z)\hat{\mathbf{y}} + 80\pi\sin(\omega t - z)\hat{\mathbf{x}}.$$

◁

Example 7.9 A plane wave propagating in a medium has

$$\mathscr{E}(\mathbf{r}, t) = e^{-\alpha z}\cos(\omega t - kz)\,\hat{\mathbf{y}},$$

where $\omega = 3 \times 10^8$. The medium is characterized by $\varepsilon_r = 1$, $\mu_r = 1$, and $\sigma = 10$. Obtain (a) α and k, and (b) $\mathscr{H}$.

Solution (a) The medium is a good conductor since

$$\frac{\sigma}{\omega\varepsilon} = 3769.91 >> 1.$$

Then,

$$\alpha = k = \sqrt{\frac{\omega\mu\sigma}{2}} = 43.41.$$

(b) We have

$$\eta = \sqrt{\frac{\omega\mu}{\sigma}}e^{j\pi/4} = 6.14e^{j\pi/4}.$$

Then,

$$\mathbf{H}(\mathbf{r}) = \frac{1}{\eta}\hat{\mathbf{k}} \times \mathbf{E}(\mathbf{r}) = -0.162e^{-\alpha z}e^{-jkz}e^{-j\pi/4}\hat{\mathbf{x}}.$$

Finally,

$$\mathscr{H}(\mathbf{r}, t) = -0.162e^{-\alpha z}\cos(\omega t - kz - \pi/4)\hat{\mathbf{x}}.$$

◁

7.4 Polarization State

Recall that for a plane wave of the form

$$\mathbf{E}(\mathbf{r}) = \mathbf{E}_o e^{-\gamma \hat{\mathbf{k}} \bullet \mathbf{r}}, \tag{7.76}$$

the constant, complex-valued vector $\mathbf{E}_o$ can be called the polarization-state of the wave. In this section, we discuss the various types of polarization states.

Polarization state is the trajectory of the time-domain electric field on the plane normal to the propagation direction. Without a loss of generality, let $\gamma = jk$ and $\hat{\mathbf{k}} = \hat{\mathbf{z}}$. Then,

$$\mathbf{E}(\mathbf{r}) = \mathbf{E}_o e^{-jkz}, \tag{7.77}$$

where $\mathbf{E}_o$ is, in general,

$$\mathbf{E}_o = E_x \,\hat{\mathbf{x}} + E_y \,\hat{\mathbf{y}}, \tag{7.78}$$

where $E_x = Ae^{ja}$ and $E_y = Be^{jb}$. Let us define $\Delta = b - a$ for later use. Depending on A, B, a, and b, polarization state can be (i) linear, (ii) circular, or (iii) elliptical.

7.4.1 Linear Polarization State

If $\Delta = n\pi$ $(n = 0, \pm 1, \pm 2, \ldots)$ (say, $a = b = 0$), then

$$\mathbf{E}(\mathbf{r}) = (A\,\hat{\mathbf{x}} + B\,\hat{\mathbf{y}})e^{-jkz}. \tag{7.79}$$

In time domain,

$$\mathscr{E}(z, t) = (A\,\hat{\mathbf{x}} + B\,\hat{\mathbf{y}}) \cos(\omega t - kz). \tag{7.80}$$

Without a loss of generality, on the plane $z = 0$, we have

$$\mathscr{E}(0, t) = (A\,\hat{\mathbf{x}} + B\,\hat{\mathbf{y}}) \cos \omega t. \tag{7.81}$$

The trajectory of $\mathscr{E}(0, t)$ versus $\omega t \in [0, 3\pi/2]$ is depicted in Fig. 7.4. As ωt varies, $\mathscr{E}(0, t)$ traces a tilted straight line. This is called linear polarization state (LP). LP can also be obtained even when $\Delta \neq \pm n\pi$, but if either $B = 0$, or $A = 0$. If $B = 0$, the wave is said to be x polarized, and if $A = 0$, it is said to be y polarized.

7.4.2 Circular Polarization State

If $A = B$, and $\Delta = -\pi/2$ (say, $a = \pi/2$ and $b = 0$), then

$$\mathbf{E}(\mathbf{r}) = A(e^{j\pi/2}\,\hat{\mathbf{x}} + \hat{\mathbf{y}})e^{-jkz}. \tag{7.82}$$

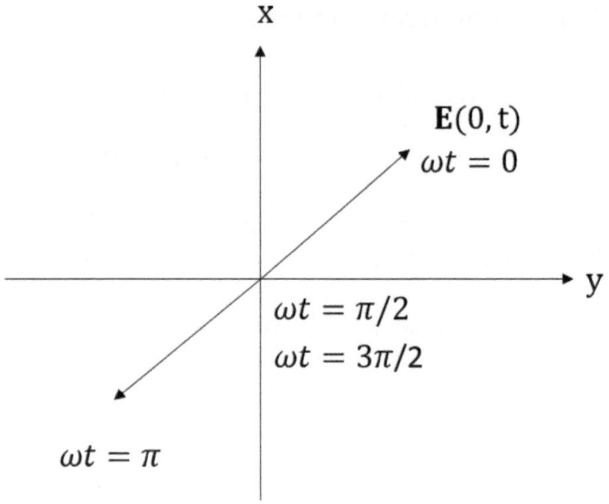

Fig. 7.4 Linear polarization state

In time domain,

$$\mathscr{E}(z,t) = A[-\sin(\omega t - kz)\,\hat{\mathbf{x}} + \cos(\omega t - kz)\,\hat{\mathbf{y}}]. \tag{7.83}$$

On the plane $z = 0$, we have

$$\mathscr{E}(0,t) = A(-\sin\omega t\,\hat{\mathbf{x}} + \cos\omega t\,\hat{\mathbf{y}}). \tag{7.84}$$

The trajectory of $\mathscr{E}(0,t)$ versus $\omega t \in [0, 3\pi/2]$ is depicted in Fig. 7.5a. As ωt varies, $\mathscr{E}(0,t)$ traces a circle. Such a polarization state is called circular polarization (CP). Also, notice that the circle has a clockwise orientation. Clockwise orientation is called right circular polarization (RCP), because if the thump points along the direction of propagation (z axis in our case), the fingers will point along the polarization state direction in the clockwise sense.

If $\Delta = \pi/2$ (say, $a = 0$ and $b = \pi/2$), then we get

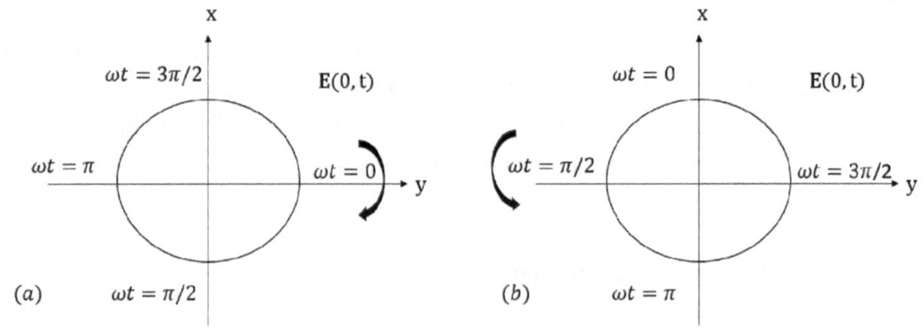

Fig. 7.5 Circular polarization state. **a** RCP, and **b** LCP

$$\mathscr{E}(0, t) = A(\cos \omega t \,\hat{\mathbf{x}} - \sin \omega t \,\hat{\mathbf{y}}), \tag{7.85}$$

the trajectory of which is shown in Fig. 7.5b. This is the left-circular polarization (LCP).

7.4.3 Elliptical Polarization State

If either the LP or CP conditions are not met, then the electric field vector on the $z = 0$ plane will trace an ellipse. Such a polarization state is called elliptical polarization (EP) (see Fig. 7.6). If $A \neq B$, but $\Delta = \pm\pi/2$, we can have right-circular elliptical polarization (REP) when $\Delta = -\pi/2$, and left-circular elliptical polarization (LEP) when $\Delta = \pi/2$.

The following summarizes polarization states.

1. Linear polarization state: When $A = 0$ or $B = 0$, or if $\Delta = n\pi$ when $A \neq 0$ and $B \neq 0$.
2. Circular polarization state: When $A = B$ and $\Delta = \pm\pi/2$. If $\Delta = +\pi/2$, it will be LCP, and if $\Delta = -\pi/2$, it will be RCP.
3. Elliptical polarization state: If the above conditions are not met.

Example 7.10 Determine the polarization state of (a) $\mathbf{E}(\mathbf{r}) = E_o(\hat{\mathbf{x}} - j\hat{\mathbf{y}})e^{-jkz}$, (b) $\mathscr{E}(\mathbf{r}, t) = e^{-0.25z}\cos(\omega t - 0.8z)\,\hat{\mathbf{x}} + 2e^{-0.25z}\sin(\omega t - 0.8z)\,\hat{\mathbf{y}}$, and (c) $\mathbf{H}(\mathbf{r}) = (\hat{\mathbf{x}} + 2\,\hat{\mathbf{y}})e^{-jkz}$.

Solution (a) After conversion to polar form, we have $A = E_o$, $B = E_o$, $a = 0$, and $b = -\pi/2$. Since $A = B$, and $\Delta = -\pi/2$, it is an RCP wave.
(b) In phasor domain, we have

$$\mathbf{E}(\mathbf{r}) = e^{-0.25z}e^{-j0.8z}\,\hat{\mathbf{x}} + 2e^{-j\pi/2}e^{-0.25z}e^{-j0.8z}\,\hat{\mathbf{y}}.$$

We have $A = 1$, $B = 2$, $a = 0$, and $b = -\pi/2$. Since $\Delta \neq \pm n\pi$, it is an LP wave. Also, since $\Delta = -\pi/2$, it is an REP wave.

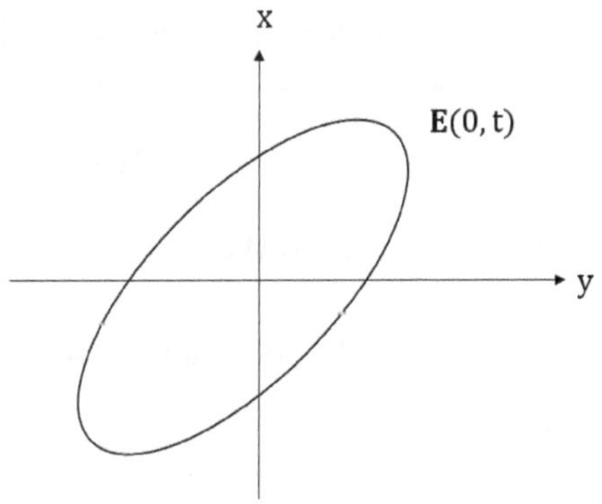

Fig. 7.6 Elliptical polarization state

(c) Using Maxwell equations, we find $\mathbf{E}(\mathbf{r})$ as

$$\mathbf{E}(\mathbf{r}) = \eta(2\,\hat{\mathbf{x}} - \hat{\mathbf{y}})e^{-jkz}.$$

After conversion to polar form, we have $A = 2\eta$, $B = \eta$, $a = 0$, and $b = \pi$. Since $\Delta = \pi$, the wave is an LP wave. ◁

Problems

7.1 Show that the magnetic field satisfies

$$\nabla^2\mathcal{H} - \mu_0\varepsilon_0\frac{\partial^2\mathcal{H}}{\partial t^2} = \mathbf{0}.$$

7.2 Show that

$$\mathbf{E}\bullet\mathbf{H} = 0$$

holds true in any unbounded isotropic dielectric-magnetic medium.

7.3 A plane wave in free space has

$$\mathcal{E} = \hat{\mathbf{y}}\,e^{-j(x+z)}.$$

Obtain $\mathcal{H}$.

7.4 A plane wave in free space has

$$\mathcal{E} = (\hat{\mathbf{x}} - j\hat{\mathbf{y}})\,e^{-jz}.$$

Obtain (a) $\mathcal{H}$, and (b) $\mathbf{W}_{\text{ave}}$.

7.5 A plane wave propagating at $\omega = 10^8$ in a good conductor with $\varepsilon_r = 1$, $\mu_r = 20$, and $\sigma = 3$. The magnetic field phasor is given by

$$\mathbf{H} = (\hat{\mathbf{x}} - j\hat{\mathbf{y}})\,e^{-61.4z}e^{-j61.4z}.$$

Obtain (a) $\mathbf{E}$, and (b) $\mathbf{W}_{\text{ave}}$.

7.6 Determine the polarization state of the following wave

$$\mathbf{H} = -2je^{-jz}\,\hat{\mathbf{x}} + 2e^{-jz}\,\hat{\mathbf{y}}.$$

7.7 A plane wave propagating in a medium has

$$\mathbf{H} = 5e^{-\alpha z}e^{-jkz}\,\hat{\mathbf{y}},$$

where $\omega = 10^9$. The medium has $\varepsilon_r = 10$, $\mu_r = 2$, and $\sigma = 3$. Obtain (a) α and k, (b) $\mathbf{E}$, and (c) the polarization state of the wave.

7.8 A plane wave with

$$\mathscr{H} = e^{-2z}\cos(\omega t - 16z)\,\hat{\mathbf{x}} + 0.5e^{-2z}\sin(\omega t - 16z)\,\hat{\mathbf{y}},$$

is a propagating in a medium with $\varepsilon_r = 1000$, $\mu_r = 1$, and $\sigma = 3$. Given that $v_p = 0.2\,c$, determine (a) ω, (b) η, (c) $\mathbf{E}$, and (d) the polarization state of the wave.

7.9 The permittivity scalar of a non-magnetized plasma [1, 2] is

$$\varepsilon(\omega) = \varepsilon_0\left[1 + \frac{\omega_p^2}{j\omega(\nu + j\omega)}\right],$$

where ω_p is called the plasma frequency, and ν is called the damping frequency. (a) Obtain expressions for the real part (ε') and the imaginary part (ε''), (b) obtain an expression for the wave number k when $\nu = 0$, and (c) given that

$$\mathbf{E} = \hat{\mathbf{x}}\,e^{-jkz},$$

write $\mathbf{E}$ when $\omega < \omega_p$ and when $\omega > \omega_p$. Discuss the effect of ω_p on the wave propagation in plasmas.

References

1. H.C. Chen, *Theory of Electromagnetic Waves: A Coordinate-free Approach* (McGraw-Hill, 1985)
2. U.S. Inan, M. Golkowski, *Principles of Plasma Physics for Engineering and Scientists* (Cambridge University Press, 2011)

Plane-Wave Reflection and Refraction

8

We discussed the propagation of plane waves in unbounded space in Chap. 7. In this chapter, we consider the situation where a plane wave encounters an object in the shape of a flat infinite-extent surface (as shown in Fig. 8.1). The entire space is then divided into two regions; a region with medium 1, and a region with medium 2. The surface separating the two regions is called an interface. In this case, part of the incident wave gets reflected by the object, and part of it gets transmitted.

The wave existing in region 1 is composed of two parts: an incident wave $\{\mathbf{E}_i(\mathbf{r}), \mathbf{H}_i(\mathbf{r}), \mathbf{k}_i\}$ that acts as a source, and a reflected wave $\{\mathbf{E}_r(\mathbf{r}), \mathbf{H}_r(\mathbf{r}), \mathbf{k}_r\}$ that results from the interaction. In region 2, the wave that exists is called the transmitted wave $\{\mathbf{E}_t(\mathbf{r}), \mathbf{H}_t(\mathbf{r}), \mathbf{k}_t\}$ that also results from the interaction. We discuss two types of incidence: normal incidence, and oblique incidence. In normal-incidence case, the incident wave propagates in a direction normal to the interface (e.g., Fig. 8.1). In oblique-incidence case, the incident wave propagates in an arbitrary direction relative to the interface. These are discussed, respectively, in Sects. 8.1 and 8.2.

8.1 Normal Incidence

Let region 1 be occupied by a medium with $(\varepsilon_1, \mu_1, \sigma_1)$, and region 2 be occupied by a medium with $(\varepsilon_2, \mu_2, \sigma_2)$. The interface separating the two regions is the plane $z = 0$. Since the plane is infinite in extent, this problem can be approached in two dimensions instead of three dimensions. Thus, we introduce the plane of incidence, which contains (i) unit vector normal to the interface $\hat{\mathbf{n}}$, and (ii) all propagation vectors $\mathbf{k}_i$, $\mathbf{k}_r$, and $\mathbf{k}_t$. In normal-incidence case, and since $\hat{\mathbf{n}} = \hat{\mathbf{z}}$, the plane of incidence could be the xz plane or the yz plane, but not the xy plane. We will choose the xz plane here (see Fig. 8.2).

H. M. Alkhoori, *Concise Introduction to Electromagnetic Fields*, Synthesis Lectures on Electromagnetics, https://doi.org/10.1007/978-3-031-60331-0_8

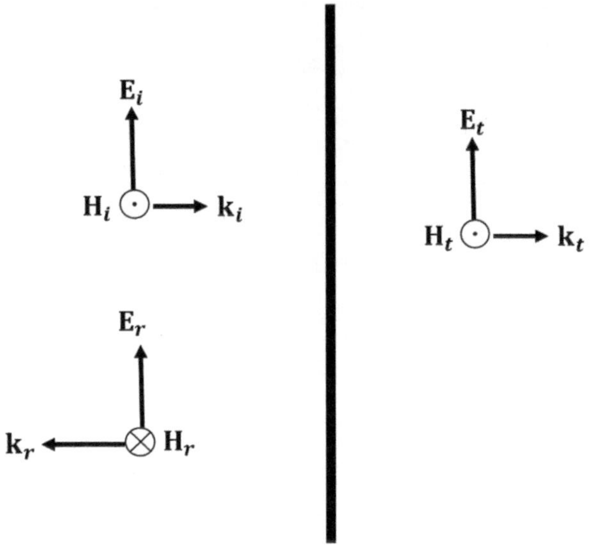

Fig. 8.1 Reflection and refraction problem

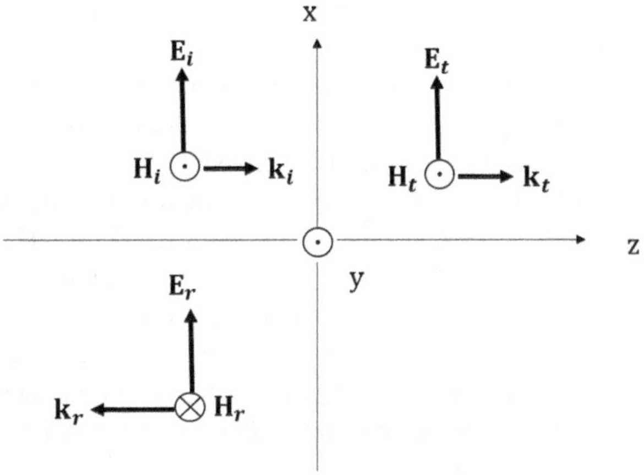

Fig. 8.2 Plane of incidence

From Fig. 8.2, we see that the incident wave has

$$\left.\begin{aligned} \mathbf{k}_i &= \gamma_1 \,\hat{\mathbf{z}} \\ \mathbf{E}_i(\mathbf{r}) &= E_o e^{-\gamma_1 z}\,\hat{\mathbf{x}} \\ \mathbf{H}_i(\mathbf{r}) &= \frac{1}{\eta_1}\hat{\mathbf{k}}_i \times \mathbf{E}_i(\mathbf{r}) = \frac{E_o}{\eta_1} e^{-\gamma_1 z}\,\hat{\mathbf{y}} \end{aligned}\right\}. \tag{8.1}$$

Also, the reflected wave has

$$\left.\begin{aligned} \mathbf{k}_r &= -\gamma_1 \hat{\mathbf{z}} \\ \mathbf{E}_r(\mathbf{r}) &= \Gamma E_o e^{\gamma_1 z}\, \hat{\mathbf{x}} \\ \mathbf{H}_r(\mathbf{r}) &= \frac{1}{\eta_1}\hat{\mathbf{k}}_r \times \mathbf{E}_r(\mathbf{r}) = -\frac{\Gamma E_o}{\eta_1} e^{\gamma_1 z}\, \hat{\mathbf{y}} \end{aligned}\right\}, \tag{8.2}$$

where Γ (called the reflection coefficient) is an unknown that has to be determined. Finally, the transmitted wave has

$$\left.\begin{aligned} \mathbf{k}_t &= \gamma_2 \hat{\mathbf{z}} \\ \mathbf{E}_t(\mathbf{r}) &= T E_o e^{-\gamma_2 z}\, \hat{\mathbf{x}} \\ \mathbf{H}_t(\mathbf{r}) &= \frac{1}{\eta_2}\hat{\mathbf{k}}_t \times \mathbf{E}_t(\mathbf{r}) = \frac{T E_o}{\eta_2} e^{-\gamma_2 z}\, \hat{\mathbf{y}} \end{aligned}\right\}, \tag{8.3}$$

where T (called the transmission coefficient) is an unknown that has to be determined. So, the total field in region 1 is

$$\left.\begin{aligned} \mathbf{E}_1(\mathbf{r}) &= \mathbf{E}_i(\mathbf{r}) + \mathbf{E}_r(\mathbf{r}) \\ \mathbf{H}_1(\mathbf{r}) &= \mathbf{H}_i(\mathbf{r}) + \mathbf{H}_r(\mathbf{r}) \end{aligned}\right\}, \tag{8.4}$$

and the total field in region 2 is

$$\left.\begin{aligned} \mathbf{E}_2(\mathbf{r}) &= \mathbf{E}_t(\mathbf{r}) \\ \mathbf{H}_2(\mathbf{r}) &= \mathbf{H}_t(\mathbf{r}) \end{aligned}\right\}. \tag{8.5}$$

Next, our task is to find Γ and T. This can be done using boundary conditions. That is,

$$\left.\begin{aligned} \hat{\mathbf{n}} \times (\mathbf{E}_1 - \mathbf{E}_2) = \mathbf{0} &\Rightarrow \mathbf{E}_1^{\|} = \mathbf{E}_2^{\|} \\ \hat{\mathbf{n}} \times (\mathbf{H}_1 - \mathbf{H}_2) = \mathbf{0} &\Rightarrow \mathbf{H}_1^{\|} = \mathbf{H}_2^{\|} \end{aligned}\right\}. \tag{8.6}$$

Since all field quantities are already tangential to the surface, we have

$$\left.\begin{aligned} \mathbf{E}_1(\mathbf{r})\Big|_{z=0} &= \mathbf{E}_2(\mathbf{r})\Big|_{z=0} \\ \mathbf{H}_1(\mathbf{r})\Big|_{z=0} &= \mathbf{H}_2(\mathbf{r})\Big|_{z=0} \end{aligned}\right\}, \tag{8.7}$$

which become

$$\left.\begin{aligned} 1 + \Gamma &= T \\ \frac{1}{\eta_1}(1 - \Gamma) &= \frac{1}{\eta_2} T \end{aligned}\right\}. \tag{8.8}$$

Solving for Γ and T, we find

$$\Gamma = \frac{\eta_2 - \eta_1}{\eta_2 + \eta_1}, \tag{8.9}$$

and

$$T = 1 + \Gamma = \frac{2\eta_2}{\eta_2 + \eta_1}. \tag{8.10}$$

Remarks

- Γ and T are complex quantities in general that depend on the mediums, as well as on the frequency of the incident wave. Therefore, the reflected and transmitted waves are, in general, not in phase.
- The magnitude of the reflection coefficient satisfies $0 \le |\Gamma| \le 1$.
- When $\Gamma = 0$ (i.e., $T = 1$), there will be no reflected wave, and thus, the incident wave will entirely be transmitted into medium 2. In the current situation, this can be obtained when $\eta_1 = \eta_2$, a condition known as impedance matching.
- When $\Gamma = -1$ (i.e., $T = 0$), there will be no transmitted wave, and thus, the incident wave will entirely be reflected by medium 2. This is useful, for example, in protecting medium 2 from heating up.

Standing Waves

In region 1, the total wave is

$$\mathbf{E}_1(\mathbf{r}) = E_o(e^{-\gamma_1 z} + \Gamma e^{\gamma_1 z})\,\hat{\mathbf{x}}. \tag{8.11}$$

Assuming non dissipative case, this becomes

$$\mathbf{E}_1(\mathbf{r}) = E_o(e^{-jk_1 z} + \Gamma e^{jk_1 z})\,\hat{\mathbf{x}}. \tag{8.12}$$

In time domain, the wave becomes

$$\mathscr{E}_1(\mathbf{r}, t) = E_o\,[\cos(\omega t - k_1 z) + \Gamma\cos(\omega t + k_1 z)]\,\hat{\mathbf{x}}. \tag{8.13}$$

We next examine the total electric field behaviour in region 1 as Γ changes from zero to unity. Without loss of generality, we let $E_o = 1$. Figure 8.3 shows plots of $E_1(z, t)$ vs. $z \in [-2, 0]\lambda$ and $t \in \{0, 1/8, 1/4, 1/2\}T$ for the cases $\Gamma \in \{0, 0.4, 0.8, 1\}$. When $\Gamma = 0$, the wave in region 1 is a traveling wave. When $\Gamma = 1$, the wave in region 1 is a standing wave. Finally, when $0 < \Gamma < 1$, the wave is partially-standing, partially-traveling wave. Physically, when $\Gamma = 0$, the total wave in region 1 is a traveling wave propagating along the z axis. As Γ increases, there will be another traveling wave propagating along the $-z$ axis. This second traveling wave will interfere with the first traveling wave, forming what is called a partially standing wave. When $\Gamma = 1$, the second traveling wave will cancel the first traveling wave, forming a purely standing wave. The standing wave ratio (S) is defined as

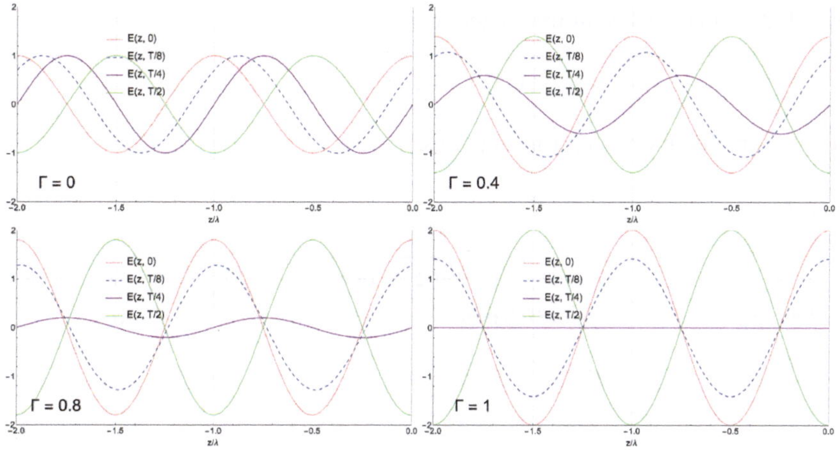

Fig. 8.3 $E_1(z, t)$ vs. $z \in [-2, 0]\lambda$ and $t \in \{0, 1/8, 1/4, 1/2\}T$ when $\Gamma \in \{0, 0.4, 0.8, 1\}$

$$S = \frac{|\mathbf{E}_1|_{\text{max}}}{|\mathbf{E}_1|_{\text{min}}} = \frac{1 + |\Gamma|}{1 - |\Gamma|}. \tag{8.14}$$

Notice that since $0 \leq |\Gamma| \leq 1$, this yields $1 \leq S < \infty$.

Example 8.1 A plane wave in free space

$$\mathbf{E}_i(\mathbf{r}) = \hat{\mathbf{x}}\, e^{-jk_o z}$$

is incident normally upon a semi-infinite non dissipative medium with $\varepsilon_r = 2.56$. (a) Determine Γ, T, S, and (b) the incident, reflected, and transmitted power.

Solution (a) We have $\eta_1 = 120\pi$ and $\eta_2 = \eta_1/1.6$. Hence, $\Gamma = \frac{1/1.6-1}{1/1.6+1} = -0.231$, $T = 1 + \Gamma = 0.769$, and $S = \frac{1+|\Gamma|}{1-|\Gamma|} = 1.6$.

(b) The incident, reflected, and transmitted power can be found, respectively, as

$$\left.\begin{aligned} \mathbf{W}^i_{\text{ave}} &= \frac{E_o^2}{2\eta_1}\,\hat{\mathbf{z}} = 1.327 \times 10^{-3}\,\hat{\mathbf{z}} \\ \mathbf{W}^r_{\text{ave}} &= -\frac{(\Gamma E_o)^2}{2\eta_1}\,\hat{\mathbf{z}} = -0.071 \times 10^{-3}\,\hat{\mathbf{z}} \\ \mathbf{W}^t_{\text{ave}} &= \frac{(T E_o)^2}{2\eta_2}\,\hat{\mathbf{z}} = 1.256 \times 10^{-3}\,\hat{\mathbf{z}} \end{aligned}\right\}.$$

Notice that $|\mathbf{W}^i_{\text{ave}}| = |\mathbf{W}^r_{\text{ave}}| + |\mathbf{W}^t_{\text{ave}}|$.

◁

Example 8.2 A plane wave in free space with

$$\mathbf{E}_i(\mathbf{r}) = E_o\,\hat{\mathbf{x}}\,e^{-jk_oz}$$

is incident normally upon a semi-infinite PEC (i.e., $\sigma \to \infty$). Obtain $\mathbf{E}_1(\mathbf{r})$ and $\mathbf{E}_2(\mathbf{r})$.

Solution From

$$\eta_2 = \sqrt{\frac{\mu_2}{\sigma_2 + j\omega\varepsilon_2}},$$

we see that since $\sigma_2 \to \infty$, $\eta_2 = 0$. Then, $\Gamma = -1$, and $T = 0$. Therefore,

$$\mathbf{E}_1(\mathbf{r}) = \mathbf{E}_i(\mathbf{r}) + \mathbf{E}_r(\mathbf{r}) = E_o\,\hat{\mathbf{x}}\,e^{-jk_oz} - E_o\,\hat{\mathbf{x}}\,e^{jk_oz} = -2jE_o \sin k_oz,$$

and

$$\mathbf{E}_2(\mathbf{r}) = \mathbf{E}_t(\mathbf{r}) = \mathbf{0}.$$

Remarks

- This example shows that, when a plane wave encounters a PEC, it will be totally reflected. This means that the wave outside the PEC is a purely standing wave.
- Whereas the terms $e^{\pm jkz}$ represent travelling waves, the term $\sin kz$ (or $\cos kz$) represents a standing wave.
- A simple radar (Radio Detection and Ranging) system is merely composed of two antennas [1, 2]. One antenna sends an incident wave to an object. The object interacts by sending a reflected wave back to the other antenna in the radar. The radar will be able to detect the object if the reflected wave has a high power.
- Vehicles are made of metals, and metals act somewhat close to PECs. Since reflection from metals is very high, vehicles can easily be detected by radars. This is not desired, for example, in some military applications.
- Reflection from vehicles can be reduced to lower levels by either coating them with some materials that are capable of manipulating reflection, or by properly designing the shape of the vehicles so as to minimize reflection.

◁

8.2 Oblique Incidence

We next consider the situation where a plane wave is incident obliquely upon an object. For simplicity, we will limit our discussion to the non dissipative case. We will let $\mathbf{k}_i$ have two components; one along the x axis, and one along the z axis. Since the interface is the plane $z = 0$ with $\hat{\mathbf{n}} = \hat{\mathbf{z}}$, the plane of incidence must be the xz plane. When the electric field phasor is perpendicular to the plane of incidence, we call such a case perpendicular polarization

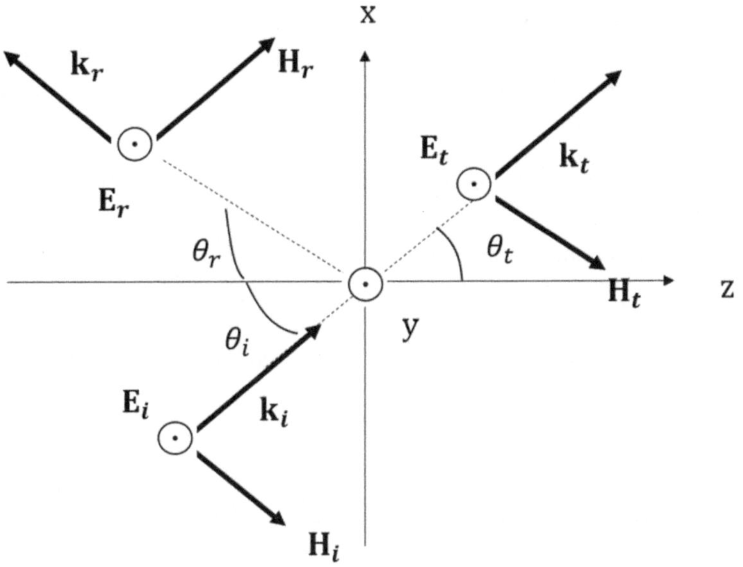

Fig. 8.4 Plane of incidence

(horizontal or E polarization), and when the electric field phasor is parallel to the plane of incidence, we call such a case parallel polarization (vertical or H polarization).

8.2.1 Perpendicular Polarization

The perpendicular-polarization case is shown in Fig. 8.4. From Fig. 8.4, we see that the incident wave has

$$\left.\begin{aligned} \mathbf{k}_i &= k_1(\sin\theta_i\,\hat{\mathbf{x}} + \cos\theta_i\,\hat{\mathbf{z}}) \\ \mathbf{E}_i(\mathbf{r}) &= \hat{\mathbf{y}}\,E_o e^{-jk_1(x\sin\theta_i + z\cos\theta_i)} \\ \mathbf{H}_i(\mathbf{r}) &= \frac{1}{\eta_1}\hat{\mathbf{k}}_i \times \mathbf{E}_i(\mathbf{r}) = (-\cos\theta_i\,\hat{\mathbf{x}} + \sin\theta_i\,\hat{\mathbf{z}})\,\frac{E_o}{\eta_1} e^{-jk_1(x\sin\theta_i + z\cos\theta_i)} \end{aligned}\right\}. \tag{8.15}$$

Also, the reflected wave has

$$\left.\begin{aligned} \mathbf{k}_r &= k_1(\sin\theta_r\,\hat{\mathbf{x}} - \cos\theta_r\,\hat{\mathbf{z}}) \\ \mathbf{E}_r(\mathbf{r}) &= \hat{\mathbf{y}}\,\Gamma_\perp E_o e^{-jk_1(x\sin\theta_r - z\cos\theta_r)} \\ \mathbf{H}_r(\mathbf{r}) &= \frac{1}{\eta_1}\hat{\mathbf{k}}_r \times \mathbf{E}_r(\mathbf{r}) = (\cos\theta_r\,\hat{\mathbf{x}} + \sin\theta_r\,\hat{\mathbf{z}})\,\frac{\Gamma_\perp E_o}{\eta_1} e^{-jk_1(x\sin\theta_r - z\cos\theta_r)} \end{aligned}\right\}. \tag{8.16}$$

Finally, the transmitted wave has

$$\left.\begin{aligned}\mathbf{k}_t &= k_2(\sin\theta_t\,\hat{\mathbf{x}} + \cos\theta_t\,\hat{\mathbf{z}})\\ \mathbf{E}_t(\mathbf{r}) &= \hat{\mathbf{y}}\,T_\perp E_o e^{-jk_2(x\sin\theta_t + z\cos\theta_t)}\\ \mathbf{H}_t(\mathbf{r}) &= \frac{1}{\eta_2}\hat{\mathbf{k}}_t \times \mathbf{E}_t(\mathbf{r}) = (-\cos\theta_t\,\hat{\mathbf{x}} + \sin\theta_t\,\hat{\mathbf{z}})\,\frac{T_\perp E_o}{\eta_2} e^{-jk_2(x\sin\theta_t + z\cos\theta_t)}\end{aligned}\right\}. \tag{8.17}$$

Now, the total fields in both regions are

$$\left.\begin{aligned}\mathbf{E}_1(\mathbf{r}) &= \hat{\mathbf{y}}\,E_o[e^{-jk_1(x\sin\theta_i + z\cos\theta_i)} + \Gamma_\perp e^{-jk_1(x\sin\theta_r - z\cos\theta_r)}]\\ \mathbf{H}_1(\mathbf{r}) &= (-\cos\theta_i\,\hat{\mathbf{x}} + \sin\theta_i\,\hat{\mathbf{z}})\,\frac{E_o}{\eta_1} e^{-jk_1(x\sin\theta_i + z\cos\theta_i)} + (\cos\theta_r\,\hat{\mathbf{x}} + \sin\theta_r\,\hat{\mathbf{z}})\,\frac{\Gamma_\perp E_o}{\eta_1} e^{-jk_1(x\sin\theta_r - z\cos\theta_r)}\end{aligned}\right\}, \tag{8.18}$$

and

$$\left.\begin{aligned}\mathbf{E}_2(\mathbf{r}) &= \hat{\mathbf{y}}\,T_\perp E_o e^{-jk_2(x\sin\theta_t + z\cos\theta_t)}\\ \mathbf{H}_2(\mathbf{r}) &= (-\cos\theta_t\,\hat{\mathbf{x}} + \sin\theta_t\,\hat{\mathbf{z}})\,\frac{T_\perp E_o}{\eta_2} e^{-jk_2(x\sin\theta_t + z\cos\theta_t)}\end{aligned}\right\}. \tag{8.19}$$

To find θ_r, θ_t, $\Gamma_\perp$ and $T_\perp$, we apply the boundary conditions

$$\left.\begin{aligned}\mathbf{E}_1^{\|}\Big|_{z=0} &= \mathbf{E}_2^{\|}\Big|_{z=0}\\ \mathbf{H}_1^{\|}\Big|_{z=0} &= \mathbf{H}_2^{\|}\Big|_{z=0}\end{aligned}\right\}. \tag{8.20}$$

After noting that $\mathbf{E}_1(\mathbf{r})$ and $\mathbf{E}_2(\mathbf{r})$ are already tangential to $z = 0$, and that only the x components of $\mathbf{H}_1(\mathbf{r})$ and $\mathbf{H}_2(\mathbf{r})$ are tangential to $z = 0$, we get

$$\left.\begin{aligned}e^{-jk_1x\sin\theta_i} + \Gamma_\perp e^{-jk_1x\sin\theta_r} &= T_\perp e^{-jk_2x\sin\theta_t}\\ \cos\theta_i\frac{1}{\eta_1}e^{-jk_1x\sin\theta_i} - \cos\theta_r\frac{\Gamma_\perp}{\eta_1}e^{-jk_1x\sin\theta_r} &= \cos\theta_t\frac{T_\perp}{\eta_2}e^{-jk_2x\sin\theta_t}\end{aligned}\right\}. \tag{8.21}$$

These can be satisfied only if

$$\theta_r = \theta_i, \tag{8.22}$$

and

$$k_1\sin\theta_i = k_2\sin\theta_t. \tag{8.23}$$

Equations (8.22) and (8.23) are called Snell laws of reflection and refraction, respectively. With these, we get

$$\left.\begin{aligned}(1 + \Gamma_\perp)e^{-jk_1x\sin\theta_i} &= T_\perp e^{-jk_2x\sin\theta_t}\\ \frac{\cos\theta_i}{\eta_1}(1 - \Gamma_\perp)e^{-jk_1x\sin\theta_i} &= \frac{\cos\theta_t}{\eta_2}T_\perp e^{-jk_2x\sin\theta_t}\end{aligned}\right\}, \tag{8.24}$$

which become

$$\left.\begin{aligned}(1+\Gamma_{\perp}) &= T_{\perp}\\ \frac{\cos\theta_i}{\eta_1}(1-\Gamma_{\perp}) &= \frac{\cos\theta_t}{\eta_2}T_{\perp}\end{aligned}\right\}, \tag{8.25}$$

and hence,

$$\Gamma_{\perp} = \frac{\eta_2\cos\theta_i - \eta_1\cos\theta_t}{\eta_2\cos\theta_i + \eta_1\cos\theta_t}, \tag{8.26}$$

and

$$T_{\perp} = \frac{2\eta_2\cos\theta_i}{\eta_2\cos\theta_i + \eta_1\cos\theta_t}. \tag{8.27}$$

8.2.2 Parallel Polarization

The parallel-polarization case is shown in Fig. 8.5. For the sake of convenience, we will express the magnetic field phasor first, then we will express the electric field phasor using Maxwell equations. From Fig. 8.5, we see that the incident wave has

$$\left.\begin{aligned}\mathbf{k}_i &= k_1(\sin\theta_i\,\hat{\mathbf{x}} + \cos\theta_i\,\hat{\mathbf{z}})\\ \mathbf{H}_i(\mathbf{r}) &= \hat{\mathbf{y}}\,\frac{E_o}{\eta_1}e^{-jk_1(x\sin\theta_i + z\cos\theta_i)}\\ \mathbf{E}_i(\mathbf{r}) &= -\eta_1\hat{\mathbf{k}}_i \times \mathbf{H}_i(\mathbf{r}) = (\cos\theta_i\,\hat{\mathbf{x}} - \sin\theta_i\,\hat{\mathbf{z}})\,E_oe^{-jk_1(x\sin\theta_i + z\cos\theta_i)}\end{aligned}\right\}. \tag{8.28}$$

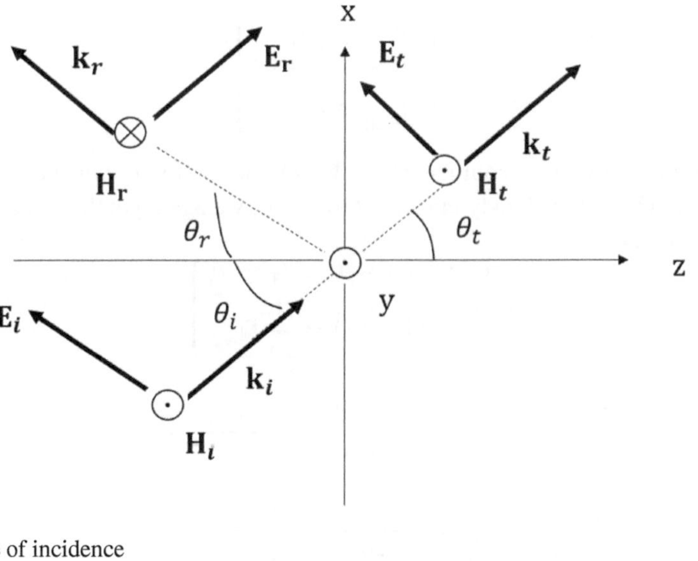

Fig. 8.5 Plane of incidence

Also, the reflected wave has

$$\left.\begin{aligned} \mathbf{k}_r &= k_1(\sin\theta_r\,\hat{\mathbf{x}} - \cos\theta_r\,\hat{\mathbf{z}}) \\ \mathbf{H}_r(\mathbf{r}) &= -\hat{\mathbf{y}}\,\frac{\Gamma_\parallel E_o}{\eta_1} e^{-jk_1(x\sin\theta_r - z\cos\theta_r)} \\ \mathbf{E}_r(\mathbf{r}) &= -\eta_1\hat{\mathbf{k}}_r \times \mathbf{H}_r(\mathbf{r}) = (\cos\theta_r\,\hat{\mathbf{x}} + \sin\theta_r\,\hat{\mathbf{z}})\,\Gamma_\parallel E_o e^{-jk_1(x\sin\theta_r - z\cos\theta_r)} \end{aligned}\right\}. \tag{8.29}$$

Finally, the transmitted wave has

$$\left.\begin{aligned} \mathbf{k}_t &= k_2(\sin\theta_t\,\hat{\mathbf{x}} + \cos\theta_t\,\hat{\mathbf{z}}) \\ \mathbf{H}_t(\mathbf{r}) &= \hat{\mathbf{y}}\,\frac{T_\parallel E_o}{\eta_2} e^{-jk_2(x\sin\theta_t + z\cos\theta_t)} \\ \mathbf{E}_t(\mathbf{r}) &= -\eta_2\hat{\mathbf{k}}_t \times \mathbf{H}_t(\mathbf{r}) = (\cos\theta_t\,\hat{\mathbf{x}} - \sin\theta_t\,\hat{\mathbf{z}})\,T_\parallel E_o e^{-jk_2(x\sin\theta_t + z\cos\theta_t)} \end{aligned}\right\}. \tag{8.30}$$

Now, the total fields in both regions are

$$\left.\begin{aligned} \mathbf{E}_1(\mathbf{r}) &= (\cos\theta_i\,\hat{\mathbf{x}} - \sin\theta_i\,\hat{\mathbf{z}})\,E_o e^{-jk_1(x\sin\theta_i + z\cos\theta_i)} + (\cos\theta_r\,\hat{\mathbf{x}} + \sin\theta_r\,\hat{\mathbf{z}})\,\Gamma_\parallel E_o e^{-jk_1(x\sin\theta_r - z\cos\theta_r)} \\ \mathbf{H}_1(\mathbf{r}) &= \hat{\mathbf{y}}\,\frac{E_o}{\eta_1}[e^{-jk_1(x\sin\theta_i + z\cos\theta_i)} - \Gamma_\parallel e^{-jk_1(x\sin\theta_r - z\cos\theta_r)}] \end{aligned}\right\}, \tag{8.31}$$

and

$$\left.\begin{aligned} \mathbf{E}_2(\mathbf{r}) &= (\cos\theta_t\,\hat{\mathbf{x}} - \sin\theta_t\,\hat{\mathbf{z}})\,T_\parallel E_o e^{-jk_2(x\sin\theta_t + z\cos\theta_t)} \\ \mathbf{H}_2(\mathbf{r}) &= \hat{\mathbf{y}}\,\frac{T_\parallel E_o}{\eta_2} e^{-jk_2(x\sin\theta_t + z\cos\theta_t)} \end{aligned}\right\}. \tag{8.32}$$

To find θ_r, θ_t, $\Gamma_\parallel$ and $T_\parallel$, we apply the boundary conditions

$$\left.\begin{aligned} \mathbf{E}_1^\parallel\Big|_{z=0} &= \mathbf{E}_2^\parallel\Big|_{z=0} \\ \mathbf{H}_1^\parallel\Big|_{z=0} &= \mathbf{H}_2^\parallel\Big|_{z=0} \end{aligned}\right\}. \tag{8.33}$$

After noting that only the x components of $\mathbf{E}_1(\mathbf{r})$ and $\mathbf{E}_2(\mathbf{r})$ are tangential to $z = 0$, and that $\mathbf{H}_1(\mathbf{r})$ and $\mathbf{H}_2(\mathbf{r})$ are already tangential to $z = 0$, we again get the same Snell laws we obtained before, and

$$\left.\begin{aligned} (1 + \Gamma_\parallel)\cos\theta_i &= T_\parallel \cos\theta_t \\ \frac{1}{\eta_1}(1 - \Gamma_\parallel) &= \frac{1}{\eta_2}T_\parallel \end{aligned}\right\}, \tag{8.34}$$

and hence,

$$\Gamma_\parallel = \frac{\eta_2\cos\theta_t - \eta_1\cos\theta_i}{\eta_2\cos\theta_t + \eta_1\cos\theta_i}, \tag{8.35}$$

and

$$T_\parallel = \frac{2\eta_2\cos\theta_i}{\eta_2\cos\theta_t + \eta_1\cos\theta_i}. \tag{8.36}$$

Zero Reflection

Unlike normal incidence, we see that $\Gamma_\perp$ and $\Gamma_\parallel$ can be made zero even when $\eta_1 \neq \eta_2$. This can be done by adjusting the angle of incidence θ_i. Such an angle can be called the polarizing angle θ_p. We next find this angle for both polarization cases.

For perpendicular-polarization case, the polarizing angle can obtained by setting $\Gamma_\perp = 0$, which becomes

$$\eta_2 \cos\theta_p = \eta_1 \cos\theta_t \Rightarrow \eta_2^2(1 - \sin^2\theta_p) = \eta_1^2(1 - \sin^2\theta_t). \tag{8.37}$$

But since

$$k_1 \sin\theta_p = k_2 \sin\theta_t \Rightarrow \sin\theta_t = \frac{k_1}{k_2}\sin\theta_p = \sqrt{\frac{\mu_1\varepsilon_1}{\mu_2\varepsilon_2}}\sin\theta_p, \tag{8.38}$$

we get

$$\sin^2\theta_p = \frac{1 - \mu_2\varepsilon_1/\mu_1\varepsilon_2}{1 - (\varepsilon_1/\varepsilon_2)^2}. \tag{8.39}$$

In practice, nonmagnetic materials are most commonly used in radio frequencies. Therefore, $\mu_1 = \mu_2 = \mu_0$, and thus,

$$\sin^2\theta_p = \frac{1}{1 + \varepsilon_1/\varepsilon_2}. \tag{8.40}$$

Since the right side is always less than unity, a polarizing angle exists for nonmagnetic materials in the perpendicular-polarization case.

For parallel-polarization case, the polarizing angle can obtained by setting $\Gamma_\parallel = 0$, which becomes

$$\eta_2 \cos\theta_t = \eta_1 \cos\theta_p. \tag{8.41}$$

Using

$$\sin\theta_t = \sqrt{\frac{\mu_1\varepsilon_1}{\mu_2\varepsilon_2}}\sin\theta_p, \tag{8.42}$$

we get

$$\sin^2\theta_p = \frac{1 - \mu_1\varepsilon_2/\mu_2\varepsilon_1}{1 - (\mu_1/\mu_2)^2}. \tag{8.43}$$

For nonmagnetic materials, the right side is infinite, indicating that a polarizing angle does not exist for nonmagnetic materials in the parallel-polarization case.

Example 8.3 A plane wave in free space with

$$\mathbf{E}_i(\mathbf{r}) = \hat{\mathbf{y}}\, e^{-j(x+3z)}$$

is incident upon a semi-infinite non dissipative medium $z \geq 0$ with $\varepsilon_r = 4$. Determine (a) the polarization of the wave (e.g., parallel or perpendicular), (b) the angle of incidence and the frequency, (c) $\mathbf{E}_r(\mathbf{r})$, and (d) $\mathbf{E}_t(\mathbf{r})$.

Solution (a) Since the unit normal is $\hat{\mathbf{z}}$, and the propagation vector contains an x and a z component, the plane of incidence is the xz plane. Since $\mathbf{E}_i(\mathbf{r})$ has a y component, this is perpendicular-polarization case.

(b) We have $\hat{\mathbf{n}} \cdot \mathbf{k}_i = k_1 \cos\theta_i$, $\hat{\mathbf{n}} = \hat{\mathbf{z}}$, $\mathbf{k}_i = \hat{\mathbf{x}} + 3\,\hat{\mathbf{z}}$, and $k_1 = \sqrt{1^2 + 3^2} = 3.16$. Hence, $\theta_i = 0.321\,(18.43°)$. Also, $\omega = k_1 c = 9.48 \times 10^8$.

(c) We have $\eta_1 = 120\pi$, $\eta_2 = 120\pi/2 = 60\pi$, $\theta_i = 0.321$, and θ_t can be found from Snell law

$$k_1 \sin\theta_i = k_2 \sin\theta_t \Rightarrow \sin\theta_t = \frac{k_1}{k_2}\sin\theta_i = \sqrt{\frac{\mu_1\varepsilon_1}{\mu_2\varepsilon_2}}\sin\theta_i \Rightarrow \theta_t = 0.158\,(9.076°).$$

Since we have perpendicular polarization, we find $\Gamma_\perp$ from

$$\Gamma_\perp = \frac{\eta_2\cos\theta_i - \eta_1\cos\theta_t}{\eta_2\cos\theta_i + \eta_1\cos\theta_t} = -0.351.$$

Also, $\mathbf{k}_r = k_1(\sin\theta_r\,\hat{\mathbf{x}} - \cos\theta_r\,\hat{\mathbf{z}}) = k_1(\sin\theta_i\,\hat{\mathbf{x}} - \cos\theta_i\,\hat{\mathbf{z}}) = 0.997\,\hat{\mathbf{x}} - 2.99\,\hat{\mathbf{z}}$. Therefore,

$$\mathbf{E}_r(\mathbf{r}) = -0.351\,\hat{\mathbf{y}}\,e^{-j(0.997x - 2.99z)}.$$

(d) We find $T_\perp$ from $T_\perp = 1 + \Gamma_\perp = 0.65$. Also, we find k_2 from $k_2 = \omega\sqrt{\mu_2\varepsilon_2} = 6.32$. Then, $\mathbf{k}_t = k_2(\sin\theta_t\,\hat{\mathbf{x}} + \cos\theta_t\,\hat{\mathbf{z}}) = 0.99\,\hat{\mathbf{x}} + 6.24\,\hat{\mathbf{z}}$. Therefore,

$$\mathbf{E}_t(\mathbf{r}) = 0.65\,\hat{\mathbf{y}}\,e^{-j(0.99x + 6.24z)}.$$

◁

Problems

8.1 A plane wave in free space with

$$\mathbf{E}_i = 4\,\hat{\mathbf{y}}\,e^{-j(5x+4z)}$$

is incident upon a semi-infinite non dissipative medium $x \geq 0$ with $\varepsilon_r = 2$. Determine (a) the polarization of the wave (e.g., parallel or perpendicular), (b) the angle of incidence, and (c) the transmitted angle.

8.2 A plane wave in free space

$$\mathbf{E}_i = \hat{\mathbf{x}}\,e^{-j0.25\,z}$$

is incident normally upon a semi-infinite medium with $\varepsilon_r = 1000$, $\mu_r = 2$, and $\sigma = 3$. The frequency is $\omega = 10^9$. (a) Determine Γ, T, S, and (b) the incident, reflected, and transmitted power.

8.3 Medium 1 is free space and medium 2 has $\varepsilon_r = 3$. An perpendicular-polarized incident plane wave with $\theta_i = 50°$ resulted in a transmitted wave with $\theta_t = 30°$. Determine $\Gamma_{\perp}$.

References

1. M.A. Richards, J.A. Scheer, W.A. Holam, *Principles of Modern Radar: Basic Principles* (SciTech Publishing, 2010)
2. C.A. Balanis, *Engineering Electromagnetics* (Wiley, 2012)

Transmission Lines 9

A real-life application of propagation of electromagnetic waves in an unbounded region is in wireless transmission of information. As another approach, information can be transmitted by means of wires. This brings us to the problem of propagation of electromagnetic waves in bounded regions. In particular, electromagnetic waves will be guided by a structure, called a waveguide, which can take many forms. A simple form can be realized as an infinitely-long hollow conducting tube of a certain cross-sectional shape (e.g., rectangular, circular, elliptical, etc.), aligned along a certain direction, and composed of a certain medium (see Fig. 9.1a for a rectangular waveguide aligned along the z axis).

Another form is constructed by making two conducting tubes occupied by a certain medium, better known as a transmission line (see Fig. 9.1b for a rectangular transmission line aligned along the z axis). Even though a transmission line consists of two conductors, whereas a waveguide consists of one, the analysis of the former is much simpler than that of the later under the assumption that the medium bounded by the two conductors in the transmission line is isotropic dielectric magnetic and that the conductors are PECs. This is attributed to the fact that satisfaction of the boundary conditions on the two conductors leads to TEM mode of propagation. That is, the propagation of electromagnetic fields in a transmission line is similar to that in an unbounded region. Moreover, due to the TEM mode of propagation, the analysis of transmission lines can be made using circuit theory instead of field theory. In this chapter, we first consider the transition from field theory to circuit theory in transmission lines in Sect. 9.1. We then discuss the governing equations of transmission lines and their time-harmonic solutions, which yields transmission lines' characteristics in Sect. 9.2. Input impedance is then discussed in Sect. 9.3, and finally, transmission lines transients is discussed in Sect. 9.4.

H. M. Alkhoori, *Concise Introduction to Electromagnetic Fields*, Synthesis Lectures on Electromagnetics, https://doi.org/10.1007/978-3-031-60331-0_9

9.1 From Field Theory to Circuit Theory

9.1.1 Voltage and Current Waves

In this section, we give a brief overview of conversion from field theory to circuit theory in transmission line theory. With reference to Fig. 9.1b, let us denote one conductor by C_1 and the other by C_2. Both conductors are assumed to be PECs. Assuming the medium occupied by the region between C_1 and C_2 to be isotropic dielectric magnetic characterized by ε and μ, satisfaction of boundary conditions on C_1 and C_2 is permitted only for TEM mode [1]. Then, from frequency-domain Maxwell equations, on writing

$$\nabla = \nabla_t + \hat{\mathbf{z}}\frac{\partial}{\partial z}, \tag{9.1}$$

where $\nabla_t = \hat{\mathbf{x}}\,\partial_x + \hat{\mathbf{y}}\,\partial_y$ is the transverse del operator, one can show that the electric field and magnetic field phasors can be written, respectively, as [1, 2]

$$\mathbf{E}(\mathbf{r}) = -\nabla_t V(x, y)e^{\mp jkz}, \tag{9.2}$$

and

$$\mathbf{H}(\mathbf{r}) = \pm\frac{1}{\eta}\hat{\mathbf{z}} \times \nabla_t V(x, y)e^{\mp jkz}, \tag{9.3}$$

where $k = \omega\sqrt{\mu}\sqrt{\varepsilon}$, $\eta = \sqrt{\mu/\varepsilon}$, and $V(x, y)$ is the solution of

$$\nabla_t^2 V(x, y) = 0 \tag{9.4}$$

in the region between the two PECs, subject to the condition $V(x, y) = 0$ on the PECs C_1 and C_2. Figure 9.2 shows streamlines plots of $\mathbf{E}$ and $\mathbf{H}$ inside a coaxial transmission line on the plane $z = 0$. Notice that both of $\mathbf{E}$ and $\mathbf{H}$ are similar to those of the static regime.

The potential difference between C_1 and C_2 on any plane is

$$V(z) = -\int_{C_1}^{C_2} \mathbf{E}(\mathbf{r}) \bullet d\mathbf{l} = e^{\mp jkz} \int_{C_1}^{C_2} \nabla_t V(x, y) \bullet d\mathbf{l} = V_o e^{\mp jkz}, \tag{9.5}$$

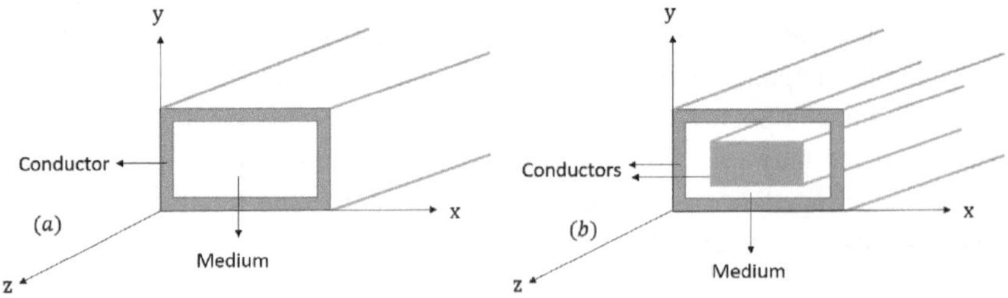

Fig. 9.1 **a** A rectangular waveguide, and **b** a transmission line

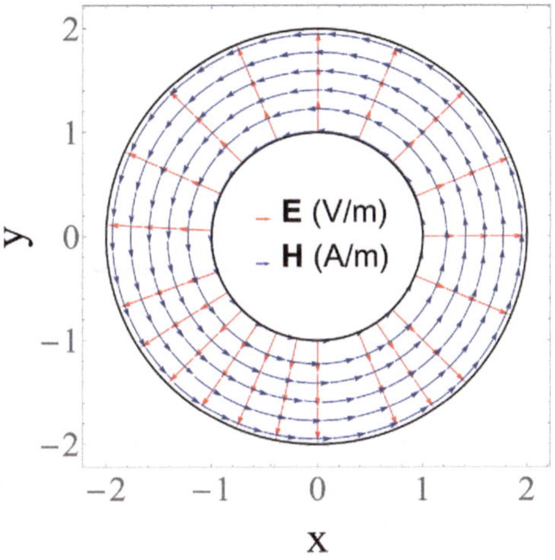

Fig. 9.2 Streamline plots of **E** and **H** for a coaxial transmission line on $z = 0$

where V_o is the potential difference between the two PECs on the plane $z = 0$. Notice that the potential difference here is a wave function. Therefore, we can call $V(z)$ in Eq. (9.5) the voltage wave phasor.

On either PECs, the surface charge density phasor $\rho_s(\mathbf{r})$ is found as

$$\rho_s(\mathbf{r}) = \hat{\mathbf{n}} \cdot \mathbf{D} = -\varepsilon e^{\mp jkz} \hat{\mathbf{n}} \cdot \nabla_t V(x, y). \tag{9.6}$$

Similarly, the surface current density phasor $\mathbf{K}(\mathbf{r})$ is found as

$$\mathbf{K}(\mathbf{r}) = \hat{\mathbf{n}} \times \mathbf{H} = \mp \frac{1}{\eta} \varepsilon e^{\mp jkz} \hat{\mathbf{n}} \times \hat{\mathbf{z}} \times \nabla_t V(x_s, y_s) = \mp \hat{\mathbf{z}} \frac{\varepsilon}{\eta} e^{\mp jkz} \hat{\mathbf{n}} \cdot \nabla_t V(x_s, y_s). \tag{9.7}$$

Now, the total current $I(z)$ around a closed curve $\mathcal{L}$ representing the PEC with positive charge is given by

$$I(z) = \oint_{\mathcal{L}} \mathbf{K}(\mathbf{r}) \cdot (\hat{\mathbf{n}} \times d\mathbf{l}) = \oint_{\mathcal{L}} \mathbf{K}(\mathbf{r}) \cdot \hat{\mathbf{z}}\, d\mathcal{L} = \pm \frac{1}{\varepsilon \eta} \oint_{\mathcal{L}} \rho_s(\mathbf{r})\, d\mathcal{L} = I_o e^{\mp jkz}, \tag{9.8}$$

where I_o is obtained on setting $z = 0$ in $\rho_s(\mathbf{r})$. Notice that the total current here is a wave function. Therefore, we can call $I(z)$ in Eq. (9.8) the current wave phasor. Therefore, the quantities associated with field theory of transmission lines are **E** and **H**, while the quantities associated with circuit theory of transmission lines are V and I.

9.1.2 Transmission Line Parameters

In field theory, the medium between the two PECs in a transmission line is described by ε and μ. From the definition of ρ_s, one can define the total charge per unit length l of a PEC. Then,

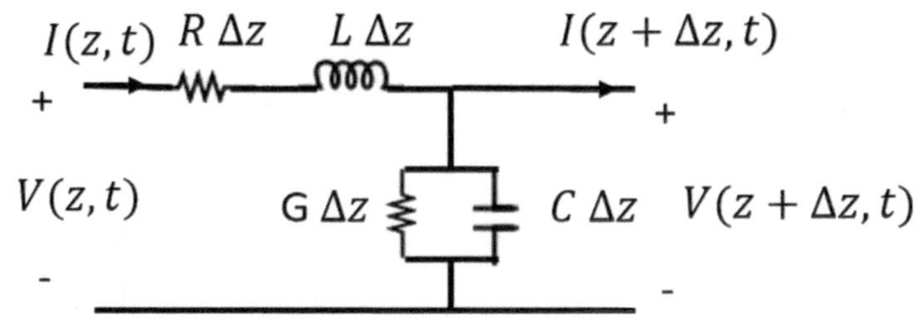

Fig. 9.3 Equivalent circuit of a transmission line for a length Δz

the capacitance per unit length C of transmission line can be found accordingly [3]. Similarly, from the definition of $\mathbf{K}$, one can define the total magnetic flux per unit length l of a PEC. Then, the inductance L per unit length of transmission line can be found accordingly [3]. In addition to ε and μ, the medium between the two conductors can be dissipative described by a nozero conductivity σ, and the conductors can be slightly non perfect described by a finite conductivity σ_c. In circuit theory, the nozero conductivity σ is translated into a resistance per unit length R, while the finite conductivity σ_c is translated into a conductance per unit length G [1]. An equivalent circuit for a length Δz of a transmission line is shown in Fig. 9.3. Therefore, the parameters associated with field theory of transmission lines are ε, μ, σ, and σ_c, while those associated with circuit theory of transmission lines are C, L, R, and G.

9.2 Telegrapher Equations

So far, we saw that the circuit theory of a transmission line consists of voltage and current wave phasors, as well as resistance, capacitance, inductance, and conductance. We next derive equations governing the voltage and current in a transmission line in time domain. Hence, we switch to time domain by $V(z) \rightarrow \mathscr{V}(z,t)$ and $I(z) \rightarrow \mathscr{I}(z,t)$. With reference to Fig. 9.3, using Kirchhoff voltage law, we get

$$\mathscr{V}(z,t) = R\Delta z \mathscr{I}(z,t) + L\Delta z \frac{\partial \mathscr{I}(z,t)}{\partial t} + \mathscr{V}(z+\Delta z,t), \tag{9.9}$$

which becomes

$$-\frac{\mathscr{V}(z+\Delta z,t) - \mathscr{V}(z,t)}{\Delta z} = R\mathscr{I}(z,t) + L\frac{\partial \mathscr{I}(z,t)}{\partial t}. \tag{9.10}$$

Taking the limit $\Delta z \rightarrow 0$, we get

$$-\frac{\partial \mathscr{V}(z,t)}{\partial z} = R\mathscr{I}(z,t) + L\frac{\partial \mathscr{I}(z,t)}{\partial t}. \tag{9.11}$$

This is a coupled, first order partial differential equation. In order to solve for $\mathscr{V}$ and $\mathscr{I}$, we need another equation. The other equation can be obtained using Kirchhoff current law as

$$\mathscr{I}(z,t) = G\Delta z\mathscr{V}(z,t) + C\Delta z\frac{\partial \mathscr{V}(z,t)}{\partial t} + \mathscr{I}(z+\Delta z,t), \tag{9.12}$$

which becomes

$$-\frac{\mathscr{I}(z+\Delta z,t) - \mathscr{I}(z,t)}{\Delta z} = G\mathscr{V}(z,t) + C\frac{\partial \mathscr{V}(z,t)}{\partial t} \tag{9.13}$$

Taking the limit $\Delta z \to 0$, we get

$$-\frac{\partial \mathscr{I}(z,t)}{\partial z} = G\mathscr{V}(z,t) + C\frac{\partial \mathscr{V}(z,t)}{\partial t}. \tag{9.14}$$

Therefore, the transmission line is governed by

$$\left.\begin{aligned} -\frac{\partial \mathscr{V}(z,t)}{\partial z} &= R\mathscr{I}(z,t) + L\frac{\partial \mathscr{I}(z,t)}{\partial t} \\ -\frac{\partial \mathscr{I}(z,t)}{\partial z} &= G\mathscr{V}(z,t) + C\frac{\partial \mathscr{V}(z,t)}{\partial t} \end{aligned}\right\}. \tag{9.15}$$

This coupled set of differential equations is called the Telegrapher equations. To solve, we need to decouple these equations by differentiating each one with respect to z and t, and substituting one into the other. Then we get

$$\frac{\partial^2 \mathscr{V}(z,t)}{\partial z^2} = LC\,\frac{\partial^2 \mathscr{V}(z,t)}{\partial t^2} + (RC+GL)\frac{\partial \mathscr{V}(z,t)}{\partial t} + RG\,\mathscr{V}(z,t), \tag{9.16}$$

and

$$\frac{\partial^2 \mathscr{I}(z,t)}{\partial z^2} = LC\,\frac{\partial^2 \mathscr{I}(z,t)}{\partial t^2} + (RC+GL)\frac{\partial \mathscr{I}(z,t)}{\partial t} + RG\,\mathscr{I}(z,t). \tag{9.17}$$

Each one of $\mathscr{V}(z,t)$ and $\mathscr{I}(z,t)$ satisfies the wave equation. This clearly indicates that the voltage and current are waves. Under time harmonic assumption, in phasor domain, solutions can be found as

$$V(z) = V_0^+ e^{-\gamma z} + V_0^- e^{\gamma z}, \tag{9.18}$$

and

$$I(z) = \frac{1}{Z_o}\left(V_0^+ e^{-\gamma z} - V_0^- e^{\gamma z}\right), \tag{9.19}$$

where

$$\gamma = \alpha + jk = \sqrt{(R+j\omega L)(G+j\omega C)} \tag{9.20}$$

is the propagation constant, and

$$Z_o = \frac{R + j\omega L}{\gamma} = \frac{\gamma}{G + j\omega C} = \sqrt{\frac{R + j\omega L}{G + j\omega C}} \tag{9.21}$$

is the characteristic impedance. The characteristic impedance is simply the ratio between the positive voltage wave amplitude to the positive current wave amplitude, or negative the ratio between the negative voltage wave amplitude to the negative current wave amplitude.

9.3 Input Impedance

Consider the situation where a transmission line of length l, characterized by γ and Z_o, connected to a source with V_s and Z_s from one side, and connected to a load with Z_L from the other side (see Fig. 9.4). When designing a circuit, the designer needs to know the input impedance Z_{in} of the circuit so that the designer knows, for instance, the amount of current needed for the load. Our task next is to find the input impedance Z_{in} seen by the source.

Originally, an incident voltage wave propagates from the source toward the load (i.e., to the right). Then, when it encounters the load, there will be a reflection, resulting in a reflected voltage wave propagating from the load toward the source. Therefore, the voltage and current wave phasors become

$$V(z) = V_0^+ e^{-\gamma z} + V_0^- e^{\gamma z} = V_o^+ \left(e^{-\gamma z} + \Gamma e^{\gamma z}\right), \tag{9.22}$$

where $V_o^- = \Gamma V_o^+$, and

$$I(z) = \frac{1}{Z_o}\left(V_0^+ e^{-\gamma z} - V_0^- e^{\gamma z}\right) = \frac{V_o^+}{Z_o}\left(e^{-\gamma z} - \Gamma e^{\gamma z}\right). \tag{9.23}$$

The impedance function $Z(z)$ can be defined as

$$Z(z) = \frac{V(z)}{I(z)} = Z_o \frac{e^{-\gamma z} + \Gamma e^{\gamma z}}{e^{-\gamma z} - \Gamma e^{\gamma z}}. \tag{9.24}$$

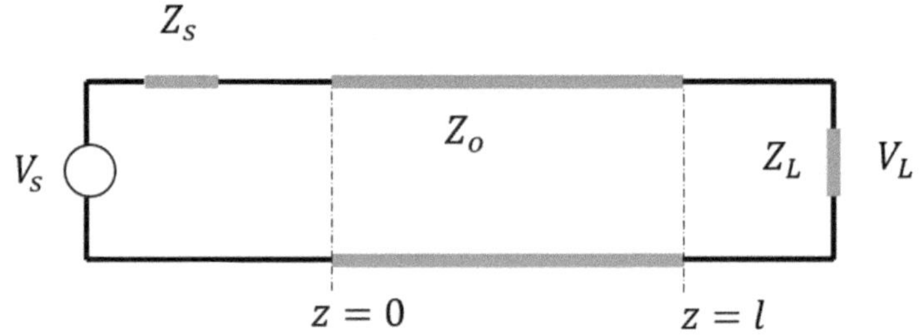

Fig. 9.4 Circuit for input impedance calculation

This impedance function is the input impedance Z_{in} at $z = 0$, and it is the load impedance Z_L at $z = l$. Hence,

$$Z_{\text{in}} = Z(z=0) = \frac{V(z=0)}{I(z=0)} = \frac{1+\Gamma}{1-\Gamma}. \tag{9.25}$$

However, Γ is unknown. To find it, we note that at the load, $Z(z = l) = Z_L$. Thus,

$$Z(z=l) = \frac{V(z=l)}{I(z=l)} = Z_o \frac{e^{-\gamma l} + \Gamma e^{\gamma l}}{e^{-\gamma l} - \Gamma e^{\gamma l}} = Z_o \frac{1 + \Gamma e^{2\gamma l}}{1 - \Gamma e^{2\gamma l}} = Z_L. \tag{9.26}$$

Solving for Γ, we get

$$\Gamma = \frac{Z_L - Z_o}{Z_L + Z_o} e^{-2\gamma l}. \tag{9.27}$$

Now, the input impedance Z_{in} becomes

$$Z_{\text{in}} = \frac{1+\Gamma}{1-\Gamma} = \frac{1 + \dfrac{Z_L - Z_o}{Z_L + Z_o} e^{-2\gamma l}}{1 - \dfrac{Z_L - Z_o}{Z_L + Z_o} e^{-2\gamma l}}, \tag{9.28}$$

which becomes

$$Z_{\text{in}} = \frac{(Z_L + Z_o) + (Z_L - Z_o)e^{-2\gamma l}}{(Z_L + Z_o) - (Z_L - Z_o)e^{-2\gamma l}} = \frac{(Z_L + Z_o)e^{\gamma l} + (Z_L - Z_o)e^{-\gamma l}}{(Z_L + Z_o)e^{\gamma l} - (Z_L - Z_o)e^{-\gamma l}}. \tag{9.29}$$

This can be written as

$$Z_{\text{in}} = Z_o \frac{Z_L \cosh \gamma l + Z_o \sinh \gamma l}{Z_o \cosh \gamma l + Z_L \sinh \gamma l}, \tag{9.30}$$

which finally becomes

$$Z_{\text{in}} = Z_o \frac{Z_L + Z_o \tanh \gamma l}{Z_o + Z_L \tanh \gamma l}. \tag{9.31}$$

To compute $\tanh(x \pm jy)$, one can use

$$\tanh(x \pm jy) = \frac{\sinh 2x \pm j \sin 2y}{\cosh 2x + \cos 2y}. \tag{9.32}$$

It can be shown that

$$|Z_{\text{in}}|\big|_{\max} = S Z_o, \tag{9.33}$$

and

$$|Z_{\text{in}}|\big|_{\min} = \frac{Z_o}{S}. \tag{9.34}$$

In practice, it is more desired to design a non dissipative transmission line. Then, $\gamma = jk$, and using $\tanh jk = j \tan k$, the input impedance becomes

$$Z_{\text{in}} = Z_o \frac{Z_L + j Z_o \tan kl}{Z_o + j Z_L \tan kl}. \tag{9.35}$$

In literature, the quantity kl is well known as the electrical length. The following special cases can be obtained as follows.

- Shorted line ($Z_L = 0$): In this case, the input impedance becomes

$$Z_{\text{in}}\Big|_{Z_L=0} = jZ_o \tan kl \equiv Z_{\text{sc}}. \tag{9.36}$$

- Open-circuited line ($Z_L = \infty$): In this case, the input impedance becomes

$$Z_{\text{in}}\Big|_{Z_L=\infty} = -jZ_o \cot kl \equiv Z_{\text{oc}}. \tag{9.37}$$

- Matched line ($Z_L = Z_o$): In this case, the input impedance becomes

$$Z_{\text{in}}\Big|_{Z_L=Z_o} = Z_o. \tag{9.38}$$

Note that

$$Z_{\text{sc}} Z_{\text{oc}} = Z_o^2. \tag{9.39}$$

9.3.1 Load Power

Assuming a non dissipative transmission line, the average power at the load is given by

$$\begin{aligned} P_{\text{ave}} &= \frac{1}{2}\text{Re}\{V(l)I(l)^*\} \\ &= \frac{1}{2}\text{Re}\{V_o^+\left(e^{-jkz} + \Gamma e^{jkz}\right)\frac{V_o^{+*}}{Z_o}\left(e^{jkz} - \Gamma^* e^{-jkz}\right)\}, \end{aligned} \tag{9.40}$$

which becomes

$$P_{\text{ave}} = \frac{|V_o^+|^2}{2Z_o}(1 - |\Gamma|^2). \tag{9.41}$$

This result shows that the transmitted power is the difference between the incident power, and the reflected power. This indicates that for an optimum design, the reflection coefficient has to be minimized.

Example 9.1 A transmission line with $\gamma = 0.5 + j\ \text{m}^{-1}$, $l = 2\ \text{m}$, and $Z_o = 50 + j40\ \Omega$. The source has $V_s = 10\ V$, $Z_s = 30\ \Omega$, and the load has $Z_L = 30 + j50\ \Omega$. Determine (a) the input impedance Z_{in}, (b) the input current at the source [i.e., $I(z = 0)$], (c) the current at the middle of the line [i.e., $I(z = l/2)$], and (d) the current at the load [i.e., $I(z = l)$].

Solution (a) $\gamma l = 1 + j2$. Then,

$$\tanh \gamma l = 1.1667 - j0.243.$$

Hence,

$$Z_{\text{in}} = Z_o \frac{Z_L + Z_o \tanh \gamma l}{Z_o + Z_L \tanh \gamma l} = 50.18 + j\,36.83\,\Omega.$$

(b)

$$I(z = 0) = \frac{V_s}{Z_{\text{in}} + Z_s} = 0.113 e^{-j24.67^\circ}\ A.$$

(c) At any point,

$$I(z) = \frac{V_o^+}{Z_o}\left(e^{-\gamma z} - \Gamma e^{\gamma z}\right).$$

At $z = l/2 = 1$, this becomes

$$I(z = 1) = \frac{V_o^+}{Z_o}\left(e^{-\gamma} - \Gamma e^{\gamma}\right).$$

So, we need Γ and V_o^+. We find Γ from

$$\Gamma = \frac{Z_L - Z_o}{Z_L + Z_o} e^{-2\gamma l} = 0.068 e^{-j9.52^\circ}.$$

To find V_o^+, we notice that

$$I(z = 0) = \frac{V_o^+}{Z_o}(1 - \Gamma).$$

Hence,

$$V_o^+ = \frac{Z_o I(z = 0)}{1 - \Gamma} = 7.78 e^{j13.29^\circ}\ V.$$

Therefore,

$$I(z = 1) = 0.078 e^{-j92.36^\circ}\ A.$$

(d) At $z = l = 2$, the current is

$$I(z = 2) = \frac{V_o^+}{Z_o}\left(e^{-2\gamma} - \Gamma e^{2\gamma}\right) = 0.063 e^{-j126.8^\circ}\ A.$$

◁

Remarks

We saw that voltage and current are waves. The question is, when does the wave nature of voltages and currents become significant? This question can be answered in two ways. First of all, let us note that the electrical length of any device

$$kl = 2\pi \frac{l}{\lambda} \tag{9.42}$$

is proportional to the ratio l/λ. Without loss of generality, let us consider non dissipative case only.

- The time-domain voltage wave can be written as

$$\mathscr{V}(z,t) = V_o^+ \cos(\omega t - kz) = V_o^+(\cos\omega t \cos kz + \sin\omega t \sin kz). \tag{9.43}$$

  Notice that z is at most l. Hence, the maximum of kz is kl. At 60 Hz, for instance, we note that $\lambda = c/f = 5 \times 10^6$ m. Thus, if the length of the transmission line is small (e.g., l is in order of few meters), the quantity kz becomes small, and thus, $\sin kz \approx 0$ and $\cos kz \approx 1$. Therefore, the time-domain voltage wave

$$\mathscr{V}(z,t) \approx V_o^+ \cos\omega t \equiv \mathscr{V}(t) \tag{9.44}$$

  reduces to the conventional voltage function used in elementary circuit courses.
- As another way, the input impedance of a transmission line is

$$Z_{\text{in}} = Z_o \frac{Z_L + jZ_o \tan kl}{Z_o + jZ_L \tan kl}. \tag{9.45}$$

  If kl is small, then $\tan kl \approx 0$, and thus, the input impedance becomes the load impedance,

$$Z_{\text{in}} = Z_L. \tag{9.46}$$

  In other words, the transmission line has no effect.

We conclude that when the electrical length of the device under consideration is not small (i.e., the length scale is comparable to the wavelength), the wave nature has to be taken into account. In power systems, and because low frequency is used, voltage and current are not regarded as waves when the length scale of the transmission line is small. When the length scale is large (e.g., in long transmission line systems), the wave nature has to be taken into account. However, since substations are extensively placed in long transmission line systems for secure power transmission, the long transmission line is divided into segments, and each segment becomes a short line, in which the wave nature is not significant. Therefore, it is rare to speak of wave nature in power systems. In communication systems, most current applications belong either to the radio wave spectral regime (i.e., f between 10 kHz to 0.1 GHz, or λ between 1 km to 1 cm), or to the mirowave spectral regime (i.e., f between 0.1 GHz to 1 THz, or λ between 1 cm to 1 10 μm). In radio wave applications, it is satisfactory to dismiss the wave nature and use some approximation-based techniques instead. On the other hand, the wave nature is more significant in the mirowave applications. This brings an important discipline in applied electromagnetics, known as microwave engineering [2].

9.3.2 Impedance Matching

We have seen that reflection in transmission lines is an undesired phenomenon. Reflection occurs because of the impedance mismatch between the transmission line and the load. Many approaches have been adopted to overcome impedance mismatch issue. Here we discuss one of these approaches, called quarter-wave transformer. Suppose that a load of impedance Z_L is connected to a transmission line with an impedance Z_o. There will be a reflection Γ between the line and the load of an amount

$$\Gamma = \frac{Z_o - Z_L}{Z_o + Z_L}. \tag{9.47}$$

To overcome this issue, we insert an additional line of an unknown impedance Z_o' of length $l = \lambda/4$ (i.e., $kl = \pi/2$) between the actual transmission line and the load. Then, the impedance in front of the additional line becomes

$$Z_{\text{in}} = Z_o' \frac{Z_L + jZ_o' \tan kl}{Z_o' + jZ_L \tan kl} = \frac{Z_o'^2}{Z_L}. \tag{9.48}$$

In order to eliminate reflection, we want $Z_{\text{in}} = Z_o$. Then, Z_o' must be

$$Z_o' = \sqrt{Z_o Z_L}. \tag{9.49}$$

In other words, we insert a line of length $\lambda/4$ between the line of Z_o and the load of Z_L. Then, we choose the impedance of the inserted line to be $\sqrt{Z_o Z_L}$. One of the drawbacks of using a quarter-wave transformer is when the load impedance is complex. Then, Z_o' must be complex, which adds undesired attenuation. Also, even if the load is purely resistive, when the operating frequency changes, the wavelength will change, and hence, the already chosen length l will no more be beneficial. Therefore, other techniques have been employed for impedance matching.

Example 9.2 A dipole antenna of impedance $Z_L = 73\,\Omega$ at $f = 400$ MHz is connected to a coaxial cable of impedance $Z_o = 50\,\Omega$. Determine l and Z_o' of the quarter-wave transformer.

Solution Since $\lambda = c/f = 75$ cm, we get $l = \lambda/4 = 18.75$ cm. So, $Z_o' = \sqrt{73 \times 50} = 60.42\,\Omega$. ◁

9.4 Transmission Line Transients

We next examine a situation where a non sinusoidal wave propagates along a transmission line. In such a case, time-domain solution of the voltage and current wave equations has to be implemented. Alternatively, we will adopt a different approach applicable only for

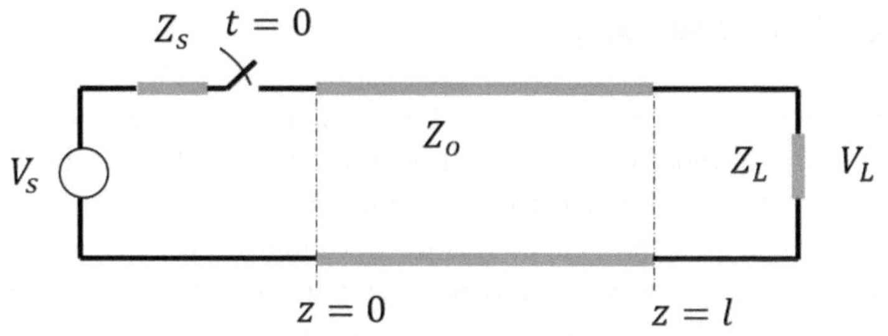

Fig. 9.5 Circuit with a switch

step and rectangular signals. Consider a circuit where a source V_s with real impedance Z_s connected to a non dissipative transmission line of impedance Z_o and length l through a switch that closes at time $t = 0$. The line is terminated by a purely resistive load Z_L (see Fig. 9.5). At $t = 0^+$, the source sees the transmission line only. So, the equivalent circuit consists of V_s, Z_s, and Z_o only. Therefore, at the source, the voltage and current are

$$\mathscr{V}(0, 0^+) = \frac{Z_o}{Z_o + Z_s} V_s = V_o, \tag{9.50}$$

and

$$\mathscr{I}(0, 0^+) = \frac{V_s}{Z_o + Z_s} = I_o. \tag{9.51}$$

At the load, these are

$$\mathscr{V}(l, 0^+) = 0, \tag{9.52}$$

and

$$\mathscr{I}(l, 0^+) = 0. \tag{9.53}$$

Then, after time $t_1 = l/v$, where $v = 1/\sqrt{LC}$, the voltage and current at the load become

$$\mathscr{V}(l, t_1) = V_o + \Gamma_L V_o, \tag{9.54}$$

and

$$\mathscr{I}(l, t_1) = I_o - \Gamma_L I_o, \tag{9.55}$$

where

$$\Gamma_L = \frac{Z_L - Z_o}{Z_L + Z_o}. \tag{9.56}$$

Then, after time $t_2 = 2t_1$, the voltage and current at the source become

$$\mathscr{V}(0, 2t_1) = V_o + \Gamma_L V_o + \Gamma_s \Gamma_L V_o, \tag{9.57}$$

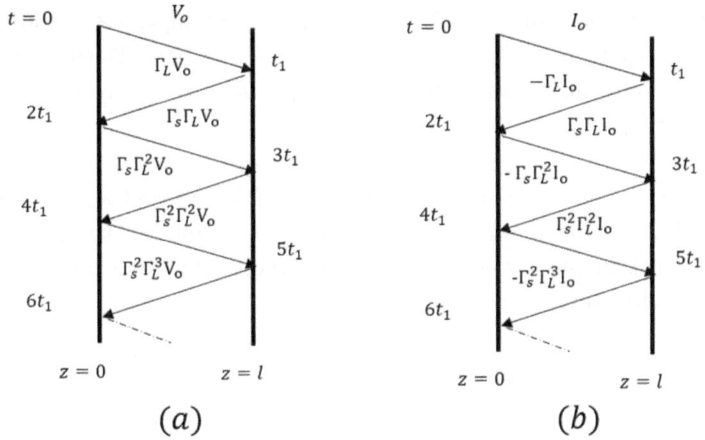

Fig. 9.6 **a** Voltage bounce diagram, and **b** current bounce diagram

and

$$\mathcal{I}(0, 2t_1) = I_o - \Gamma_L I_o + \Gamma_s \Gamma_L I_o, \tag{9.58}$$

where

$$\Gamma_s = \frac{Z_s - Z_o}{Z_s + Z_o}. \tag{9.59}$$

The process continues until steady state is reached. The overall process can be described by the so-called bounce diagrams shown in Fig. 9.6.

Problems

9.1 A coaxial transmission line with radii a and b $(b > a)$ has the following field expressions in $a < \rho < b$.

$$\left.\begin{aligned} \mathbf{E} &= \frac{V_o}{\rho \ln(b/a)} e^{-jkz}\, \hat{\boldsymbol{\rho}} \\ \mathbf{H} &= \frac{I_o}{2\pi\rho} e^{-jkz}\, \hat{\boldsymbol{\phi}} \end{aligned}\right\},$$

where V_o is the potential difference between $\rho = a$ and $\rho = b$, and I_o is the current flowing on $\rho = a$ and $\rho = b$. Obtain expressions for the capacitance per unit length and inductance per unit length.

9.2 A transmission line operating at $f = 1$ MHz has $R = 30\ \Omega$/km, $L = 100$ mH/km, $G = 0$, and $C = 20\ \mu$F/km. Determine γ and Z_0.

9.3 A transmission line with $\gamma = 0.25 + j0.25\ \mathrm{m}^{-1}$, $l = 2$ m, and $Z_o = 50 + j40\ \Omega$. The source has $V_s = 10$ V, $Z_s = 0\ \Omega$, and the load has $Z_L = 10\ \Omega$. Determine (a) the input impedance Z_{in}, (b) the input current at the source [i.e., $I_s(z = 0)$], and (c) the current at the load [i.e., $I_s(z = l)$].

9.4 A transmission line with $\gamma = 0.15 + j0.5\ \mathrm{m}^{-1}$, $l = 2$ m, and $Z_o = 50\ \Omega$. The source has $V_s = 10$ V, $Z_s = 10\ \Omega$, and the load has $Z_L = 60 - j40\ \Omega$. Determine (a) the input impedance Z_{in}, (b) the input voltage at the source [i.e., $V_s(z = 0)$], and (c) the voltage at the middle of the line [i.e., $V_s(z = l/2)$].

References

1. R.E. Collin, *Field Theory of Guided Waves* (IEEE Press, 1991)
2. D.M. Pozar, *Microwave Engineering* (Wiley, 2011)
3. E.J. Rothwell, M.J. Cloud, *Electromagnetics*, 3rd edn. (CRC Press, 2018)

Waveguides and Cavities

10

Transmission lines can be used to transmit waves effectively below the microwave regime. In the microwave regime and beyond, they become less efficient due to high attenuation. As an alternative approach, waveguides can be used to transmit waves. Unlike transmission lines, waveguides do not support TEM mode. Consequently, circuit theory is no more valid and restoration to field theory is a must. Recall that a waveguide can be realized as an infinitely-long hollow conducting tube of a certain cross-sectional shape aligned along a certain direction, and composed of a certain medium. When the ends of a waveguide are closed, it becomes an energy-storing device called a resonator (or a cavity). In this chapter, we limit our discussion to the case where the cross sectional shape of the waveguide is rectangular, and the medium is isotropic dielectric magnetic. Also, the alignment axis will taken to be the z axis. First, we give a short review on some required mathematical background in Sect. 10.1. This includes boundary-value problems, from which eigenvalue problems are defined. Then, we discuss the solution of Maxwell equations inside a waveguide in Sect. 10.2. As to be shown, three modes result, called transverse electric (TE), transverse magnetic (TM) modes, and hybrid (HE) modes. The TM and TE are discussed, respectively, in in Sects. 10.3 and 10.4. We then discuss power transmission and attenuation in waveguides in Sect. 10.5. Finally, we give a brief overview of rectangular cavities in Sect. 10.6. A useful computer program is given in the appendix at the end of the chapter. For a shorthand notation, let us denote $\partial_u \equiv \frac{\partial}{\partial u}$, where $u = \{x, y, z\}$, in this chapter.

H. M. Alkhoori, *Concise Introduction to Electromagnetic Fields*, Synthesis Lectures on Electromagnetics, https://doi.org/10.1007/978-3-031-60331-0_10

10.1 Boundary-Value Problems

The equation

$$y'' + k^2 y = 0,$$

where k^2 is a real constant, is a second-order ordinary differential equation (ODE). To solve it, we let

$$y(x) = ce^{mx}, \tag{10.1}$$

where c is a constant, and m is an unknown constant that has to be determined. After we substitute $y(x)$ into the ODE, we get

$$m^2 + k^2 = 0. \tag{10.2}$$

Then, we get two solutions as

$$y = c_1 e^{jkx} + c_2 e^{-jkx}. \tag{10.3}$$

Using Euler identity, solution can be written as

$$y(x) = c_1 \sin kx + c_2 \cos kx. \tag{10.4}$$

An ODE with boundary conditions is called a boundary-value problem [1–4]. An example is

$$\left.\begin{array}{ll} y'' + k^2 y = 0 & a < x < b \\ y(a) = K_1 & \\ y(b) = K_2 & \end{array}\right\}, \tag{10.5}$$

where K_1 and K_2 are given constants.
Example 10.1 Solve

$$\left.\begin{array}{ll} y'' + 25y = 0 & 0 < x < \pi/10 \\ y(0) = 1 & \\ y(\pi/10) = 1 & \end{array}\right\}. \tag{10.6}$$

Solution. $y = \cos 5x + \sin 5x$. ◁

An eigenvalue problem is a boundary-value problem of the form

$$\left.\begin{array}{ll} \phi'' + \lambda\phi = 0, & a < x < b \\ \phi(a) = 0 & \\ \phi(b) = 0 & \end{array}\right\},$$

where a and b are finite, λ are unknown real constants that have to be determined, and $\phi(x)$ are unknown nontrivial functions that have to be determined. The constants λ are called eigenvalues, and the constant $\phi(x)$ are called eigenfunctions.

Example 10.2 Solve the eigenvalue problem (i.e., find the eigenvalues and the eigenfunctions)

$$\left.\begin{array}{ll}\phi'' + \lambda\phi = 0 & 0 < x < L \\ \phi(0) = 0 & \\ \phi(L) = 0 & \end{array}\right\} .$$

Solution. We begin with solving the ODE. Notice that this a constant-coefficient ODE with an unknown parameter λ. So, we let

$$\phi(x) = ce^{mx}.$$

Then, we get

$$(m^2 + \lambda)\phi(x) = 0.$$

For a non-trivial $\phi(x)$, we get

$$m^2 + \lambda = 0 \Rightarrow m = \sqrt{-\lambda}.$$

At this point, λ is unknown. But since it is real, it can either be positive, negative, or zero. For each case, we write the corresponding solution. That is,

$$\phi(x) = \begin{cases} c_1 + c_2 x, & \lambda = 0, \\ c_1 \cosh\sqrt{\lambda}x + c_2 \sinh\sqrt{\lambda}x, & \lambda < 0, \\ c_1 \cos\sqrt{\lambda}x + c_2 \sin\sqrt{\lambda}x, & \lambda > 0. \end{cases}$$

If $\lambda = 0$, then application of boundary conditions on the associated solution gives $\phi(0) = 0 = c_1$ and $\phi(L) = c_2 L = 0 \Rightarrow c_2 = 0$. Therefore, $\phi(x) = 0$. But, $\phi(x) = 0$ is the trivial solution; hence, $\lambda = 0$ is not an eigenvalue. If $\lambda < 0$, then application of boundary conditions on the associated solution gives $\phi(0) = 0 = c_1$ and $\phi(L) = c_2 \sinh\sqrt{\lambda}L = 0 \Rightarrow c_2 = 0$ since $\sinh\sqrt{\lambda}L \neq 0$ for $\lambda \neq 0$. Therefore, $\phi(x) = 0$. Again, since $\phi(x) = 0$ is the trivial solution, $\lambda < 0$ are not eigenvalues. If $\lambda > 0$, then application of boundary conditions on the associated solution gives $\phi(0) = 0 = c_1$ and $\phi(L) = c_2 \sin\sqrt{\lambda}L = 0 \Rightarrow \sin\sqrt{\lambda}L = 0 \Rightarrow \lambda_n = \left(\frac{n\pi}{L}\right)^2$, $n = 1, 2, 3, \ldots$. Therefore, $\phi_n(x) = \sin\frac{n\pi x}{L}$, $n = 1, 2, 3, \ldots$. Thus, the eigenvalues are $\lambda_n = \left(\frac{n\pi}{L}\right)^2$, $n = 1, 2, 3, \ldots$, and the eigenfunctions are $\phi_n(x) = \sin\frac{n\pi x}{L}$. Note that we have not considered $n = -1, -2, \ldots$ because they yield the same eigenfunctions. ◁

Example 10.3 Solve the eigenvalue problem

$$\left.\begin{array}{ll}\phi'' + \lambda\phi = 0 & 0 < x < L\\ \phi'(0) = 0 & \\ \phi'(L) = 0 & \end{array}\right\}.$$

Solution. In general,

$$\phi(x) = \begin{cases} c_1 + c_2 x, & \lambda = 0,\\ c_1 \cosh\sqrt{\lambda}x + c_2 \sinh\sqrt{\lambda}x, & \lambda < 0,\\ c_1 \cos\sqrt{\lambda}x + c_2 \sin\sqrt{\lambda}x, & \lambda > 0. \end{cases}$$

If $\lambda = 0$, boundary conditions give $\phi'(0) = 0 \Rightarrow c_2 = 0$ and $\phi'(L) = 0 \Rightarrow 0 = 0$. Hence, $\phi(x) = c_1 = 1$, where we made $c_1 = 1$ for convenience. As $\phi(x) \neq 0$, $\lambda = 0$ is an eigenvalue. If $\lambda < 0$, boundary conditions give $\phi'(0) = 0 \Rightarrow c_2 = 0$ and $\phi'(L) = 0 \Rightarrow c_1 = 0$. Hence, $\phi(x) = 0$, and $\lambda < 0$ are not eigenvalues. If $\lambda > 0$, boundary conditions give $\phi'(0) = 0 \Rightarrow c_2 = 0$ and $\phi'(L) = 0 \Rightarrow c_1 \sin\sqrt{\lambda}L = 0 \Rightarrow \sin\sqrt{\lambda}L = 0 \Rightarrow \lambda_n = \left(\frac{n\pi}{L}\right)^2$, $n = 1, 2, 3, \ldots$. Therefore, $\phi_n(x) = \cos\frac{n\pi x}{L}$, $n = 1, 2, 3, \ldots$. Thus, the eigenvalues are $\left(\frac{n\pi}{L}\right)^2$, $n = 0, 1, 2, \ldots$, and the eigenfunctions are $\phi_n(x) = \cos\dfrac{n\pi x}{L}$. ◁

10.2 Rectangular Waveguides

Consider a waveguide aligned along the z axis, and with dimensions a and b along the x axis and y axis, respectively. The medium inside is isotropic dielectric magnetic (see Fig. 10.1). Our target is to find $\mathbf{E}$ and $\mathbf{H}$ when the domain is bounded along the x and y axes, but unbounded along the z axis. Frequency-domain Maxwell curl equations are

$$\left.\begin{array}{l}\nabla \times \mathbf{E} = -j\omega\mu\mathbf{H}\\ \nabla \times \mathbf{H} = j\omega\varepsilon\mathbf{E}\end{array}\right\}. \tag{10.7}$$

In addition to these, we impose the following boundary condition on the boundary

$$\hat{\mathbf{n}} \times \mathbf{E} = \mathbf{0}. \tag{10.8}$$

This boundary condition is derived as follows. In general, at the interface between medium 1 and medium 2, we have

$$\hat{\mathbf{n}} \times (\mathbf{E}_1 - \mathbf{E}_2) = \mathbf{0}. \tag{10.9}$$

In our case, medium 2 is a PEC. Hence, $\mathbf{E}_2 = 0$, and $\mathbf{E}_1$ is simply our unknown field $\mathbf{E}$. Thus, we solve

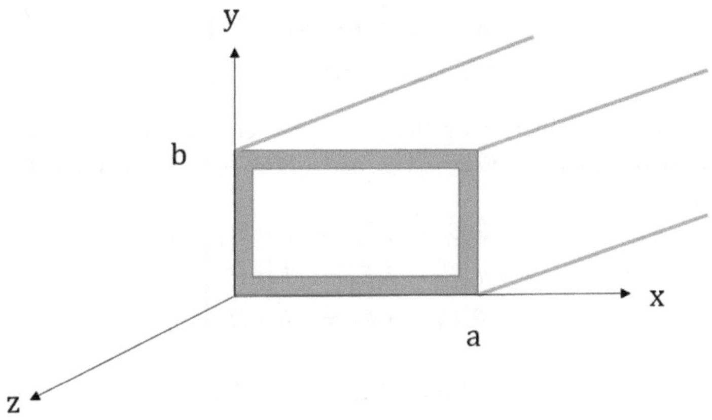

Fig. 10.1 Waveguide geometry

$$\left.\begin{aligned} \nabla \times \mathbf{E} &= -j\omega\mu\mathbf{H} \\ \nabla \times \mathbf{H} &= j\omega\varepsilon\mathbf{E} \\ \hat{\mathbf{n}} \times \mathbf{E} &= \mathbf{0}. \end{aligned}\right\}. \tag{10.10}$$

This is a set of vector partial differential equations. Here, we solve these vector partial differential equations by solving only for two scalar partial differential equations as follows. We will express the transverse components $\{E_x, E_y, H_x, H_y\}$ into the axial (i.e., longitudinal) components $\{E_z, H_z\}$. Then, we will derive the axial components' equations.

10.2.1 Decomposition into Axial Components

Each vector equation is indeed three scalar equations. So, from Eq. (10.10)$_1$, we get

$$\left.\begin{aligned} \partial_y E_z - \partial_z E_y &= -j\omega\mu H_x \\ \partial_z E_x - \partial_x E_z &= -j\omega\mu H_y \\ \partial_x E_y - \partial_y E_x &= -j\omega\mu H_z \end{aligned}\right\}. \tag{10.11}$$

Likewise, from Eq. (10.10)$_2$, we get

$$\left.\begin{aligned} \partial_y H_z - \partial_z H_y &= j\omega\varepsilon E_x \\ \partial_z H_x - \partial_x H_z &= j\omega\varepsilon E_y \\ \partial_x H_y - \partial_y H_x &= j\omega\varepsilon E_z \end{aligned}\right\}. \tag{10.12}$$

Now, since the domain is infinite along the z axis, we note that $\mathbf{E}$ and $\mathbf{H}$ will have the form

$$\left.\begin{aligned}\mathbf{E}(x,y,z)&=\mathbf{E}(x,y)e^{-jk_z z}\\ \mathbf{H}(x,y,z)&=\mathbf{H}(x,y)e^{-jk_z z}\end{aligned}\right\},\tag{10.13}$$

where k_z is the wave number along the z axis. Notice that we considered a positively-propagating wave only. Thus, $\partial_z\mathbf{E}=-jk_z\mathbf{E}$, and $\partial_z\mathbf{H}=-jk_z\mathbf{H}$. Therefore, Eqs. (10.11) and (10.12) become

$$\left.\begin{aligned}\partial_y E_z+jk_z E_y&=-j\omega\mu H_x\\ -jk_z E_x-\partial_x E_z&=-j\omega\mu H_y\\ \partial_x E_y-\partial_y E_x&=-j\omega\mu H_z\end{aligned}\right\},\tag{10.14}$$

and

$$\left.\begin{aligned}\partial_y H_z+jk_z H_y&=j\omega\varepsilon E_x\\ -jk_z H_x-\partial_x H_z&=j\omega\varepsilon E_y\\ \partial_x H_y-\partial_y H_x&=j\omega\varepsilon E_z\end{aligned}\right\}.\tag{10.15}$$

Now, if we combine Eqs. $(10.15)_1$ with $(10.14)_2$, we get

$$\left.\begin{aligned}E_x&=\frac{1}{j\omega\varepsilon}\left(\partial_y H_z+jk_z H_y\right)\\ &=\frac{1}{j\omega\varepsilon}\left[\partial_y H_z+\frac{jk_z}{j\omega\mu}\left(jk_z E_x+\partial_x E_z\right)\right]\\ &=\frac{1}{j\omega\varepsilon}\partial_y H_z+\frac{k_z^2}{k^2}E_x-j\frac{k_z}{k^2}\partial_x E_z\end{aligned}\right\},\tag{10.16}$$

where $k^2=\omega^2\mu\varepsilon$. Then, on multiplying by k^2, we get

$$\left.\begin{aligned}k^2E_x&=-j\frac{k^2}{\omega\varepsilon}\partial_y H_z+k_z^2E_x-jk_z\partial_x E_z\\ (k^2-k_z^2)E_x&=-j\frac{k^2}{\omega\varepsilon}\partial_y H_z-jk_z\partial_x E_z\\ (k^2-k_z^2)E_x&=-j\omega\mu\partial_y H_z-jk_z\partial_x E_z\\ E_x&=\frac{1}{k_t^2}\left(-j\omega\mu\partial_y H_z-jk_z\partial_x E_z\right)\end{aligned}\right\},\tag{10.17}$$

where $k_t^2=k^2-k_z^2$. By the same token, we can express E_y in terms of E_z and H_z, H_x in terms of E_z and H_z, and H_y in terms of E_z and H_z. Thus,

$$\left.\begin{aligned} E_x &= \frac{1}{k_t^2}\left(-jk_z\partial_x E_z - j\omega\mu\partial_y H_z\right) \\ E_y &= \frac{1}{k_t^2}\left(-jk_z\partial_y E_z + j\omega\mu\partial_x H_z\right) \\ H_x &= \frac{1}{k_t^2}\left(j\omega\varepsilon\partial_y E_z - jk_z\partial_x H_z\right) \\ H_y &= \frac{1}{k_t^2}\left(-j\omega\varepsilon\partial_x E_z - jk_z\partial_y H_z\right) \end{aligned}\right\}. \tag{10.18}$$

Equation (10.18) can also be written more compactly in vector notation as

$$\left.\begin{aligned} \mathbf{E}_t &= -\frac{jk_z}{k_t^2}\nabla_t E_z + \frac{j\omega\mu}{k_t^2}\hat{\mathbf{z}}\times\nabla_t H_z \\ \mathbf{H}_t &= -\frac{j\omega\varepsilon}{k_t^2}\hat{\mathbf{z}}\times\nabla_t E_z - \frac{jk_z}{k_t^2}\nabla_t H_z \end{aligned}\right\}, \tag{10.19}$$

where $\mathbf{E}_t = E_x\,\hat{\mathbf{x}} + E_y\,\hat{\mathbf{y}}$, $\mathbf{H}_t = H_x\,\hat{\mathbf{x}} + H_y\,\hat{\mathbf{y}}$, and $\nabla_t = \hat{\mathbf{x}}\,\partial_x + \hat{\mathbf{y}}\,\partial_y$ is the transverse del operator.

Remarks

- Notice that since k_z is unknown, then $k_t = \sqrt{k^2 - k_z^2}$ is also unknown. As to be shown, k_t will be infinitely-many numbers found from the boundary conditions, and these will be nothing but the eigenvalues of the problem.
- Since the transverse components $\mathbf{E}_t$ and $\mathbf{H}_t$ can be found from the axial components E_z and H_z, we only need to solve for the axial components.
- Notice that in waveguides, we can either have (i) $E_z \neq 0, H_z = 0$, (ii) $E_z = 0, H_z \neq 0$, or (iii) $E_z \neq 0, H_z \neq 0$.
- Case (i) is called transverse magnetic (TM) mode since **H** will be transverse (i.e., normal) to the direction of propagation, case (ii) is called transverse electric (TE) mode since **E** will be transverse (i.e., normal) to the direction of propagation, and case (iii) is called hybrid (HE) mode.
- Notice that transverse electromagnetic (TEM) mode (i.e., $E_z = H_z = 0$) is not possible in waveguides. This is a major difference between transmission lines and waveguides.
- Finally, notice that extension to the case where the medium is dissipative is straightforward by making ε complex.

10.2.2 Axial Components Equations

TM Case

In isotropic dielectric-magnetic medium, we get from Maxwell equations

$$\nabla^2 \mathbf{E} + k^2 \mathbf{E} = \mathbf{0}. \tag{10.20}$$

Since we are dealing with Cartesian system, this means that

$$\nabla^2 E_z + k^2 E_z = 0, \tag{10.21}$$

which is the three-dimensional scalar Helmholtz equation. Now,

$$\nabla^2 = \partial_{xx} + \partial_{yy} + \partial_{zz} = \nabla_t^2 + \partial_{zz}, \tag{10.22}$$

where $\nabla_t^2 = \partial_{xx} + \partial_{yy}$ is the transverse Laplacian. Then, since $E_z \propto e^{-jk_z z}$, it follows that $\partial_{zz} E_z = -k_z^2 E_z$, which implies that the scalar Helmholtz equation becomes

$$\nabla_t^2 E_z + k_t^2 E_z = 0, \tag{10.23}$$

which is a two-dimensional scalar Helmholtz equation. Since we know in advance that $E_z \propto e^{-jk_z z}$, let us make it as

$$E_z(x, y, z) = \phi(x, y) e^{-jk_z z}. \tag{10.24}$$

In addition to E_z, we have the boundary condition

$$\hat{\mathbf{n}} \times \mathbf{E} = \mathbf{0}, \tag{10.25}$$

which implies that the tangential electric field component is zero at the boundary. The boundaries are $x = 0$, $x = a$, $y = 0$, and $y = b$. On $x = \{0, a\}$, we see that the tangential components are E_y and E_z. Since we are concerned with E_z only, then $E_z = 0$ on $x = \{0, a\}$. Likewise, On $y = \{0, b\}$, we see that the tangential components are E_x and E_z. Since we are concerned with E_z only, then $E_z = 0$ on $y = \{0, b\}$. Therefore, we conclude that the axial electric field equation is

$$\left.\begin{array}{lll} \nabla_t^2 \phi + k_t^2 \phi = 0 & 0 < x < a & 0 < x < b \\ \phi(0, y) = 0 & \phi(a, y) = 0 & \\ \phi(x, 0) = 0 & \phi(x, b) = 0 & \end{array}\right\}, \tag{10.26}$$

which is nothing but a two-dimensional eigenvalue problem.

TE Case

Similarly, we get from Maxwell equations

$$\nabla_t^2 H_z + k_t^2 H_z = 0. \tag{10.27}$$

Since we know in advance that $H_z \propto e^{-jk_z z}$, let us make it as

$$H_z(x, y, z) = \psi(x, y) e^{-jk_z z}. \tag{10.28}$$

For the magnetic field boundary conditions, we note that

$$\left.\begin{aligned} E_x &= \frac{1}{k_t^2}\left(-jk_z \partial_x E_z - j\omega\mu \partial_y H_z\right) \\ E_y &= \frac{1}{k_t^2}\left(-jk_z \partial_y E_z + j\omega\mu \partial_x H_z\right) \end{aligned}\right\}. \tag{10.29}$$

In TE mode, $E_z = 0$. Hence,

$$\left.\begin{aligned} E_x &= \frac{-j\omega\mu}{k_t^2}\partial_y H_z \\ E_y &= \frac{j\omega\mu}{k_t^2}\partial_x H_z \end{aligned}\right\}. \tag{10.30}$$

On $x = \{0, a\}$, we have $E_y = 0$, which means $\partial_x H_z = 0$. On $y = \{0, b\}$, we have $E_x = 0$, which means $\partial_y H_z = 0$. Therefore, we conclude that the axial magnetic field equation is

$$\left.\begin{array}{llll} \nabla_t^2 \psi + k_t^2 \psi = 0 & & 0 < x < a & 0 < x < b \\ \partial_x \psi(0, y) = 0 & \partial_x \psi(a, y) = 0 & & \\ \partial_y \psi(x, 0) = 0 & \partial_y \psi(x, b) = 0 & & \end{array}\right\}. \tag{10.31}$$

Summary

Therefore, for rectangular waveguide made of an isotropic dielectric-magnetic medium with perfect conducting boundary, we solve

$$\left.\begin{array}{llll} \nabla_t^2 \phi + k_t^2 \phi = 0 & & 0 < x < a & 0 < x < b \\ \phi(0, y) = 0 & \phi(a, y) = 0 & & \\ \phi(x, 0) = 0 & \phi(x, b) = 0 & & \end{array}\right\}. \tag{10.32}$$

and

$$\left.\begin{array}{llll} \nabla_t^2 \psi + k_t^2 \psi = 0 & & 0 < x < a & 0 < x < b \\ \partial_x \psi(0, y) = 0 & \partial_x \psi(a, y) = 0 & & \\ \partial_y \psi(x, 0) = 0 & \partial_y \psi(x, b) = 0 & & \end{array}\right\}. \tag{10.33}$$

Then,

$$\left.E_z(x,y,z) = \phi(x,y)e^{-jk_z z} \qquad H_z(x,y,z) = \psi(x,y)e^{-jk_z z}\right\}. \tag{10.34}$$

Finally, we find the transverse components from

$$\left.\begin{aligned}
E_x &= \frac{1}{k_t^2}\left(-jk_z\partial_x E_z - j\omega\mu\partial_y H_z\right)\\
E_y &= \frac{1}{k_t^2}\left(-jk_z\partial_y E_z + j\omega\mu\partial_x H_z\right)\\
H_x &= \frac{1}{k_t^2}\left(j\omega\varepsilon\partial_y E_z - jk_z\partial_x H_z\right)\\
H_y &= \frac{1}{k_t^2}\left(-j\omega\varepsilon\partial_x E_z - jk_z\partial_y H_z\right)
\end{aligned}\right\}. \tag{10.35}$$

10.2.3 Some Aspects

We saw that the wave number is

$$k_z = \sqrt{k^2 - k_t^2} = \sqrt{\omega^2\mu\varepsilon - k_t^2}. \tag{10.36}$$

Depending on the frequency of operation, we can have the following three cases

$$k_z = \begin{cases} \sqrt{\omega^2\mu\varepsilon - k_t^2}, & \omega\sqrt{\mu\varepsilon} > k_t,\\ 0, & \omega\sqrt{\mu\varepsilon} = k_t,\\ -j\sqrt{k_t^2 - \omega^2\mu\varepsilon}, & \omega\sqrt{\mu\varepsilon} < k_t. \end{cases} \tag{10.37}$$

- When $\omega\sqrt{\mu\varepsilon} > k_t$, $e^{-j\sqrt{\omega^2\mu\varepsilon - k_t^2}z}$ will be complex, and thus, the mode will propagate in the waveguide. Such a mode is called a propagating mode.
- When $\omega\sqrt{\mu\varepsilon} < k_t$, $e^{-\sqrt{k_t^2 - \omega^2\mu\varepsilon}z}$ will be real, and thus, the mode will be attenuated in the waveguide. Such a mode is called an evanescent mode.
- The point where the mode changes from propagation to attenuation is called cutoff.

So, we define the cutoff wave number $k_c = k_t$, the cutoff wavelength $\lambda_c = \dfrac{2\pi}{k_c}$, and the cutoff frequency $\omega_c = \dfrac{k_c}{\sqrt{\mu\varepsilon}}$ (or $f_c = \dfrac{k_c}{2\pi\sqrt{\mu\varepsilon}}$). Thus, we see that, unlike transmission lines, only frequencies above the cutoff frequency can propagate in waveguides, indicating that a waveguide acts like a high-pass filter. So, we can write the wave number as

$$k_z = k\sqrt{1 - \left(\frac{k_c}{k}\right)^2}. \tag{10.38}$$

If we examine the phase velocity in a waveguide made of free space, we find that it is

$$v_p = \frac{\omega}{k_z} = \frac{c}{\sqrt{1 - \left(\frac{k_c}{k}\right)^2}}, \tag{10.39}$$

which is larger than the speed of light. However, it can be shown that information does not propagate at the phase velocity v_p, but at the energy or group velocity v_g, which is given by

$$v_g = \left(\frac{dk_z}{d\omega}\right)^{-1} = c\sqrt{1 - \left(\frac{k_c}{k}\right)^2}, \tag{10.40}$$

which is less than the speed of light.

The wave impedance in a waveguide can be found from

$$\eta = \frac{E_x}{H_y} = -\frac{E_y}{H_x}. \tag{10.41}$$

This can be verified from Maxwell equations. For TM mode, from

$$\left.\begin{aligned} E_x &= \frac{-jk_z}{k_t^2}\partial_x E_z \\ E_y &= \frac{-jk_z}{k_t^2}\partial_y E_z \\ H_x &= \frac{j\omega\varepsilon}{k_t^2}\partial_y E_z \\ H_y &= \frac{-j\omega\varepsilon}{k_t^2}\partial_x E_z \end{aligned}\right\}, \tag{10.42}$$

we get

$$\eta_{TM} = \frac{E_x}{H_y} = \frac{k_z}{\omega\varepsilon} = \eta'\sqrt{1 - \left(\frac{k_c}{k}\right)^2}, \tag{10.43}$$

where $\eta' = \sqrt{\mu/\varepsilon}$. Therefore, we see that

$$\eta_{TM} = \begin{cases} \text{Resistive}, & f > f_c, \\ 0, & f = f_c, \\ \text{Capacitive}, & f < f_c. \end{cases} \tag{10.44}$$

For TE mode, from

$$\left.\begin{aligned} E_x &= \frac{-j\omega\mu}{k_t^2}\partial_y H_z \\ E_y &= \frac{j\omega\mu}{k_t^2}\partial_x H_z \\ H_x &= \frac{-jk_z}{k_t^2}\partial_x H_z \\ H_y &= \frac{-jk_z}{k_t^2}\partial_y H_z \end{aligned}\right\}, \tag{10.45}$$

we get

$$\eta_{TE} = \frac{E_x}{H_y} = \frac{\omega\mu}{k_z} = \frac{\eta'}{\sqrt{1-\left(\frac{k_c}{k}\right)^2}}. \tag{10.46}$$

Therefore, we see that

$$\eta_{TE} = \begin{cases} \text{Resistive}, & f > f_c, \\ \infty, & f = f_c, \\ \text{Inductive}, & f < f_c. \end{cases} \tag{10.47}$$

10.3 Transverse Magnetic (TM) Mode

TM mode is characterized by $H_z = 0$. Therefore, we solve

$$\left.\begin{array}{lll} \nabla_t^2\phi + k_t^2\phi = 0 & 0 < x < a & 0 < x < b \\ \phi(0, y) = 0 & \phi(a, y) = 0 & \\ \phi(x, 0) = 0 & \phi(x, b) = 0 & \end{array}\right\}. \tag{10.48}$$

Then, after finding

$$E_z(x, y, z) = \phi(x, y)e^{-jk_z z}, \tag{10.49}$$

we find the remaining components from

$$\left.\begin{aligned} E_x &= \frac{-jk_z}{k_t^2}\partial_x E_z \\ E_y &= \frac{-jk_z}{k_t^2}\partial_y E_z \\ H_x &= \frac{j\omega\varepsilon}{k_t^2}\partial_y E_z \\ H_y &= \frac{-j\omega\varepsilon}{k_t^2}\partial_x E_z \end{aligned}\right\}. \tag{10.50}$$

Example 10.4 Solve Eq. (10.48).

Solution. We let

$$\phi(x, y) = X(x)Y(y).$$

Then,

$$X''Y + XY'' + k_t^2 XY = 0 \Rightarrow \frac{X''}{X} + \frac{Y''}{Y} = -k_t^2.$$

At this stage, due to the boundary conditions, we let

$$\frac{X''}{X} = -\lambda_x,$$

and

$$\frac{Y''}{Y} = -\lambda_y.$$

Hence, we get

$$X'' + \lambda_x X = 0,$$

and

$$Y'' + \lambda_y Y = 0.$$

We have two homogeneous boundary conditions for X, which are

$$\phi(0, y) = X(0)Y(y) = 0 \Rightarrow X(0) = 0,$$

and

$$\phi(a, y) = X(a)Y(y) = 0 \Rightarrow X(a) = 0.$$

Also, we have two homogeneous boundary conditions for Y, which are

$$\phi(x, 0) = X(x)Y(0) = 0 \Rightarrow Y(0) = 0,$$

and

$$\phi(x, b) = X(x)Y(b) = 0 \Rightarrow Y(b) = 0.$$

Hence, the PDE becomes

$$\left.\begin{array}{ll} X'' + \lambda_x X = 0 & 0 < x < a \\ X(0) = 0 & \\ X(a) = 0 & \end{array}\right\},$$

and

$$\left.\begin{array}{ll} Y'' + \lambda_y Y = 0 & 0 < y < b \\ Y(0) = 0 & \\ Y(b) = 0 & \end{array}\right\}.$$

Then, we get

$$X_m(x) = A_m \sin\frac{m\pi x}{a}, \qquad \lambda_x^m = \left(\frac{m\pi}{a}\right)^2, m \geq 1,$$

and

$$Y_n(y) = B_n \sin \frac{n\pi y}{b}, \qquad \lambda_y^n = \left(\frac{n\pi}{b}\right)^2, n \geq 1.$$

Notice that we have attached the constants A_m and B_n for later convenience. Their values are unimportant in source-free regions. Therefore, the eigenvalues are

$$k_t^{mn} = \sqrt{\lambda_x^m + \lambda_y^n} = \sqrt{\left(\frac{m\pi}{a}\right)^2 + \left(\frac{n\pi}{b}\right)^2},$$

and the eigenfunctions are

$$\phi_{nm}(x, y) = E_{mn} \sin \frac{m\pi x}{a} \sin \frac{n\pi y}{b},$$

where $E_{mn} = A_m B_n$. ◁

Now that $\phi_{nm}(x, y)$ has been found, we can find E_z as

$$E_z = E_{mn} \sin \frac{m\pi x}{a} \sin \frac{n\pi y}{b} e^{-jk_z^{mn} z}, \tag{10.51}$$

where

$$k_z^{mn} = \sqrt{\omega^2 \mu\varepsilon - \left(\frac{m\pi}{a}\right)^2 - \left(\frac{n\pi}{b}\right)^2}\Bigg|_{m\geq 1, n\geq 1}. \tag{10.52}$$

The remaining components are found as

$$\left.\begin{aligned} E_x &= -E_{mn} \frac{jk_z^{mn}}{k_t^{mn2}} \frac{m\pi}{a} \cos \frac{n\pi x}{a} \sin \frac{n\pi y}{b} e^{-jk_z^{mn} z} \\ E_y &= -E_{mn} \frac{jk_z^{mn}}{k_t^{mn2}} \frac{n\pi}{b} \sin \frac{m\pi x}{a} \cos \frac{n\pi y}{b} e^{-jk_z^{mn} z} \\ H_x &= E_{mn} \frac{j\omega\varepsilon}{k_t^{mn2}} \frac{n\pi}{b} \sin \frac{m\pi x}{a} \cos \frac{n\pi y}{b} e^{-jk_z^{mn} z} \\ H_y &= -E_{mn} \frac{j\omega\varepsilon}{k_t^{mn2}} \frac{m\pi}{a} \cos \frac{m\pi x}{a} \sin \frac{n\pi y}{b} e^{-jk_z^{mn} z} \end{aligned}\right\}. \tag{10.53}$$

Each pair $\{m, n\}$ represents a mode TM_{mn} with an associated cutoff frequency, given by

$$f_c^{mn} = \frac{1}{2\sqrt{\mu\varepsilon}} \sqrt{\left(\frac{m}{a}\right)^2 + \left(\frac{n}{b}\right)^2}\Bigg|_{m\geq 1, n\geq 1}. \tag{10.54}$$

The mode with the smallest cutoff frequency is called the dominant mode. If two modes have the same cutoff frequency, they are called degenerate modes. The first mode in the TM case is TM_{11}, which has a cutoff frequency

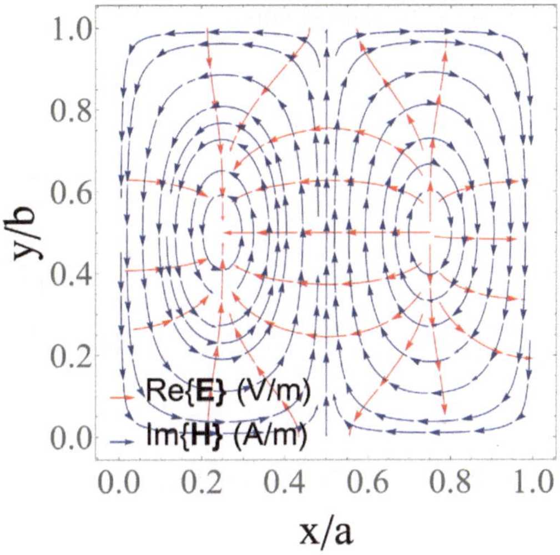

Fig. 10.2 Streamline plots of Re{**E**} and Im{**H**} on $z = 0$ for the mode TM_{21}

$$f_c^{11} = \frac{1}{2\sqrt{\mu\varepsilon}}\sqrt{\left(\frac{1}{a}\right)^2 + \left(\frac{1}{b}\right)^2}. \tag{10.55}$$

In other words, when the frequency of operation f exceeds f_c^{11}, the first mode will get excited. The second mode is TM_{21} (if $a > b$), TM_{12} (if $a < b$), or both (if $a = b$), will have a higher cutoff frequency.

Figure 10.2 shows two-dimensional streamline plots of Re{**E**} and Im{**H**} on the $z = 0$ plane for the mode TM_{21} when the waveguide is occupied by free space and $a = b = \lambda_o$.

10.4 Transverse Electric (TE) Mode

TE mode is characterized by $E_z = 0$. Therefore, we solve

$$\left.\begin{array}{lll} \nabla_t^2\psi + k_t^2\psi = 0 & 0 < x < a & 0 < y < b \\ \partial_x\psi(0, y) = 0 & \partial_x\psi(a, y) = 0 & \\ \partial_y\psi(x, 0) = 0 & \partial_y\psi(x, b) = 0 & \end{array}\right\}. \tag{10.56}$$

Then, after finding

$$H_z(x, y, z) = \psi(x, y)e^{-jk_z z}, \tag{10.57}$$

we find the remaining components from

$$\left.\begin{aligned} E_x &= \frac{-j\omega\mu}{k_t^2}\partial_y H_z \\ E_y &= \frac{j\omega\mu}{k_t^2}\partial_x H_z \\ H_x &= \frac{-jk_z}{k_t^2}\partial_x H_z \\ H_y &= \frac{-jk_z}{k_t^2}\partial_y H_z \end{aligned}\right\}. \tag{10.58}$$

Example 10.5 Solve Eq. (10.56).

Solution. The only difference appears in the boundary conditions. That is,

$$\partial_x \psi(0, y) = X'(0)Y(y) = 0 \Rightarrow X'(0) = 0,$$

$$\partial_x \psi(a, y) = X'(a)Y(y) = 0 \Rightarrow X'(a) = 0,$$

$$\partial_y \psi(0, y) = X(x)Y'(0) = 0 \Rightarrow Y'(0) = 0,$$

and

$$\partial_y \psi(x, b) = X(x)Y'(b) = 0 \Rightarrow Y'(b) = 0.$$

Hence, the PDE becomes

$$\left.\begin{aligned} &X'' + \lambda_x X = 0 \qquad 0 < x < a \\ &X'(0) = 0 \\ &X'(a) = 0 \end{aligned}\right\},$$

and

$$\left.\begin{aligned} &Y'' + \lambda_y Y = 0 \qquad 0 < y < b \\ &Y'(0) = 0 \\ &Y'(b) = 0 \end{aligned}\right\}.$$

Then, we get

$$X_m(x) = A_m \cos\frac{m\pi x}{a}, \qquad \lambda_x^m = \left(\frac{m\pi}{a}\right)^2, m \geq 0,$$

and

$$Y_n(y) = B_n \cos\frac{n\pi y}{b}, \qquad \lambda_y^n = \left(\frac{n\pi}{b}\right)^2, n \geq 0.$$

Therefore, the eigenvalues are

$$k_t^{mn} = \sqrt{\lambda_x^m + \lambda_y^n} = \sqrt{\left(\frac{m\pi}{a}\right)^2 + \left(\frac{n\pi}{b}\right)^2},$$

and the eigenfunction are

$$\psi_{nm}(x, y) = H_{mn} \cos \frac{m\pi x}{a} \cos \frac{n\pi y}{b},$$

where $H_{mn} = A_m B_n$. ◁

Therefore, we find H_z as

$$H_z = H_{mn} \cos \frac{m\pi x}{a} \cos \frac{n\pi y}{b} e^{-jk_z^{mn} z}, \tag{10.59}$$

where

$$k_z^{mn} = \sqrt{\omega^2 \mu\varepsilon - \left(\frac{m\pi}{a}\right)^2 - \left(\frac{n\pi}{b}\right)^2}\Bigg|_{m\geq 0, n\geq 0}. \tag{10.60}$$

The remaining components are found as

$$\left.\begin{aligned}
E_x &= H_{mn} \frac{j\omega\mu}{k_t^{mn2}} \frac{n\pi}{b} \cos \frac{m\pi x}{a} \sin \frac{n\pi y}{b} e^{-jk_z^{mn} z} \\
E_y &= -H_{mn} \frac{j\omega\mu}{k_t^{mn2}} \frac{m\pi}{a} \sin \frac{m\pi x}{a} \cos \frac{n\pi y}{b} e^{-jk_z^{mn} z} \\
H_x &= H_{mn} \frac{jk_z^{mn}}{k_t^{mn2}} \frac{m\pi}{a} \sin \frac{m\pi x}{a} \cos \frac{n\pi y}{b} e^{-jk_z^{mn} z} \\
H_y &= H_{mn} \frac{jk_z^{mn}}{k_t^{mn2}} \frac{n\pi}{b} \cos \frac{m\pi x}{a} \sin \frac{n\pi y}{b} e^{-jk_z^{mn} z}
\end{aligned}\right\}. \tag{10.61}$$

Notice that we can't have $m = n = 0$ because then $\mathbf{E} = \mathbf{0}$, indicating that an electromagnetic wave will not exist in the waveguide. Each pair $\{m, n\}$ represents a mode TE_{mn} with an associated cutoff frequency, given by

$$f_c^{mn} = \frac{1}{2\sqrt{\mu\varepsilon}} \sqrt{\left(\frac{m}{a}\right)^2 + \left(\frac{n}{b}\right)^2}\Bigg|_{m\geq 0, n\geq 0, m+n>0}. \tag{10.62}$$

The first mode in the TE case is TE_{10} (if $a > b$), TE_{01} (if $a < b$), or both (if $a = b$). For any a/b, it can be seen that the dominant mode is always TE_{01} or TE_{10}. So, TE modes are excited before TM modes.

Figure 10.3 shows two-dimensional streamline plots of Im{**E**} and Re{**H**} on the $z = 0$ plane for the mode TE_{21} when the waveguide is occupied by free space and $a = b = \lambda_o$.

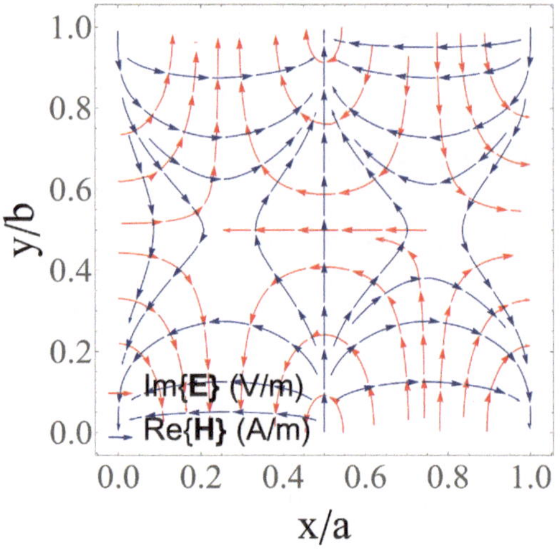

Fig. 10.3 Streamline plots of Im{**E**} and Re{**H**} on $z = 0$ for the mode TE_{21}

10.5 Power and Attenuation in Waveguides

10.5.1 Power

Given a surface S with a unit normal oriented along the z axis, the average power crossing that surface is found as

$$P_{\text{ave}} = \int_S \mathbf{W}_{\text{ave}} \cdot d\mathbf{S} = \int_0^a dx \int_0^b dy\, \mathbf{W}_{\text{ave}} \cdot \hat{\mathbf{z}}, \tag{10.63}$$

where

$$\mathbf{W}_{\text{ave}} \cdot \hat{\mathbf{z}} = \frac{1}{2}\text{Re}\{\mathbf{E} \times \mathbf{H}^*\} \cdot \hat{\mathbf{z}} = \frac{1}{2}\text{Re}\{E_x H_y^* - E_y H_x^*\}. \tag{10.64}$$

Notice that each mode has its own power. That is,

$$P_{\text{ave}} \equiv P_{mn}. \tag{10.65}$$

Example 10.6 Obtain the average power for the TE_{10} mode, assuming a non dissipative medium.

Solution. For TE_{10}, the field components are

$$\left.\begin{aligned} E_x &= H_y = 0 \\ E_y &= -H_{10}\frac{j\omega\mu}{k_t^2}\frac{\pi}{a}\sin\frac{\pi x}{a}e^{-jk_z z} \\ H_x &= H_{10}\frac{jk_z}{k_t^2}\frac{\pi}{a}\sin\frac{\pi x}{a}e^{-jk_z z} \end{aligned}\right\},$$

where

$$k_t \equiv k_t^{10} = \pi/a, \qquad k_z \equiv k_z^{10} = k\sqrt{1-\left(\frac{k_t}{k}\right)^2}.$$

So,

$$\left.\begin{aligned} E_y &= -H_{10}\frac{j\omega\mu a}{\pi}\sin\frac{\pi x}{a}e^{-jk_z z} \\ H_x &= H_{10}\frac{jk_z a}{\pi}\sin\frac{\pi x}{a}e^{-jk_z z} \end{aligned}\right\}.$$

Then,

$$\mathbf{W}_{\text{ave}} = \hat{\mathbf{z}}\frac{\omega\mu k_z a^2|H_{10}|^2}{2\pi^2}\sin^2\frac{\pi x}{a}.$$

Finally,

$$P_{10}^{TE} = \frac{\omega\mu a^2|H_{10}|^2}{2\pi^2}k_z\int_0^a dx\int_0^b dy\,\sin^2\frac{\pi x}{a} = \frac{\omega\mu k_z|H_{10}|^2a^3b}{4\pi^2}.$$

Note that the power P_{ave} in this example is obtained at a fixed surface $z =$ constant. However, the condition $z =$ constant does not have any effect on the power because $\mathbf{W}_{\text{ave}}$ itself is z independent. This holds only in non dissipative case. ◁

10.5.2 Attenuation

For a non dissipative waveguide, a propagating wave (i.e., $f > f_c$) will travel without attenuation. When dissipation presents, attenuation will take place. There are two dissipation mechanisms in a waveguide. These are (i) the dissipation stemming from the medium (called dielectric dissipation if the medium is non magnetic), and (ii) the dissipation stemming from the finite conductivity of the walls (called conduction dissipation).

A. Dielectric Dissipation

To take the dielectric dissipation into account, we replace k_z by $-j\gamma_z$. Then, in the non dissipative case, since

$$k_z = \sqrt{k^2 - k_t^2}, \tag{10.66}$$

in the dissipative case, we have

$$\gamma_z = \sqrt{k_t^2 - k^2} = \alpha_d + jk_z, \tag{10.67}$$

where the unknown α_d is the attenuation stemming from the dielectric dissipation, and the unknown k_z is the new wave number. To determine these, on substituting

$$k_t = \sqrt{\left(\frac{m\pi}{a}\right)^2 + \left(\frac{n\pi}{b}\right)^2}, \tag{10.68}$$

and

$$k^2 = \omega^2\mu\varepsilon_c = \omega^2\mu(\varepsilon' - j\varepsilon''), \tag{10.69}$$

with the assumption that $\alpha_d << k_z$, we get

$$k_z = \sqrt{\omega^2\mu\varepsilon' - \left(\frac{m\pi}{a}\right)^2 - \left(\frac{n\pi}{b}\right)^2} = \omega\sqrt{\mu\varepsilon'}\sqrt{1 - \left(\frac{f_c}{f}\right)^2}, \tag{10.70}$$

which is same as that of the non dissipative case, and

$$\alpha_d = \frac{\omega\varepsilon''\eta'}{2\sqrt{1 - \left(\frac{f_c}{f}\right)^2}}, \tag{10.71}$$

where $\eta' = \sqrt{\mu/\varepsilon'}$.

B. Conduction Dissipation

Suppose that the walls are made of non perfect conductors of conductivity σ_c. Let the associated field phasors be denoted by $\mathbf{E}_c$ and $\mathbf{H}_c$. Since the walls are no more PECs, the condition

$$\hat{\mathbf{n}} \times \mathbf{E}_c = \mathbf{0} \tag{10.72}$$

will not hold anymore, but is replaced by

$$\hat{\mathbf{n}} \times \mathbf{E}_c = \eta\mathbf{H}_c^{\parallel}, \tag{10.73}$$

where η is the conductor impedance, and $\mathbf{H}_c^{\parallel}$ is the tangential component of $\mathbf{H}_c$ at the conductor. Therefore, to get exact solutions of $\mathbf{E}_c$ and $\mathbf{H}_c$, we solve

$$\left.\begin{aligned} \nabla \times \mathbf{E}_c &= -j\omega\mu\mathbf{H}_c \\ \nabla \times \mathbf{H}_c &= j\omega\varepsilon\mathbf{E}_c \\ \hat{\mathbf{n}} \times \mathbf{E}_c &= \eta\mathbf{H}_c^{\parallel} \end{aligned}\right\}. \tag{10.74}$$

However, solving this set is somewhat more involved and is subject to numerical analysis [5, 6]. Alternatively, when dissipation is small, we may consider the walls being made of good conductors. Then, we let

$$\left.\begin{aligned}\mathbf{E}_c &= \mathbf{E}\, e^{-\alpha_c z}\\ \mathbf{H}_c &= \mathbf{H}\, e^{-\alpha_c z}\end{aligned}\right\}, \tag{10.75}$$

where **E** and **H** are electric and magnetic field phasors associated with the non dissipative case, and the unknown α_c is the attenuation constant due to the wall dissipation. To find α_c, we use conservation of energy as follows. For a good conductor, the impedance is given by

$$\eta = \sqrt{\frac{\omega\mu}{2\sigma_c}}(1+j) = R_s + jX_s, \tag{10.76}$$

with

$$R_s = X_s = \sqrt{\frac{\omega\mu}{2\sigma_c}} = \sqrt{\frac{\pi f \mu}{\sigma_c}}. \tag{10.77}$$

Then, at the boundaries (i.e., $x = \{0, a\}, y = \{0, b\}$), there will be an induced surface current $\mathbf{K}_c$, which gives rise to a dissipated power P_c given by

$$P_c(z) = \frac{1}{2}\mathrm{Re}\left\{\eta \oint_S \mathbf{K}_c \bullet \mathbf{K}_c^*\, d\mathcal{S}\right\} = \frac{R_s}{2}\oint_S \mathbf{K}_c \bullet \mathbf{K}_c^*\, d\mathcal{S}, \tag{10.78}$$

where

$$\mathbf{K}_c = \hat{\mathbf{n}} \times \mathbf{H}_c, \tag{10.79}$$

and $d\mathcal{S}$ is the differential surface element normal to the surface. Notice that the dissipated power will always have the form

$$P_c(z) = P_c^o e^{-2\alpha_c z}, \tag{10.80}$$

where P_c^o is obtained using **E** and **H**. In other words,

$$P_c^o = \frac{R_s}{2}\oint_S \mathbf{K} \bullet \mathbf{K}^*\, d\mathcal{S}, \tag{10.81}$$

where

$$\mathbf{K} = \hat{\mathbf{n}} \times \mathbf{H}. \tag{10.82}$$

The average Poynting vector

$$\mathbf{W}_{\text{ave}} = \frac{1}{2}\mathrm{Re}\{\mathbf{E}_c \times \mathbf{H}_c^*\} = \frac{1}{2}\mathrm{Re}\{\mathbf{E} \times \mathbf{H}^*\}e^{-2\alpha_c z} \tag{10.83}$$

is clearly z dependent. Therefore, the total power P_{ave} must also depend on z. This power can be obtained as

$$P_{\text{ave}}(z) = P_{\text{ave}}^o e^{-2\alpha_c z}, \tag{10.84}$$

where P^o_{ave} is obtained using $\mathbf{E}$ and $\mathbf{H}$. That is,

$$P^o_{\text{ave}} = \frac{1}{2}\int_S \text{Re}\{\mathbf{E}\times\mathbf{H}^*\}\cdot\hat{\mathbf{z}}\,d\mathcal{S}. \tag{10.85}$$

Due to conservation of energy, over a length l of the waveguide, the dissipated power must be equal to the rate of decrease of the power multiplied by that length. That is

$$P_c(l) = -l\frac{dP_{\text{ave}}(z)}{dz}\bigg|_{z=l} = 2\alpha_c l P_{\text{ave}}(l). \tag{10.86}$$

Therefore,

$$\alpha_c = \frac{P_c(l)}{2lP_{\text{ave}}(l)} = \frac{P^o_c}{2lP^o_{\text{ave}}}. \tag{10.87}$$

Example 10.7 Obtain the attenuation due to the conduction dissipation for TE_{10} mode.

Solution. For TE_{10}, the nonzero field components are

$$\left.\begin{aligned} E_y &= -H_{10}\frac{j\omega\mu}{k_t{}^2}\frac{\pi}{a}\sin\frac{\pi x}{a}e^{-jk_z z} \\ H_x &= H_{10}\frac{jk_z}{k_t{}^2}\frac{\pi}{a}\sin\frac{\pi x}{a}e^{-jk_z z} \\ H_z &= H_{10}\cos\frac{\pi x}{a}e^{-jk_z z} \end{aligned}\right\},$$

where

$$k_t = \pi/a,$$

and

$$k_z = k\sqrt{1-\left(\frac{k_t}{k}\right)^2}.$$

Therefore,

$$\left.\begin{aligned} E_y &= -H_{10}\frac{j\omega\mu a}{\pi}\sin\frac{\pi x}{a}e^{-jk_z z} \\ H_x &= H_{10}\frac{jk_z a}{\pi}\sin\frac{\pi x}{a}e^{-jk_z z} \\ H_z &= H_{10}\cos\frac{\pi x}{a}e^{-jk_z z} \end{aligned}\right\},$$

We have already found P^o_{ave} as

$$P^o_{\text{ave}} \equiv P^{TE}_{10} = \frac{\omega\mu k_z|H_{10}|^2 a^3 b}{4\pi^2}.$$

To find P^o_c, we notice that we have 4 surfaces. Therefore,

$$P_c^o = \frac{R_s}{2}\oint_S \mathbf{K}\cdot\mathbf{K}^*\,d\mathcal{S} = \frac{R_s}{2}\left(\int_{①}\mathbf{K}\cdot\mathbf{K}^*\,d\mathcal{S}_1 + \int_{②}\mathbf{K}\cdot\mathbf{K}^*\,d\mathcal{S}_2 + \int_{③}\mathbf{K}\cdot\mathbf{K}^*\,d\mathcal{S}_3 + \int_{④}\mathbf{K}\cdot\mathbf{K}^*\,d\mathcal{S}_4\right).$$

Here, ① is the surface $x = 0$ with $\hat{\mathbf{n}} = -\hat{\mathbf{x}}$, ② is the surface $x = a$ with $\hat{\mathbf{n}} = \hat{\mathbf{x}}$, ③ is the surface $y = 0$ with $\hat{\mathbf{n}} = -\hat{\mathbf{y}}$, and ④ is the surface $y = b$ with $\hat{\mathbf{n}} = \hat{\mathbf{y}}$. At ①, we have

$$\left.\begin{aligned}
\mathbf{K} = \hat{\mathbf{n}}\times\mathbf{H}\Big|_{x=0} &= -\hat{\mathbf{x}}\times(H_x\,\hat{\mathbf{x}} + H_z\,\hat{\mathbf{z}})\Big|_{x=0} = \hat{\mathbf{y}}\,H_z\Big|_{x=0}\\
\Rightarrow \mathbf{K}\cdot\mathbf{K}^* = |H_z|^2\Big|_{x=0} &= |H_{10}|^2\cos^2\frac{\pi x}{a}\Big|_{x=0} = |H_{10}|^2\\
\Rightarrow \int_{①}\mathbf{K}\cdot\mathbf{K}^*\,d\mathcal{S}_1 &= \int_0^b dy\int_0^l dz\,|H_{10}|^2 = |H_{10}|^2 bl
\end{aligned}\right\}.$$

At ②, the only difference is that we evaluate at $x = a$. Then,

$$\left.\begin{aligned}
\mathbf{K}\cdot\mathbf{K}^* &= |H_z|^2\Big|_{x=a} = |H_{10}|^2\\
\Rightarrow \int_{②}\mathbf{K}\cdot\mathbf{K}^*\,d\mathcal{S}_2 &= \int_0^b dy\int_0^l dz\,|H_{10}|^2 = |H_{10}|^2 bl
\end{aligned}\right\}.$$

At ③, we have

$$\left.\begin{aligned}
\mathbf{K} = \hat{\mathbf{n}}\times\mathbf{H}\Big|_{y=0} &= -\hat{\mathbf{y}}\times(H_x\,\hat{\mathbf{x}} + H_z\,\hat{\mathbf{z}})\Big|_{y=0} = (-H_z\,\hat{\mathbf{x}} + H_x\,\hat{\mathbf{z}})\Big|_{y=0}\\
\Rightarrow \mathbf{K}\cdot\mathbf{K}^* = \left(|H_x|^2 + |H_z|^2\right)\Big|_{y=0} &= \frac{k_z^2a^2}{\pi^2}|H_{10}|^2\sin^2\frac{\pi x}{a} + |H_{10}|^2\cos^2\frac{\pi x}{a}\\
\Rightarrow \int_{③}\mathbf{K}\cdot\mathbf{K}^*\,d\mathcal{S}_3 &= \int_0^a dx\int_0^l dz\left(\frac{k_z^2a^2}{\pi^2}|H_{10}|^2\sin^2\frac{\pi x}{a} + |H_{10}|^2\cos^2\frac{\pi x}{a}\right)\\
&= \frac{|H_{10}|^2 al}{2}\left(1 + \frac{k_z^2a^2}{\pi^2}\right)
\end{aligned}\right\}.$$

At ④, the only difference is that we evaluate at $y = b$. Then,

$$\left.\begin{aligned}
\mathbf{K}\cdot\mathbf{K}^* = \left(|H_x|^2 + |H_z|^2\right)\Big|_{y=b} &= \frac{k_z^2a^2}{\pi^2}|H_{10}|^2\sin^2\frac{\pi x}{a} + |H_{10}|^2\cos^2\frac{\pi x}{a}\\
\Rightarrow \int_{④}\mathbf{K}\cdot\mathbf{K}^*\,d\mathcal{S}_4 &= \int_0^a dx\int_0^l dz\left(\frac{k_z^2a^2}{\pi^2}|H_{10}|^2\sin^2\frac{\pi x}{a} + |H_{10}|^2\cos^2\frac{\pi x}{a}\right)\\
&= \frac{|H_{10}|^2 al}{2}\left(1 + \frac{k_z^2a^2}{\pi^2}\right)
\end{aligned}\right\}.$$

Therefore,

$$P_c^o = R_s l|H_{10}|^2\left[b + \frac{a}{2}\left(1 + \frac{k_z^2a^2}{\pi^2}\right)\right].$$

Finally,

$$\alpha_c = \frac{P_c^o}{2lP_{\text{ave}}^o} = \frac{2\pi^2 R_s \left[b + \frac{a}{2}\left(1 + \frac{k_z^2 a^2}{\pi^2}\right)\right]}{\omega\mu k_z a^3 b}.$$

◁

Example 10.8 Obtain the attenuation due to the conduction dissipation for TE_{01} mode.

Solution. For TE_{01}, the nonzero field components are

$$\left.\begin{aligned} E_x &= H_{01}\frac{j\omega\mu}{k_t{}^2}\frac{\pi}{b}\sin\frac{\pi y}{b}e^{-jk_z z} \\ H_y &= H_{01}\frac{jk_z}{k_t^2}\frac{\pi}{b}\sin\frac{\pi y}{b}e^{-jk_z z} \\ H_z &= H_{01}\cos\frac{\pi y}{b}e^{-jk_z z} \end{aligned}\right\},$$

where

$$k_t = \pi/b,$$

and

$$k_z = k\sqrt{1 - \left(\frac{k_t}{k}\right)^2}.$$

Then,

$$\left.\begin{aligned} E_x &= H_{01}\frac{j\omega\mu b}{\pi}\sin\frac{\pi y}{b}e^{-jk_z z} \\ H_y &= H_{01}\frac{jk_z b}{\pi}\sin\frac{\pi y}{b}e^{-jk_z z} \\ H_z &= H_{01}\cos\frac{\pi y}{b}e^{-jk_z z} \end{aligned}\right\}.$$

$$\mathbf{W}_{\text{ave}} = \hat{\mathbf{z}}\frac{\omega\mu k_z b^2 |H_{01}|^2}{2\pi^2}\sin^2\frac{\pi y}{b}.$$

$$P_{\text{ave}}^o \equiv P_{01}^{TE} = \frac{\omega\mu k_z b^2 |H_{01}|^2}{2\pi^2}\int_0^a dx \int_0^b dy \sin^2\frac{\pi y}{b} = \frac{\omega\mu k_z |H_{01}|^2 a b^3}{4\pi^2}.$$

$$P_c^o = \frac{R_s}{2}\oint_S \mathbf{K}\cdot\mathbf{K}^*\,dS = \frac{R_s}{2}\left(\int_{①}\mathbf{K}\cdot\mathbf{K}^*\,dS_1 + \int_{②}\mathbf{K}\cdot\mathbf{K}^*\,dS_2 + \int_{③}\mathbf{K}\cdot\mathbf{K}^*\,dS_3 + \int_{④}\mathbf{K}\cdot\mathbf{K}^*\,dS_4\right).$$

At ①, we have

$$\left.\begin{aligned}
&\mathbf{K} = \hat{\mathbf{n}} \times \mathbf{H}\Big|_{x=0} = -\hat{\mathbf{x}} \times (H_y\,\hat{\mathbf{y}} + H_z\,\hat{\mathbf{z}})\Big|_{x=0} = (H_z\,\hat{\mathbf{y}} - H_y\,\hat{\mathbf{z}})\Big|_{x=0} \\
&\Rightarrow \mathbf{K} \bullet \mathbf{K}^* = (|H_y|^2 + |H_z|^2)\Big|_{x=0} = \left(\frac{k_z^2 b^2}{\pi^2}|H_{01}|^2 \sin^2\frac{\pi y}{b} + |H_{01}|^2\cos^2\frac{\pi y}{b}\right)\Big|_{x=0} \\
&\Rightarrow \int_{①} \mathbf{K} \bullet \mathbf{K}^*\, dS_1 = \int_0^b dy \int_0^l dz \left(\frac{k_z^2 b^2}{\pi^2}|H_{01}|^2 \sin^2\frac{\pi y}{b} + |H_{01}|^2\cos^2\frac{\pi y}{b}\right) = \frac{|H_{01}|^2 bl}{2}\left(1 + \frac{k_z^2 b^2}{\pi^2}\right)
\end{aligned}\right\}.$$

At ②, the only difference is that we evaluate at $x = a$. Then,

$$\left.\begin{aligned}
&\mathbf{K} \bullet \mathbf{K}^* = (|H_y|^2 + |H_z|^2)\Big|_{x=a} = \frac{k_z^2 b^2}{\pi^2}|H_{01}|^2 \sin^2\frac{\pi y}{b} + |H_{01}|^2\cos^2\frac{\pi y}{b} \\
&\Rightarrow \int_{②} \mathbf{K} \bullet \mathbf{K}^*\, dS_2 = \int_0^b dy \int_0^l dz \left(\frac{k_z^2 b^2}{\pi^2}|H_{01}|^2 \sin^2\frac{\pi y}{b} + |H_{01}|^2\cos^2\frac{\pi y}{b}\right) = \frac{|H_{01}|^2 bl}{2}\left(1 + \frac{k_z^2 b^2}{\pi^2}\right)
\end{aligned}\right\}.$$

At ③, we have

$$\left.\begin{aligned}
&\mathbf{K} = \hat{\mathbf{n}} \times \mathbf{H}\Big|_{y=0} = -\hat{\mathbf{y}} \times (H_y\,\hat{\mathbf{y}} + H_z\,\hat{\mathbf{z}})\Big|_{y=0} = -\hat{\mathbf{x}}\,H_z\Big|_{y=0} \\
&\Rightarrow \mathbf{K} \bullet \mathbf{K}^* = |H_z|^2\Big|_{y=0} = |H_{01}|^2\cos^2\frac{\pi y}{b}\Big|_{y=0} = |H_{01}|^2 \\
&\Rightarrow \int_{③} \mathbf{K} \bullet \mathbf{K}^*\, dS_3 = \int_0^a dx \int_0^l dz\, |H_{01}|^2 = |H_{01}|^2 al
\end{aligned}\right\}.$$

At ④, the only difference is that we evaluate at $y = b$. Then,

$$\left.\begin{aligned}
&\mathbf{K} \bullet \mathbf{K}^* = |H_z|^2\Big|_{y=b} = |H_{01}|^2\cos^2\frac{\pi y}{b}\Big|_{y=b} = |H_{01}|^2 \\
&\Rightarrow \int_{④} \mathbf{K} \bullet \mathbf{K}^*\, dS_4 = \int_0^a dx \int_0^l dz\, |H_{01}|^2 = |H_{01}|^2 al
\end{aligned}\right\}.$$

Therefore,

$$P_c^o = R_s l |H_{01}|^2 \left[a + \frac{b}{2}\left(1 + \frac{k_z^2 b^2}{\pi^2}\right)\right].$$

Finally,

$$\alpha_c = \frac{P_c^o}{2lP_{\text{ave}}^o} = \frac{2\pi^2 R_s \left[a + \frac{b}{2}\left(1 + \frac{k_z^2 b^2}{\pi^2}\right)\right]}{\omega\mu k_z b^3 a}.$$

◁

10.6 Rectangular Cavity

A cavity can be realized as a waveguide enclosed at the ends $z = 0$ and $z = c$ by a PEC. Like waveguides, electric field and magnetic field phasors in cavities can be decomposed into TE and TM modes. However, expressions of the axial field phasors E_z and H_z are slightly different from those of waveguides due to the presence of the additional boundaries $z = 0$ and $z = c$. At the boundaries $z = 0$ and $z = c$, the condition $\hat{\mathbf{n}} \times \mathbf{E} = \mathbf{0}$ implies that $E_x = E_y = 0$. However, conditions on E_z and H_z can't be obtained from $\hat{\mathbf{n}} \times \mathbf{E} = \mathbf{0}$. To remedy this issue, starting from Eq. (10.10), it can be shown that

$$\left.\begin{aligned} j\omega\varepsilon E_x &= \partial_y H_z + \frac{1}{j\omega\mu}\left(\partial_{zz}E_x - \partial_{xz}E_z\right) \\ j\omega\varepsilon E_y &= -\partial_x H_z - \frac{1}{j\omega\mu}\left(\partial_{yz}E_z - \partial_{zz}E_y\right) \end{aligned}\right\}, \tag{10.88}$$

where it should be emphasized that we don't make use of the equality $\partial_z \neq -jk_z$. Then, for TM mode (i.e., $H_z = 0$), $E_x = E_y = 0$ imply that $\partial_{xz}E_z = 0$, which, when used in separation of variables procedure, can be made as $\partial_z E_z = 0$. For TE mode (i.e., $E_z = 0$), $E_x = E_y = 0$ imply that $\partial_x H_z = \partial_y H_z = 0$, which can be made as $H_z = 0$.

10.6.1 Transverse Magnetic (TM) Mode

In TM mode, the associated eigenvalue problem becomes

$$\left.\begin{array}{llll} \nabla^2 E_z + k_r^2 E_z = 0 & 0 < x < a & 0 < y < b & 0 < z < c \\ E_z(0, y, z) = 0 & E_z(a, y, z) = 0 & & \\ E_z(x, 0, z) = 0 & E_z(x, b, z) = 0 & & \\ \partial_z E_z(x, y, 0) = 0 & \partial_z E_z(x, y, c) = 0 & & \end{array}\right\}. \tag{10.89}$$

where E_z and k_r are unknown. Solutions are

$$E_z = E_{mnp} \sin\frac{m\pi x}{a} \sin\frac{n\pi y}{b} \cos\frac{p\pi z}{c}, \tag{10.90}$$

and

$$k_r^{mnp} = \sqrt{\left(\frac{m\pi}{a}\right)^2 + \left(\frac{n\pi}{b}\right)^2 + \left(\frac{p\pi}{c}\right)^2}\Bigg|_{m\geq1,n\geq1,p\geq0}. \tag{10.91}$$

The frequency in which $k = \omega\sqrt{\mu\varepsilon} = k_r$ is satisfied is called the resonant frequency f_r. Hence,

$$f_r^{mnp} = \frac{1}{2\pi\sqrt{\mu\varepsilon}} \frac{1}{\sqrt{\left(\frac{m}{a}\right)^2 + \left(\frac{n}{b}\right)^2 + \left(\frac{p}{c}\right)^2}}\Bigg|_{m\geq1,n\geq1,p\geq0}. \tag{10.92}$$

10.6.2 Transverse Electric (TE) Mode

In TE mode, the associated eigenvalue problem becomes

$$\left.\begin{array}{llll} \nabla^2 H_z + k_r^2 H_z = 0 & 0 < x < a & 0 < y < b & 0 < z < c \\ \partial_x H_z(0, y, z) = 0 & \partial_x H_z(a, y, z) = 0 & & \\ \partial_y H_z(x, 0, z) = 0 & \partial_y H_z(x, b, z) = 0 & & \\ H_z(x, y, 0) = 0 & H_z(x, y, c) = 0 & & \end{array}\right\}. \tag{10.93}$$

where H_z and k_r are unknown. Solutions are

$$H_z = H_{mnp} \cos \frac{m\pi x}{a} \cos \frac{n\pi y}{b} \sin \frac{p\pi z}{c}, \tag{10.94}$$

and

$$k_r^{mnp} = \sqrt{\left(\frac{m\pi}{a}\right)^2 + \left(\frac{n\pi}{b}\right)^2 + \left(\frac{p\pi}{c}\right)^2}\Bigg|_{m\geq 0, n\geq 0, p\geq 1}. \tag{10.95}$$

The resonant frequency becomes

$$f_r^{mnp} = \frac{1}{2\sqrt{\mu\varepsilon}} \frac{1}{\sqrt{\left(\frac{m}{a}\right)^2 + \left(\frac{n}{b}\right)^2 + \left(\frac{p}{c}\right)^2}}\Bigg|_{m\geq 0, n\geq 0, p\geq 1}. \tag{10.96}$$

Problems

10.1 Prove Eqs. $(10.18)_2$, $(10.18)_3$, and $(10.18)_4$.

10.2 Inhomogeneously-filled waveguides are used frequently in microwave devices, such as polarizers, phase shifters, etc. For a waveguide aligned along the z axis with an inhomogeneous medium, it can be shown that the axial-field equations are [6]

$$\nabla_t \times \left(\omega\varepsilon \hat{\mathbf{z}} \times \nabla_t E_z + k_z \nabla_t H_z\right) = -\omega\varepsilon k_t^2 E_z \,\hat{\mathbf{z}},$$

and

$$\nabla_t \times \left(\omega\mu \hat{\mathbf{z}} \times \nabla_t H_z - k_z \nabla_t E_z\right) = -\omega\mu k_t^2 H_z \,\hat{\mathbf{z}}.$$

where $\varepsilon \equiv \varepsilon(x, y, z)$ and $\mu \equiv \mu(x, y, z)$. Show that for a homogeneous medium (i.e., $\varepsilon(x, y, z) = \varepsilon$ and $\mu(x, y, z) = \mu$), the axial-field equations will reduce to

$$\nabla_t^2 E_z + k_t^2 E_z = 0,$$

and

$$\nabla_t^2 H_z + k_t^2 H_z = 0.$$

10.3 A ferrite-filled waveguide is used frequently in the X-band regime due to its capability to allow wave propagation with minimal attenuation. The non-zero TE_{m0} mode fields in a certain situation are found as [7]

$$\left.\begin{aligned} E_y &= -j\frac{\omega\mu}{k_t}\left(1-\frac{\alpha^2}{\mu^2}\right)H_m \sin\frac{m\pi x}{a} e^{-jk_z z} \\ H_x &= -j\frac{k_z}{k_t}H_m\left(\sin\frac{m\pi x}{a}+\frac{\alpha k_z}{\mu k_t}\cos\frac{m\pi x}{a}\right)e^{-jk_z z} \\ H_z &= H_m\left(\cos\frac{m\pi x}{a}+\frac{\alpha k_z}{\mu k_t}\sin\frac{m\pi x}{a}\right)e^{-jk_z z} \end{aligned}\right\},$$

where α is a parameter describing the anisotropy of the medium. Obtain (a) the average power P_{ave} over the plane $\{z = 0, x \in [0, a], y \in [0, b]\}$, and (b) the power dissipated P_c^0 over a length l of the waveguide.

Appendix

The following computer program can be used to compute **E** and **H**, and to procude streamline plots similar to those in Figs. 10.2 and 10.3.

```
\[Lambda]o = 600 10^(−9); ko = 2 \[Pi]/\[Lambda]o; \[Eta]o =
 120 \[Pi]; \[Mu]o = 4 \[Pi]*10^−7; \[Omega] =
 ko *3*10^8; \[Epsilon]o = 10^−9/(36 \[Pi]);
a = \[Lambda]o; b = a; Eo = 1; Ho = 0; mmode = 2; nmode = 1; ko = \
\[Omega] Sqrt[\[Mu]o \[Epsilon]o ];
kt[m_, n_] := Sqrt[((m \[Pi] )/a)^2 + ((n \[Pi] )/b)^2];
kz[m_, n_] := Sqrt[ko^2 − kt[m, n]^2];
\[Phi]z[x_, y_, z_, m_, n_] :=
  Eo Sin[(m \[Pi] x)/a] Sin[(n \[Pi] y)/b] ;
\[Psi]z[x_, y_, z_, m_, n_] :=
  Ho Cos[(m \[Pi] x)/a] Cos[(n \[Pi] y)/b] ;
Et[x_, y_, z_, m_,
   n_] = −(( I kz[m, n])/kt[m, n]^2)
      Grad[\[Phi]z[x, y, z, m, n], {x, y, z}, "Cartesian"] + (
     I \[Omega] \[Mu]o)/
     kt[m, n]^2 Cross[{0, 0, 1},
      Grad[\[Psi]z[x, y, z, m, n], {x, y, z}, "Cartesian"]];
Ht[x_, y_, z_, m_,
   n_] = −((I \[Omega] \[Epsilon]o)/kt[m, n]^2) Cross[{0, 0, 1},
      Grad[\[Phi]z[x, y, z, m, n], {x, y, z}, "Cartesian"]] − (
     I kz[m, n])/
     kt[m, n]^2 Grad[\[Psi]z[x, y, z, m, n], {x, y, z}, "Cartesian"];
Efield[x_, y_, z_, m_,
   n_] := (Et[x, y, z, m,
      n] + \[Phi]z[x, y, z, m, n] {0, 0, 1}) Exp[− I kz[m, n] z];
Hfield[x_, y_, z_, m_,
   n_] := (Ht[x, y, z, m,
      n] + \[Psi]z[x, y, z, m, n] {0, 0, 1}) Exp[− I kz[m, n] z];
\]Ex[x_, y_, z_] := Re[Efield[x, y, z, mmode, nmode] . {1, 0, 0}];
Ey[x_, y_, z_] := Re[Efield[x, y, z, mmode, nmode] . {0, 1, 0}];
Ez[x_, y_, z_] := Re[Efield[x, y, z, mmode, nmode] . {0, 0, 1}];
```

```
Hx[x_, y_, z_] := Im[Hfield[x, y, z, mmode, nmode] . {1, 0, 0}];
Hy[x_, y_, z_] := Im[Hfield[x, y, z, mmode, nmode] . {0, 1, 0}];
Hz[x_, y_, z_] := Im[Hfield[x, y, z, mmode, nmode] . {0, 0, 1}];

StreamPlot[{{Ex[a x, b y, 0], Ey[a x, b y, 0]}, {Hx[a x, b y, 0],
   Hy[a x, b y, 0]}}, {x, 0, 1}, {y, 0, 1},
 ImageSize -> 72*9,
 AspectRatio -> 1,
 StreamColorFunction -> None,
 StreamStyle -> {Red, Blue},
 StreamScale -> Automatic,
 StreamPoints -> 30,
 FrameLabel -> {Style["x/a", 50, Black], Style["y/b", 50, Black] },
 FrameStyle -> Directive[40, Plain],
 PlotLegends -> Placed[{"Re{\!\(\*
StyleBox[\"E\",\nFontWeight->\"Bold\"]\)\!\(\*
StyleBox[\"}\",\nFontWeight->\"Bold\"]\) (V/m)", "Im{\!\(\*
StyleBox[\"H\",\nFontWeight->\"Bold\"]\)\!\(\*
StyleBox[\"}\",\nFontWeight->\"Bold\"]\) (A/m)"}, {0.2, 0.1}],
 BaseStyle -> {FontWeight -> Plain, FontSize -> 30,
   FontFamily -> "Times New Roman"}, Frame -> True]
```

References

1. D.G. Zill, *Advanced Engineering Mathematics* (Jones & Bartlett, 2018)
2. G.B. Arfken, H.J. Weber, F.E. Harris, *Mathematical Methods for Physicists* (Elsevier, 2013)
3. E. Kreyszig, *Advanced Engineering Mathematics*, 10th edn. (Wiley, 2011)
4. O'Neil, *Engineering Mathematics* (Thomson, 2003)
5. R.E. Collin, *Field Theory of Guided Waves* (IEEE Press, 1991)
6. J. Jin, *Theory and Computation of Electromagnetic Fields*, 2nd edn. (Wiley, 2015)
7. E.J. Rothwell, M.J. Cloud, *Electromagnetics*, 3rd edn. (CRC Press, 2018)

Part IV
Radiation and Antennas

The last part of the book is devoted to solution of Maxwell equations in regions containing sources, as well as to the discussion of the sources themselves. Those sources are better known as antennas, and the process in which fields are generated from their sources is conventionally known as radiation. This part consists of 4 chapters. Chapter 11 presents solution of electromagnetic fields using the method of potentials. In Chap. 12, antenna fundamentals including definition and properties are discussed. Chapter 13 is about the simplest antenna structure, called wire antennas. Finally, Chap. 14 discusses the performance of a group of antennas, called antenna arrays.

11 Electromagnetic Radiation and Auxiliary Potentials

In this chapter, we discuss solution of Maxwell equations in the presence of sources, a problem known as a radiation problem. After outlining the solution procedure and a difficulty associated with it in Sect. 11.1, we describe an alternative procedure using auxiliary functions called vector potentials in Sects. 11.2 and 11.3. Using these, Maxwell equations are solved in Sect. 11.4 and 11.5. Time-domain solution is briefly discussed in Sect. 11.5. Then, far-zone radiation using vector potentials is discussed in Sect. 11.7. Finally, current distribution is discussed in Sect. 11.8. For brevity, we make the **r** dependence of all phasors implicit. Explicit **r** dependence will be illustrated whenever is needed.

11.1 Solution Procedure

Given electric and magnetic current density phasors $\mathbf{J}_e$ and $\mathbf{J}_m$ occupying a region $\mathcal{V}'$, which is immersed an unbounded isotropic dielectric-magnetic medium characterized by ε and μ, the electric field and magnetic field phasors **E** and **H** can be found upon solving Maxwell equations

$$\left.\begin{aligned} \nabla \times \mathbf{E} + j\omega\mu\mathbf{H} &= -\mathbf{J}_m \\ \nabla \times \mathbf{H} - j\omega\varepsilon\mathbf{E} &= \mathbf{J}_e \\ \nabla \bullet \mathbf{E} &= -\frac{1}{j\omega\varepsilon}\nabla \bullet \mathbf{J}_e \\ \nabla \bullet \mathbf{H} &= -\frac{1}{j\omega\mu}\nabla \bullet \mathbf{J}_m \end{aligned}\right\}. \tag{11.1}$$

To find **E**, we take the curl of Eq. (11.1)$_1$ to get

$$\nabla \times (\nabla \times \mathbf{E}) + j\omega\mu\nabla \times \mathbf{H} = -\nabla \times \mathbf{J}_m. \tag{11.2}$$

H. M. Alkhoori, *Concise Introduction to Electromagnetic Fields*, Synthesis Lectures on Electromagnetics, https://doi.org/10.1007/978-3-031-60331-0_11

Substitution of Eq. $(11.1)_2$ into Eq. (11.2) gives

$$\nabla \times (\nabla \times \mathbf{E}) - k^2\mathbf{E} = -j\omega\mu\mathbf{J}_e - \nabla \times \mathbf{J}_m, \tag{11.3}$$

where we made $k = \omega\sqrt{\mu\varepsilon}$. By the same token, we get an equation for $\mathbf{H}$ as

$$\nabla \times (\nabla \times \mathbf{H}) - k^2\mathbf{H} = -j\omega\varepsilon\mathbf{J}_m + \nabla \times \mathbf{J}_e. \tag{11.4}$$

These are the inhomogeneous vector wave equations, whose solutions can't be written is standard form (see Chap. 2) without the use of dyadic Green function. Since dyadics is beyond the level of this introductory book, we give an alternative method whereby the electric field and magnetic field phasors can be obtained. Such a method can be called the method of potentials, or auxiliary functions method. In particular, we introduce the magnetic vector potential $\mathbf{A}$ and the electric vector potential $\mathbf{F}$ as an intermediate step which can be found from the current sources. Once these auxiliary functions are found, the electric field and magnetic field phasors can be found upon differentiating these potential functions. Another set of auxiliary functions that are commonly used consists of the vector electric Hertzian potential $\mathbf{\Pi}_e$, and the vector magnetic Hertzian potential $\mathbf{\Pi}_m$.

11.2 The Magnetic Vector Potential

11.2.1 Definition

In the absence of magnetic current source $\mathbf{J}_m$, frequency-domain Maxwell equations become

$$\left.\begin{aligned} \nabla \times \mathbf{E} + j\omega\mu\mathbf{H} &= \mathbf{0} \\ \nabla \times \mathbf{H} - j\omega\varepsilon\mathbf{E} &= \mathbf{J}_e \\ \nabla \bullet \mathbf{E} &= -\frac{1}{j\omega\varepsilon}\nabla \bullet \mathbf{J}_e \\ \nabla \bullet \mathbf{H} &= 0 \end{aligned}\right\}. \tag{11.5}$$

Since $\nabla \bullet \mathbf{H} = 0$, the magnetic field is solenoidal. That is, it can be derived from a magnetic vector potential as

$$\mathbf{H} = \frac{1}{\mu}\nabla \times \mathbf{A}, \tag{11.6}$$

where $\mathbf{A}$ is the magnetic vector potential. Next, we need to express $\mathbf{E}$ in terms of $\mathbf{A}$. Substitution of $\mathbf{H} = \frac{1}{\mu}\nabla \times \mathbf{A}$ into Eq. $(11.5)_1$ gives

$$\nabla \times \mathbf{E} + j\omega\nabla \times \mathbf{A} = \mathbf{0} \Rightarrow \nabla \times (\mathbf{E} + j\omega\mathbf{A}) = \mathbf{0}. \tag{11.7}$$

Since $\nabla \times (\mathbf{E} + j\omega\mathbf{A}) = \mathbf{0}$, the quantity $\mathbf{E} + j\omega\mathbf{A}$ is conservative. That is, it can be derived from an electric scalar potential as

$$\mathbf{E} + j\omega\mathbf{A} = -\nabla V. \tag{11.8}$$

Therefore,

$$\mathbf{E} = -\nabla V - j\omega\mathbf{A}. \tag{11.9}$$

Thus far, if the potential functions V and $\mathbf{A}$ are known, the electric field and magnetic field phasors can be found as

$$\left.\begin{aligned} \mathbf{E} &= -\nabla V - j\omega\mathbf{A} \\ \mathbf{H} &= \frac{1}{\mu}\nabla \times \mathbf{A} \end{aligned}\right\}. \tag{11.10}$$

Then, what remains is to find an equation for each of V and $\mathbf{A}$. This is done next.

11.2.2 Governing Equation

To find an equation for $\mathbf{A}$, we substitute $\mathbf{E} = -\nabla V - j\omega\mathbf{A}$ and $\mathbf{H} = \frac{1}{\mu}\nabla \times \mathbf{A}$ into Eq. $(11.5)_2$. Then,

$$\left.\begin{aligned} &\frac{1}{\mu}\nabla \times (\nabla \times \mathbf{A}) + j\omega\varepsilon\nabla V - \omega^2\varepsilon\mathbf{A} = \mathbf{J}_e \\ &\Rightarrow \nabla \times (\nabla \times \mathbf{A}) - k^2\mathbf{A} = \mu\mathbf{J}_e - j\omega\mu\varepsilon\nabla V \\ &\Rightarrow \nabla^2\mathbf{A} + k^2\mathbf{A} = -\mu\mathbf{J}_e + \nabla(\nabla \bullet \mathbf{A} + j\omega\mu\varepsilon V) \end{aligned}\right\}. \tag{11.11}$$

Now, since $\nabla \bullet \mathbf{A}$ is not specified, we let it be

$$\nabla \bullet \mathbf{A} = -j\omega\mu\varepsilon V \Rightarrow V = -\frac{1}{j\omega\mu\varepsilon}\nabla \bullet \mathbf{A} \tag{11.12}$$

a choice known as Lorentz condition. Therefore, we get

$$\nabla^2\mathbf{A} + k^2\mathbf{A} = -\mu\mathbf{J}_e, \tag{11.13}$$

which is the inhomogeneous vector Helmholtz equation. Once $\mathbf{A}$ is found, V is expressed in terms of $\mathbf{A}$ using Lorentz condition. Therefore, we can express the electric field and magnetic field phasors exclusively in terms of $\mathbf{A}$ as

$$\left.\begin{aligned} \mathbf{E} &= -j\omega\mathbf{A} - \frac{j}{\omega\mu\varepsilon}\nabla(\nabla \bullet \mathbf{A}) \\ \mathbf{H} &= \frac{1}{\mu}\nabla \times \mathbf{A} \end{aligned}\right\}. \tag{11.14}$$

11.3 The Electric Vector Potential

11.3.1 Definition

In the absence of electric current source $\mathbf{J}_e$, frequency-domain Maxwell equations become

$$\left.\begin{aligned} \nabla \times \mathbf{E} + j\omega\mu\mathbf{H} &= -\mathbf{J}_m \\ \nabla \times \mathbf{H} - j\omega\varepsilon\mathbf{E} &= \mathbf{0} \\ \nabla \bullet \mathbf{E} &= 0 \\ \nabla \bullet \mathbf{H} &= -\frac{1}{j\omega\mu}\nabla \bullet \mathbf{J}_m \end{aligned}\right\}. \tag{11.15}$$

Since $\nabla \bullet \mathbf{E} = 0$, the electric field is solenoidal. That is, it can be derived from an electric vector potential as

$$\mathbf{E} = -\frac{1}{\varepsilon}\nabla \times \mathbf{F}, \tag{11.16}$$

where $\mathbf{F}$ is the electric vector potential. Next, we need to express $\mathbf{H}$ in terms of $\mathbf{F}$. Substitution of $\mathbf{E} = -\frac{1}{\varepsilon}\nabla \times \mathbf{F}$ into Eq. $(11.15)_2$ gives

$$\nabla \times \mathbf{H} + j\omega\nabla \times \mathbf{F} = \mathbf{0} \Rightarrow \nabla \times (\mathbf{H} + j\omega\mathbf{F}) = \mathbf{0}. \tag{11.17}$$

Since $\nabla \times (\mathbf{H} + j\omega\mathbf{F}) = \mathbf{0}$, the quantity $\mathbf{H} + j\omega\mathbf{F}$ is conservative. That is, it can be derived from a magnetic scalar potential as

$$\mathbf{H} + j\omega\mathbf{F} = -\nabla V_m. \tag{11.18}$$

Therefore,

$$\mathbf{H} = -\nabla V_m - j\omega\mathbf{F}. \tag{11.19}$$

So, the electric field and magnetic field phasors can be found as

$$\left.\begin{aligned} \mathbf{E} &= -\frac{1}{\varepsilon}\nabla \times \mathbf{F} \\ \mathbf{H} &= -\nabla V_m - j\omega\mathbf{F} \end{aligned}\right\}. \tag{11.20}$$

Then, what remains is to find an equation for each of V_m and $\mathbf{F}$. This is done next.

11.3.2 Governing Equation

To find an equation for $\mathbf{F}$, we substitute $\mathbf{E} = -\frac{1}{\varepsilon}\nabla \times \mathbf{F}$ and $\mathbf{H} = -\nabla V_m - j\omega\mathbf{F}$ into Eq. $(11.15)_1$. Then,

$$\left.\begin{aligned} &-\frac{1}{\varepsilon}\nabla\times(\nabla\times\mathbf{F})+j\omega\mu\nabla V_m-\omega^2\mu\mathbf{F}=-\mathbf{J}_m\\ &\Rightarrow\nabla\times(\nabla\times\mathbf{F})-k^2\mathbf{F}=\varepsilon\mathbf{J}_m-j\omega\mu\varepsilon\nabla V_m\\ &\Rightarrow\nabla^2\mathbf{F}+k^2\mathbf{F}=-\varepsilon\mathbf{J}_m+\nabla(\nabla\bullet\mathbf{F}+j\omega\mu\varepsilon\nabla V_m) \end{aligned}\right\}. \tag{11.21}$$

Now, since $\nabla\bullet\mathbf{F}$ is not specified, we let it be

$$\nabla\bullet\mathbf{F}=-j\omega\mu\varepsilon V_m\Rightarrow V_m=-\frac{1}{j\omega\mu\varepsilon}\nabla\bullet\mathbf{F}, \tag{11.22}$$

which is again Lorentz condition. Therefore, we get

$$\nabla^2\mathbf{F}+k^2\mathbf{F}=-\varepsilon\mathbf{J}_m, \tag{11.23}$$

which is another inhomogeneous vector Helmholtz equation. Once $\mathbf{F}$ is found, V_m is expressed in terms of $\mathbf{F}$ using Lorentz condition. Therefore, we can express the electric field and magnetic field phasors exclusively in terms of $\mathbf{F}$ as

$$\left.\begin{aligned} \mathbf{E}&=-\frac{1}{\varepsilon}\nabla\times\mathbf{F}\\ \mathbf{H}&=-j\omega\mathbf{F}-\frac{j}{\omega\mu\varepsilon}\nabla(\nabla\bullet\mathbf{F}) \end{aligned}\right\}. \tag{11.24}$$

11.4 Total Electric Field and Magnetic Field Phasors

In the preceding sections, we saw that the set $\{\mathbf{J}_e\neq\mathbf{0},\mathbf{J}_m=\mathbf{0}\}$ generates a magnetic vector potential satisfying

$$\nabla^2\mathbf{A}+k^2\mathbf{A}=-\mu\mathbf{J}_e. \tag{11.25}$$

Once this is solved, the associated electric field and magnetic field phasors can be found from

$$\left.\begin{aligned} \mathbf{E}&=-j\omega\mathbf{A}-\frac{j}{\omega\mu\varepsilon}\nabla(\nabla\bullet\mathbf{A})\\ \mathbf{H}&=\frac{1}{\mu}\nabla\times\mathbf{A} \end{aligned}\right\}. \tag{11.26}$$

Likewise, the set $\{\mathbf{J}_e=\mathbf{0},\mathbf{J}_m\neq\mathbf{0}\}$ generates an electric vector potential satisfying

$$\nabla^2\mathbf{F}+k^2\mathbf{F}=-\varepsilon\mathbf{J}_m. \tag{11.27}$$

Once this is solved, the associated electric field and magnetic field phasors can be found from

$$\left.\begin{aligned} \mathbf{E}&=-\frac{1}{\varepsilon}\nabla\times\mathbf{F}\\ \mathbf{H}&=-j\omega\mathbf{F}-\frac{j}{\omega\mu\varepsilon}\nabla(\nabla\bullet\mathbf{F}) \end{aligned}\right\}. \tag{11.28}$$

Then, what if $\{\mathbf{J}_e \neq \mathbf{0}, \mathbf{J}_m \neq \mathbf{0}\}$? In this case, by virtue of superposition, the total solution is

$$\left.\begin{aligned} \mathbf{E} &= -j\omega\mathbf{A} - \frac{j}{\omega\mu\varepsilon}\nabla(\nabla\bullet\mathbf{A}) - \frac{1}{\varepsilon}\nabla\times\mathbf{F} \\ \mathbf{H} &= \frac{1}{\mu}\nabla\times\mathbf{A} - j\omega\mathbf{F} - \frac{j}{\omega\mu\varepsilon}\nabla(\nabla\bullet\mathbf{F}) \end{aligned}\right\}. \tag{11.29}$$

11.5 Solution of the Inhomogeneous Vector Helmholtz Equation

We next discuss the solution of

$$\nabla^2\mathbf{A} + k^2\mathbf{A} = -\mu\mathbf{J}_e, \tag{11.30}$$

and

$$\nabla^2\mathbf{F} + k^2\mathbf{F} = -\varepsilon\mathbf{J}_m. \tag{11.31}$$

Since both equations have the same form with the correspondences $\mathbf{A} \to \mathbf{F}$ and $\mu\mathbf{J}_e \to \varepsilon\mathbf{J}_m$, and since both equations have the same radiation condition (to be discussed), we will consider one of these two equations; namely the equation of $\mathbf{A}$.

The formal solution of Eq. (11.30) is

$$\mathbf{A}(\mathbf{r}) = \mu \int_{\mathcal{V}'} G(\mathbf{r}, \mathbf{r}')\, \mathbf{J}_e(\mathbf{r}')\, d\mathcal{V}', \tag{11.32}$$

where $G(\mathbf{r}, \mathbf{r}')$ is the Green function of the problem. The problem is solved, in principle, once $G(\mathbf{r}, \mathbf{r}')$ is found. To find it, we substitute Eq. (11.32) back into Eq. (11.30) to get

$$\int_{\mathcal{V}'} \left[\nabla^2 G(\mathbf{r}, \mathbf{r}') + k^2 G(\mathbf{r}, \mathbf{r}')\right] \mu\mathbf{J}_e(\mathbf{r}')\, d\mathcal{V}' = -\mu\mathbf{J}_e(\mathbf{r}). \tag{11.33}$$

This equation is similar to

$$\int_{\mathcal{V}'} -\delta(\mathbf{r} - \mathbf{r}')\mu\mathbf{J}_e(\mathbf{r}')\, d\mathcal{V}' = -\mu\mathbf{J}_e(\mathbf{r}), \tag{11.34}$$

where $\delta(\mathbf{r} - \mathbf{r}')$ is the three-dimensional Dirac delta function with properties

$$\delta(\mathbf{r} - \mathbf{r}') = \begin{cases} \infty, & \mathbf{r} = \mathbf{r}', \\ 0, & \mathbf{r} \neq \mathbf{r}', \end{cases}$$

and

$$\int_{\mathcal{V}'} \delta(\mathbf{r} - \mathbf{r}')\, d\mathcal{V}' = \begin{cases} 1, & \mathbf{r} \in \mathcal{V}', \\ 0, & \mathbf{r} \notin \mathcal{V}'. \end{cases}$$

Comparing Eqs. (11.33) with (11.34), we deduce that $G(\mathbf{r}, \mathbf{r}')$ is found from

$$\nabla^2 G(\mathbf{r}, \mathbf{r}') + k^2 G(\mathbf{r}, \mathbf{r}') = -\delta(\mathbf{r} - \mathbf{r}'). \tag{11.35}$$

This equations describes the response $G(\mathbf{r}, \mathbf{r}')$ of a point source $\delta(\mathbf{r} - \mathbf{r}')$ located at $\mathbf{r}'$. For simplicity, let us assume that the point source is located at the origin (i.e., $\mathbf{r}' = \mathbf{0}$). Then, the problem reduces to

$$\nabla^2 G(\mathbf{r}, \mathbf{0}) + k^2 G(\mathbf{r}, \mathbf{0}) = -\delta(\mathbf{r}). \tag{11.36}$$

We solve this equation in spherical coordinates. Due to symmetry, there will be no θ or ϕ dependence. That is, $\mathbf{r} \rightarrow r$. This means the partial differential equation will reduce to an ordinary differential equation of the form

$$\frac{1}{r^2}\frac{d}{dr}\left[r^2 \frac{dG(r,0)}{dr}\right] + k^2 G(r,0) = -\delta(r). \tag{11.37}$$

In addition to Eq. (11.37), as $r \rightarrow \infty$, the Green function must satisfy the so-called radiation condition (Sommerfeld condition)

$$\lim_{r\rightarrow\infty} r\left[\frac{dG(r,0)}{dr} + jkG(r,0)\right] = 0. \tag{11.38}$$

This condition is originally derived in vector form for $\mathbf{E}$ and $\mathbf{H}$. When transformed to scalar form, it becomes Eq. (11.38). In words, this equation states that the solution very far away from the source must include outgoing waves only, and not incoming waves. At any point $r \neq 0$, Eq. (11.37) becomes

$$\frac{1}{r^2}\frac{d}{dr}\left[r^2 \frac{dG(r,0)}{dr}\right] + k^2 G(r,0) = 0, \tag{11.39}$$

which has solution

$$G(r,0) = C_1 \frac{e^{-jkr}}{r} + C_2 \frac{e^{jkr}}{r}, \tag{11.40}$$

where C_1 and C_2 are unknown constants that must be determined. The terms $\dfrac{e^{\pm jkr}}{r}$ are called spherical waves (recall that $e^{\pm jkr}$ are plane waves). To see the role of the radiation condition, we have to understand the meaning of each term. The term $\dfrac{e^{-jkr}}{r}$ in time domain becomes $\dfrac{\cos(\omega t - kr)}{r}$, while the term $\dfrac{e^{jkr}}{r}$ in time domain becomes $\dfrac{\cos(\omega t + kr)}{r}$. If we plot each term versus time, we will see that the term $\dfrac{\cos(\omega t - kr)}{r}$ represents an outgoing spherical wave propagating radially away from the source with a speed equal to ω/k, while the term $\dfrac{\cos(\omega t + kr)}{r}$ is an incoming spherical wave propagating radially into the source with a speed equal to ω/k. Logically, the source must transmit a wave, not receive it. Therefore, we reject the incoming spherical wave term by setting $C_2 = 0$. Note that this is done when

$e^{j\omega t}$ convention is adopted. When $e^{-j\omega t}$ convention is adopted, then the converse becomes true. Therefore, the solution so far is

$$G(r, 0) = C_1 \frac{e^{-jkr}}{r}. \tag{11.41}$$

The constant C_1 is found upon applying a procedure that involves the delta function and divergence theorem, which results in $C_1 = \dfrac{1}{4\pi}$. Hence, the solution of Eq. (11.37) subject to the condition Eq. (11.38) is

$$G(r, 0) = \frac{e^{-jkr}}{4\pi r}. \tag{11.42}$$

Now, if the source is placed at any arbitrary $\mathbf{r}'$, then $r \to R = |\mathbf{r} - \mathbf{r}'|$. Therefore, the Green function of the problem is

$$G(\mathbf{r}, \mathbf{r}') = \frac{e^{-jkR}}{4\pi R}. \tag{11.43}$$

This is called the unbounded-space Green function for an isotropic dielectric-magnetic medium. Hence, the solution of the magnetic vector potential is

$$\mathbf{A}(\mathbf{r}) = \frac{\mu}{4\pi} \int_{\mathcal{V}'} \mathbf{J}_e(\mathbf{r}') \frac{e^{-jkR}}{R} \, d\mathcal{V}', \qquad R = |\mathbf{r} - \mathbf{r}'|. \tag{11.44}$$

Using $\mathbf{A} \to \mathbf{F}$ and $\mu \mathbf{J}_e \to \varepsilon \mathbf{J}_m$, we get

$$\mathbf{F}(\mathbf{r}) = \frac{\varepsilon}{4\pi} \int_{\mathcal{V}'} \mathbf{J}_m(\mathbf{r}') \frac{e^{-jkR}}{R} \, d\mathcal{V}', \qquad R = |\mathbf{r} - \mathbf{r}'|. \tag{11.45}$$

If the sources are an electric surface current $\mathbf{K}_e$ and a magnetic surface current $\mathbf{K}_m$ residing a surface $\mathcal{S}'$, the solutions become

$$\mathbf{A}(\mathbf{r}) = \frac{\mu}{4\pi} \int_{\mathcal{S}'} \mathbf{K}_e(\mathbf{r}') \frac{e^{-jkR}}{R} \, d\mathcal{S}', \tag{11.46}$$

and

$$\mathbf{F}(\mathbf{r}) = \frac{\varepsilon}{4\pi} \int_{\mathcal{S}'} \mathbf{K}_m(\mathbf{r}') \frac{e^{-jkR}}{R} \, d\mathcal{S}'. \tag{11.47}$$

Finally, if the sources are an electric line current $\mathbf{I}_e$ and a magnetic line current $\mathbf{I}_m$ residing a line $\mathcal{L}'$, the solutions become

$$\mathbf{A}(\mathbf{r}) = \frac{\mu}{4\pi} \int_{\mathcal{L}'} \mathbf{I}_e(\mathbf{r}') \frac{e^{-jkR}}{R} \, d\mathcal{L}', \tag{11.48}$$

and

$$\mathbf{F}(\mathbf{r}) = \frac{\varepsilon}{4\pi} \int_{\mathcal{L}'} \mathbf{I}_m(\mathbf{r}') \frac{e^{-jkR}}{R} \, d\mathcal{L}'. \tag{11.49}$$

11.6 Time-Domain Radiation

Even though this chapter is devoted to frequency-domain radiation, time-domain radiation provides more insights into radiation mechanism. Time-domain radiation is harder to deal with than frequency-domain radiation; yet, the former can be obtained from the later using Fourier analysis. Here, we just outline one very important consequence of time-domain radiation; namely, the finite speed of propagation.

In free space, we have

$$\left.\begin{aligned} \mathbf{A}(\mathbf{r}) &= \frac{\mu_0}{4\pi}\int_{\mathcal{V}'} \mathbf{J}_e(\mathbf{r}')\,\frac{e^{-jkR}}{R}\,d\mathcal{V}' \\ \mathbf{F}(\mathbf{r}) &= \frac{\varepsilon_0}{4\pi}\int_{\mathcal{V}'} \mathbf{J}_m(\mathbf{r}')\,\frac{e^{-jkR}}{R}\,d\mathcal{V}' \end{aligned}\right\}. \tag{11.50}$$

Using inverse Fourier transform, these expressions in time domain become

$$\left.\begin{aligned} \mathscr{A}(\mathbf{r},t) &= \frac{\mu_0}{4\pi}\int_{\mathcal{V}'} \frac{\mathscr{J}_e(\mathbf{r}',t-R/c)}{R}\,d\mathcal{V}' \\ \mathscr{F}(\mathbf{r},t) &= \frac{\varepsilon_0}{4\pi}\int_{\mathcal{V}'} \frac{\mathscr{J}_m(\mathbf{r}',t-R/c)}{R}\,d\mathcal{V}' \end{aligned}\right\}, \tag{11.51}$$

where c is the speed of light. With these, the time-domain electric field and magnetic field can be found as [1]

$$\left.\begin{aligned} \mathscr{E}(\mathbf{r},t) &= -\frac{\partial\mathscr{A}(\mathbf{r},t)}{\partial t} + \frac{1}{\mu_0\varepsilon_0}\int_0^{t-R/c} \nabla[\nabla\cdot\mathscr{A}(\mathbf{r},\tau)]\,d\tau - \frac{1}{\varepsilon_0}\nabla\times\mathscr{F}(\mathbf{r},t) \\ \mathscr{H}(\mathbf{r},t) &= \frac{1}{\mu_0}\nabla\times\mathscr{A}(\mathbf{r},t) - \frac{\partial\mathscr{F}(\mathbf{r},t)}{\partial t} + \frac{1}{\mu_0\varepsilon_0}\int_0^{t-R/c} \nabla[\nabla\cdot\mathscr{F}(\mathbf{r},\tau)]\,d\tau \end{aligned}\right\}. \tag{11.52}$$

Equations (11.51) and (11.52) show that if a disturbance (input) occurs at a certain time, then the fields (output) measured a distance R from the disturbance will not show up until a time interval equal to R/c. This represents a delay for information transmission. Since it is known theoretically that the maximum speed of energy propagation in free space is the speed of light, the smallest delay possible over a length R is simply R/c.

11.7 Far-Zone Approximation

Field calculation everywhere for simple sources is not usually that simple. On the other hand, far-zone-field calculation is easier than field calculation everywhere, although the former might be still not easy. In many antenna applications, it is the far-zone fields that we are interested to obtain. Therefore, far-zone calculation is usually adopted. In this section, we will show how potential integrals simplify when the field point is in the far zone. This will, in turn, simplify obtaining the fields. We will also discuss electric field and magnetic field phasors characteristics in the far zone region.

The potential integrals involve the distance function R. So, in order to approximate the potential in the far zone, let us examine the distance function. In general,

$$R = \sqrt{r^2 + r'^2 - 2rr'(\hat{\mathbf{r}} \bullet \hat{\mathbf{r}}')} = r\sqrt{1 - 2(\hat{\mathbf{r}} \bullet \hat{\mathbf{r}}')\frac{r'}{r} + \left(\frac{r'}{r}\right)^2}, \tag{11.53}$$

where we made $\mathbf{r} = r\,\hat{\mathbf{r}}$ and $\mathbf{r}' = r'\,\hat{\mathbf{r}}'$. In the far zone, we have $r >> r'$, which implies that $r'/r << 1$. Therefore we ignore the part $(r'/r)^2$ in favour of r'/r and 1. Then,

$$R \approx r\sqrt{1 - 2(\hat{\mathbf{r}} \bullet \hat{\mathbf{r}}')\frac{r'}{r}} \approx r\left[1 - (\hat{\mathbf{r}} \bullet \hat{\mathbf{r}}')\frac{r'}{r}\right] = r - r'(\hat{\mathbf{r}} \bullet \hat{\mathbf{r}}'), \tag{11.54}$$

where use has been made of

$$\sqrt{1-x} = 1 - \frac{x}{2} - \frac{x^2}{8} + \cdots, \tag{11.55}$$

with $x = r/r'$. Notice that we have ignored x^2 term, etc., since $r'/r << 1$. Therefore, in the far zone, we have

$$R \approx r - r'(\hat{\mathbf{r}} \bullet \hat{\mathbf{r}}') = r - \hat{\mathbf{r}} \bullet \mathbf{r}'. \tag{11.56}$$

The Green function in general is

$$G(\mathbf{r}, \mathbf{r}') = \frac{e^{-jkR}}{4\pi R}. \tag{11.57}$$

In the far zone, we use the approximation $R \approx r - \hat{\mathbf{r}} \bullet \mathbf{r}'$ in the exponential term, and the approximation $R \approx r$ in the denominator term for further simplification. Using the approximation $R \approx r$ in the exponential term can cause a significant error because of the term k. Therefore, the far-zone Green function becomes

$$G(\mathbf{r}, \mathbf{r}') \approx \frac{e^{-jkr}}{4\pi r} e^{j\mathbf{k} \bullet \mathbf{r}'}, \tag{11.58}$$

where $\mathbf{k} = k\,\hat{\mathbf{r}}$. Therefore, the potential integrals due to volume currents $\{\mathbf{J}_e, \mathbf{J}_m\}$ in the far zone become

$$\left.\begin{aligned} \mathbf{A}(\mathbf{r}) &\approx \frac{\mu}{4\pi}\frac{e^{-jkr}}{r}\int_{\mathcal{V}'} \mathbf{J}_e(\mathbf{r}')\, e^{j\mathbf{k} \bullet \mathbf{r}'}\, d\mathcal{V}' \\ \mathbf{F}(\mathbf{r}) &\approx \frac{\varepsilon}{4\pi}\frac{e^{-jkr}}{r}\int_{\mathcal{V}'} \mathbf{J}_m(\mathbf{r}')\, e^{j\mathbf{k} \bullet \mathbf{r}'}\, d\mathcal{V}' \end{aligned}\right\}. \tag{11.59}$$

Similar expressions can be obtained for surface and line currents.

Before obtaining **E** and **H**, let us note that the integrals in Eq. (11.59) are independent of r. Then, on defining

$$\mathbf{a}(\theta, \phi) = \int_{\mathcal{V}'} \mathbf{J}_e(\mathbf{r}')\, e^{j\mathbf{k} \bullet \mathbf{r}'}\, d\mathcal{V}' \tag{11.60}$$

as the electric radiation vector, and

$$\mathbf{f}(\theta, \phi) = \int_{\mathcal{V}'} \mathbf{J}_m(\mathbf{r}')\, e^{j\mathbf{k} \bullet \mathbf{r}'}\, d\mathcal{V}' \tag{11.61}$$

as the magnetic radiation vector, we see that the potentials become

$$\left.\begin{aligned} \mathbf{A}(\mathbf{r}) &\approx \frac{\mu}{4\pi}\frac{e^{-jkr}}{r}\mathbf{a}(\theta, \phi) \\ \mathbf{F}(\mathbf{r}) &\approx \frac{\varepsilon}{4\pi}\frac{e^{-jkr}}{r}\mathbf{f}(\theta, \phi) \end{aligned}\right\}. \tag{11.62}$$

Notice that $\mathbf{a}(\theta, \phi)$ is nothing but the three-dimensional Fourier transform of the electric current $\mathbf{J}_e(\mathbf{r})$. Similarly, $\mathbf{f}(\theta, \phi)$ is nothing but the three-dimensional Fourier transform of the magnetic current $\mathbf{J}_m(\mathbf{r})$. Also, this shows clearly that both **A** and **F** are dependent upon r through the term e^{-jkr}/r only.

To find **E** and **H**, we insert Eqs. (11.59) into Eqs. (11.29). This gives an expression for each of **E** and **H** that involves terms like $1/r$, as well as $1/r^2$. In the far zone, as to be shown in Chap. 12, only terms like $1/r$ count in the far zone. On ignoring the remaining terms, it can be shown that **E** and **H** will only have $\hat{\boldsymbol{\theta}}$ and $\hat{\boldsymbol{\phi}}$ components. Consequently, they can be found from

$$\left.\begin{aligned} \mathbf{E} &\approx -j\omega\mathbf{A}_\perp + \frac{jk}{\varepsilon}\hat{\mathbf{r}} \times \mathbf{F}_\perp \\ \mathbf{H} &\approx -\frac{jk}{\mu}\hat{\mathbf{r}} \times \mathbf{A}_\perp - j\omega\mathbf{F}_\perp \end{aligned}\right\}, \tag{11.63}$$

where

$$\left.\begin{aligned} \mathbf{A}_\perp &= \mathbf{A} - (\mathbf{A} \bullet \hat{\mathbf{r}})\hat{\mathbf{r}} \\ \mathbf{F}_\perp &= \mathbf{F} - (\mathbf{F} \bullet \hat{\mathbf{r}})\hat{\mathbf{r}} \end{aligned}\right\}. \tag{11.64}$$

In words, only components of **A** and **F** that are perpendicular to $\hat{\mathbf{r}}$ matter in the far zone. Notice that Eq. (11.63) immediately imply that both **E** and **H** will include no $\hat{\mathbf{r}}$ component. Thus, **E** and **H** will have the form

$$\left.\begin{aligned} \mathbf{E}(\mathbf{r}) &\approx \mathbf{E}_\perp(\theta, \phi)\frac{e^{-jkr}}{r} \\ \mathbf{H}(\mathbf{r}) &\approx \mathbf{H}_\perp(\theta, \phi)\frac{e^{-jkr}}{r} \end{aligned}\right\}, \tag{11.65}$$

with the property

$$\mathbf{E}_\perp \bullet \hat{\mathbf{r}} = \mathbf{H}_\perp \bullet \hat{\mathbf{r}} = 0. \tag{11.66}$$

The following can be deduced from the above.

- Due to the factor $\dfrac{e^{-jkr}}{r}$, each of $\mathbf{E}$ and $\mathbf{H}$ is a vector spherical wave that propagates along the r direction.
- Moreover, since $\mathbf{E} \cdot \hat{\mathbf{r}} = \mathbf{H} \cdot \hat{\mathbf{r}} = 0$, both of $\mathbf{E}$ and $\mathbf{H}$ are transverse to the direction of propagation. This implies that fields are TEM in far-zone region.
- So, an electromagnetic field in the far zone is somewhat like a plane wave with a propagation vector $\mathbf{k} = k\,\hat{\mathbf{r}}$, but with an amplitude decreasing like $1/r$. Hence, the ∇ operator in Maxwell equations is replaced by $-jk\hat{\mathbf{r}}$.
- Replacing the ∇ operator in Maxwell equations gives another way for finding the far-zone fields. Given the far-zone $\mathbf{A}$, we find $\mathbf{E}$ from

$$\mathbf{E} \approx -j\omega\mathbf{A}_{\perp}. \tag{11.67}$$

Then, $\mathbf{H}$ is found using Maxwell equation as

$$\mathbf{H} \approx \frac{\hat{\mathbf{r}}}{\eta} \times \mathbf{E}. \tag{11.68}$$

Next, given the far-zone $\mathbf{F}$, we find $\mathbf{H}$ from

$$\mathbf{H} \approx -j\omega\mathbf{F}_{\perp}. \tag{11.69}$$

Then, $\mathbf{E}$ is found using Maxwell equation as

$$\mathbf{E} \approx -\eta\hat{\mathbf{r}} \times \mathbf{H}. \tag{11.70}$$

11.8 Current Distribution

In antenna theory, an antenna carries sources $\{\mathbf{J}_e, \mathbf{J}_m\}$. Given those sources, we find the vector potentials $\{\mathbf{A}, \mathbf{F}\}$, and then the field phasors $\{\mathbf{E}, \mathbf{H}\}$. Apparently, the problem is formally solved. Once the current distribution is prescribed on the antenna, the electric field and magnetic field phasors can be found as indicated above. However, in practice, the current distribution is itself not completely known; it might be known at the feeding point of the antenna where the source is connected to. To illustrate this mathematically, we have the situation

$$\mathbf{A}(\mathbf{r}) = \int_{\mathcal{V}'} \underline{\underline{G}}(\mathbf{r}, \mathbf{r}') \cdot \mathbf{f}(\mathbf{r}')\, d\mathcal{V}', \tag{11.71}$$

where the output $\mathbf{A}$ is to be found. However, the input $\mathbf{f}$ itself is unknown. If it happens that $\mathbf{A}$ is known in a portion $\mathbf{r}_o$ of the antenna, then, on that portion, the input $\mathbf{f}$ can be found. That is, on the portion $\mathbf{r}_o$, we have

$$\mathbf{A}(\mathbf{r}_o) = \int_{\mathcal{V}'} \underline{\underline{G}}(\mathbf{r}_o, \mathbf{r}') \cdot \mathbf{f}(\mathbf{r}')\, d\mathcal{V}', \tag{11.72}$$

where $\mathbf{A}(\mathbf{r}_o)$ is known, and $\mathbf{f}(\mathbf{r}')$ is unknown. Since the unknown appears inside an integral sign, Eq. (11.72) is called an integral equation. Once Eq. (11.72) is solved for the unknown $\mathbf{f}$, insertion of $\mathbf{f}$ into Eq. (11.71) gives $\mathbf{A}$ everywhere.

The subject of integral equations is well established, and there exist methods to solve them. One of the famous numerical methods to solve integral equations is the Method of Moments (MoM) (see Chap. 13). Alternatively, electromagnetics problems can be numerically solved directly using Maxwell equations. This can be done using a vast number of numerical methods, including Finite Difference Method (FDM), Finite Difference Time Domain Method (FDTD), Finite Element Method (FEM), Extended Boundary Condition Method (EBCM), etc. [2–7]. The subject concerns with these methods is called computational electromagnetics. There exist software packages based on these numerical methods. Well known of these are FEKO [8], HFSS [9], etc.

Problems

11.1 Starting from $\mathbf{H} = j\omega\varepsilon\nabla \times \mathbf{\Pi}_e$, where $\mathbf{\Pi}_e$ is the electric Hertzian potential, (a) derive its associated vector Helmholtz equation, and (b) express $\mathbf{E}$ and $\mathbf{H}$ in terms of it.

11.2 Starting from $\mathbf{E} = -j\omega\mu\nabla \times \mathbf{\Pi}_h$, where $\mathbf{\Pi}_h$ is the magnetic Hertzian potential, (a) derive its associated vector Helmholtz equation, and (b) express $\mathbf{E}$ and $\mathbf{H}$ in terms of it.

11.3 An aperture antenna can be realized as a rectangular element $\{x \in [-a/2, a/2], y \in [-b/2, b/2]\}$ placed on the PEC $z = 0$. Using an equivalence theorem and image theory (to be discussed in Chap. 13), the equivalent electric and magnetic surface currents can be approximated by

$$\mathbf{K}_e \approx \mathbf{0}, \qquad \mathbf{K}_m \approx 2E_o\,\hat{\mathbf{x}}, \qquad x \in [-a/2, a/2], y \in [-b/2, b/2],$$

where E_o is the electric field amplitude on the element. Obtain far-zone $\mathbf{E}$ and $\mathbf{H}$ above the plane. **Hint:**

$$\int_{-c/2}^{c/2} e^{j\alpha z}\, dz = c\frac{\sin(\alpha c/2)}{\alpha c/2}.$$

References

1. J. Jin, *Theory and Computation of Electromagnetic Fields*, 2nd edn. (Wiley, 2015)
2. G.C. Hsiao, R.E. Kleinman, D.-Q. Wang, Applications of boundary integral equation methods in 3D electromagnetic scattering. J. Comput. Appl. Math. **104**, 89–110 (1999)
3. P. Ylä-Oijala, M. Taskinen, S. Järvenpää, Surface integral equation formulations for solving electromagnetic scattering problems with iterative methods. Radio Sci. **40**, RS6002 (2005)
4. K.S. Kunz, R.S. Luebbers, *The Finite Difference Time Domain Method for Electromagnetics* (CRC Press, 1993)
5. P.B. Monk, *Finite Element Methods for Maxwell's Equations* (Oxford University, 2003)
6. P.C. Waterman, Matrix formulation of electromagnetic scattering. Proc. IEEE **53**, 805–812 (1965)
7. A. Lakhtakia, The Ewald–Oseen extinction theorem and the extended boundary condition method, in *The World of Applied Electromagnetics*, ed. by A. Lakhtakia, C.M. Furse (Springer, 2018), pp. 481–513
8. Altair, Altair FEKO™ software overview. https://altairhyperworks.com/product/FEKO
9. Ansys, https://www.ansys.com/products/electronics/ansys-hfss

Antenna Fundamentals 12

An antenna is an element carrying a time-varying current that radiates an electromagnetic wave. Such an element is designed to achieve certain properties. Antenna properties can be classified into three categories, as proposed by Rothwell and Cloud [1]. These are (i) radiation properties, (ii) circuit properties, and (iii) radiation-circuit properties. After defining antenna regions in Sect. 12.1, those properties are discussed, respectively, in Sects. 12.2, 12.3, and 12.4. A useful computer program is given in the appendix at the end of the chapter.

12.1 Antenna Regions

The radiated electric field and magnetic field phasors **E** and **H** from an antenna contain reactive components that do not radiate, and radiating components that do radiate. These components exist in the space surrounding the antenna. Such a space can be divided into three regions as follows:

- Reactive near-zone region: This region is defined as the region immediately surrounding the antenna. Usually, it is valid when $r < R_1$, with $R_1 = 0.62\sqrt{D^3/\lambda}$, where D is the largest antenna dimension. In this region, the reactive components dominate over the radiating components.
- Radiating near-zone region: This region is defined as the region beyond the reactive near-zone region and the far-zone region. Usually, it is valid when $R_1 < r < R_2$, where $R_2 = 2D^2/\lambda$. In this region, the radiating components dominate.
- Far-zone region: This region is defined as the region beyond the radiating near-zone region (see Sect. 11.7). Usually, it is valid when $r > R_2$. In the far-zone region, field

H. M. Alkhoori, *Concise Introduction to Electromagnetic Fields*, Synthesis Lectures on Electromagnetics, https://doi.org/10.1007/978-3-031-60331-0_12

phasors are in the form $|\mathbf{E}| \propto \dfrac{1}{r}$ and $|\mathbf{H}| \propto \dfrac{1}{r}$. Moreover, as discussed in Sect. 11.7, the electromagnetic wave becomes TEM in such region. Since antennas serve to transmit information over great distances, the far-zone region is usually the region of interest.

12.2 Radiation Properties

In this section, we discuss radiation properties of antennas. These include radiation power, radiation intensity, directivity, and antenna polarization.

12.2.1 Radiation Power

Recall that for a time-harmonic electric field and magnetic field phasors $\{\mathbf{E}, \mathbf{H}\}$, we define the complex Poynting vector $\mathbf{W}$ as

$$\mathbf{W} = \frac{1}{2}\mathbf{E} \times \mathbf{H}^*. \tag{12.1}$$

The real part of it is the average Poynting vector given by

$$\mathbf{W}_{\text{ave}} = \frac{1}{2}\text{Re}\{\mathbf{E} \times \mathbf{H}^*\}. \tag{12.2}$$

Given a surface $\mathcal{S}$, the total average power of the vector $\mathbf{W}_{\text{ave}}$ over a surface $\mathcal{S}$ is the flux of $\mathbf{W}_{\text{ave}}$ over $\mathcal{S}$. That is,

$$P_{\text{ave}} = \int_{\mathcal{S}} \mathbf{W}_{\text{ave}} \bullet d\mathbf{S} \tag{12.3}$$

From Sect. 11.7, given far-zone $\mathbf{E}$ and $\mathbf{H}$, the far-zone average Poynting vector can be calculated as

$$\mathbf{W}_{\text{ave}} = \frac{1}{2\eta}|\mathbf{E}|^2\,\hat{\mathbf{r}}. \tag{12.4}$$

When P_{ave} is computed in the far zone region, it is called the radiation power P_{rad}. Mathematically,

$$P_{\text{rad}} = \lim_{r\to\infty} P_{\text{ave}} = \lim_{r\to\infty} \int_0^{2\pi}\int_0^{\pi} \mathbf{W}_{\text{ave}} \bullet \hat{\mathbf{r}} r^2 \sin\theta \, d\theta d\phi. \tag{12.5}$$

The quantity

$$d\Omega = \sin\theta \, d\theta d\phi \tag{12.6}$$

is often called the differential solid angle of a sphere (in Sr). Using it, the radiation power can be written as

$$P_{\text{rad}} = \lim_{r\to\infty} \int_{\Omega} r^2 \mathbf{W}_{\text{ave}} \bullet \hat{\mathbf{r}}\, d\Omega, \tag{12.7}$$

It is to be noted that the limit operator is not needed if far-zone **E** and **H** are used.

Example 12.1 Given the far-zone expressions

$$\mathbf{E} = \frac{2A_o \sin\theta e^{-jkr}}{r}\hat{\boldsymbol{\theta}},$$

and

$$\mathbf{H} = \frac{e^{-jkr}}{r}\hat{\boldsymbol{\phi}},$$

where A_o is a constant, obtain the radiation power. **Hint:** $\int_0^{\pi} \sin^2\theta\, d\theta = \pi/2$.

Solution After finding

$$\mathbf{W} = \frac{1}{2}\mathbf{E}\times\mathbf{H}^* = \frac{A_o \sin\theta}{r^2}\hat{\mathbf{r}} \Rightarrow \mathbf{W}_{\text{ave}} = \text{Re}\{\mathbf{W}\} = \frac{A_o \sin\theta}{r^2}\hat{\mathbf{r}},$$

we get

$$P_{\text{rad}} = \int_0^{2\pi}\int_0^{\pi} \frac{A_o \sin\theta}{r^2}\hat{\mathbf{r}} \bullet \hat{\mathbf{r}} r^2 \sin\theta\, d\theta d\phi = 2\pi A_o \int_0^{\pi} \sin^2\theta\, d\theta = \pi^2 A_o.$$

Notice that the insertion of the limit in this example did not have an effect. This is because far-zone fields were used. ◁

12.2.2 Radiation Intensity

The radiation intensity $U(\theta, \phi)$ of a transmitting antenna is the radiated power per unit solid angle. From Eq. (12.7), we define $U(\theta, \phi)$ as

$$U(\theta, \phi) = \lim_{r\to\infty} r^2 \mathbf{W}_{\text{ave}} \bullet \hat{\mathbf{r}}. \tag{12.8}$$

Using this, the radiation power is found as

$$P_{\text{rad}} = \int_{\Omega} U(\theta, \phi)\, d\Omega. \tag{12.9}$$

Example 12.2 Given

$$U(\theta, \phi) = \sin\theta,$$

obtain the radiation power. **Hint:** $\int_0^{\pi} \sin^2\theta \, d\theta = \pi/2$.

Solution

$$P_{\text{rad}} = \int_{\Omega} U(\theta, \phi)\, d\Omega = \int_0^{2\pi}\int_0^{\pi} \sin^2\theta = \pi^2.$$

◁

12.2.3 Antenna Pattern

An antenna pattern is a spatial distribution of an antenna property versus the angles θ and ϕ. There are two antenna pattern types: (i) field pattern, and (ii) power pattern. Field pattern can be, for instance, a plot of $|\mathbf{E}|$, $|E_r|$, $|E_\theta|$, $|E_\phi|$, etc. Radiation pattern can be, for instance, a plot of $|\mathbf{E}|^2$, $U(\theta, \phi)$, etc. Antenna pattern is a three-dimensional plot (both θ and ϕ are varied), but often, two-dimensional cuts are made (θ is varied and ϕ is fixed, or θ is fixed and ϕ is varied). An antenna pattern can be normalized with respect to its maximum. Figure 12.1a shows a three-dimensional pattern for the radiation intensity. A two-dimensional cut can be obtained by, for example, setting $\phi = 0°$, and varying $\theta \in [0, 180°]$. This gives a polar plot on the xz plane, as shown in Fig. 12.1b. Notice that the upper part of the plot represents the portion of the xz plane associated with $z > 0$, whereas the lower part of the plot represents the portion of the xz plane associated with $z < 0$.

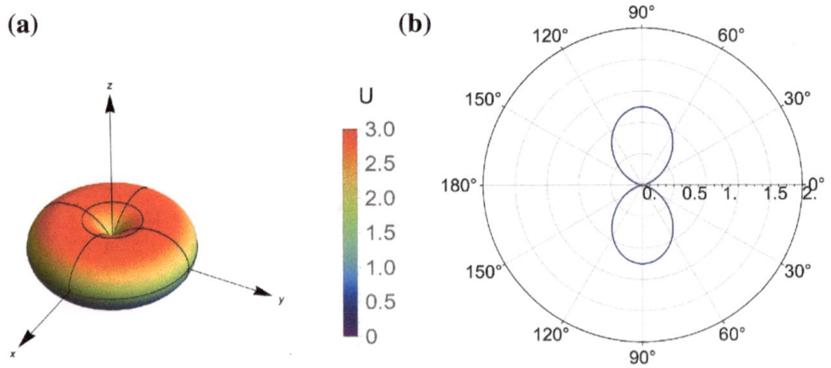

Fig. 12.1 **a** Three-dimensional antenna pattern, and **b** two-dimensional antenna pattern

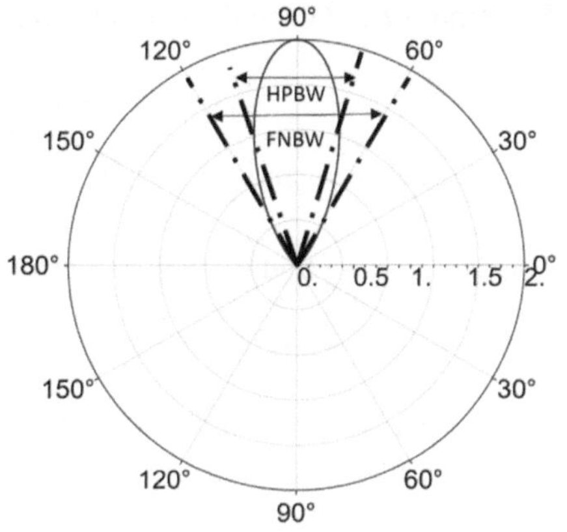

Fig. 12.2 HPBW and FNBW calculation when $U_{\max} = 2$

A pattern lobe is a portion of the radiation pattern with a local maximum. Lobes are classified as main (major) lobes, side (minor) lobes, and back lobes. A main lobe is the one containing the maximum amplitude $U_{\max}$, whereas a side lobe is the one not containing $U_{\max}$. Finally, a back lobe is a lobe in a direction opposite to the main lobe. A null is a point where the antenna pattern is zero (i.e., $U = 0$). The beamwidth is used to describe the width of the main lobe. When few nulls occur, it is more convenient to use the half-power beamwidth (HPBW), which is the angle between half-power points $U_{\max}/2$. When many nulls occur, it is more convenient to use the first-null beamwidth (FNBW), which is defined as the angle between the nulls forming the main lobe (see Fig. 12.2).

Antenna pattern can be made in linear scale or in logarithmic scale. Figure 12.3a shows a two-dimensional pattern for the radiation intensity in linear scale. Sometimes, side lobes are barely visible when the pattern is made in linear scale. To overcome this issue, logarithmic scale can be used instead of linear scale, as shown in Fig. 12.3b.

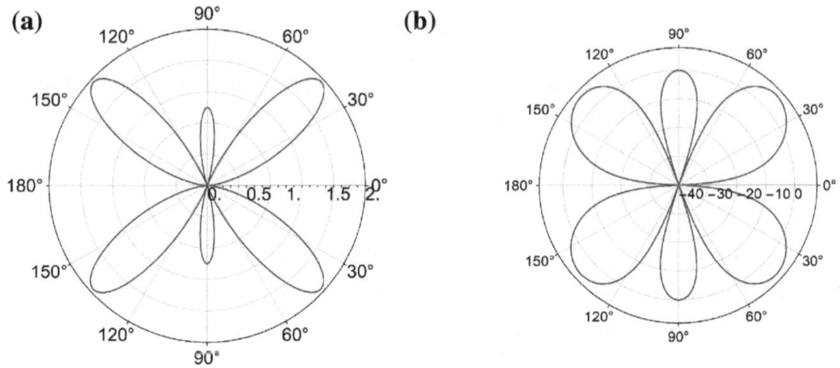

Fig. 12.3 Two-dimensional antenna pattern in **a** linear scale, and **b** in logarithmic scale

Example 12.3 Given the normalized radiation intensity

$$U(\theta) = \cos^2\theta \cos^2 2\theta,$$

obtain (a) HPBW, and (b) FNBW.

Solution (a) We find θ_h such that $U(\theta_h) = U_{\max}/2 = 1/2$. That is,

$$U(\theta_h) = \cos^2\theta_h \cos^2 2\theta_h = 1/2 \Rightarrow \cos\theta_h \cos 2\theta_h = 0.707 \Rightarrow \theta_h = 0.251.$$

Due to symmetry, we find that HPBW$= 2\theta_h = 0.502$.

(b) We find θ_n such that $U(\theta_h) = 0$. That is,

$$U(\theta_n) = \cos^2\theta_n \cos^2 2\theta_n = 0 \Rightarrow \cos\theta_n \cos 2\theta_n = 0 \Rightarrow \theta_n = \{0.78, 1.57\}.$$

We pick the smallest θ_n. Therefore, and due to symmetry, we find that FNBW$= 2\theta_n = 1.56$.

◁

12.2.4 Directivity

An isotropic antenna is a fictitious antenna radiating equally in all directions. Mathematically, $U(\theta,\phi) = U_o$, where U_o is a constant. It appears as a sphere in three-dimensional plots, and as a circle in two-dimensional plots. The radiation power of an isotropic antenna is

$$P_{\text{rad}} = \int_\Omega U(\theta,\phi)\, d\Omega = 4\pi U_o. \tag{12.10}$$

This means that if an antenna is supplied with P_{rad}, its radiation intensity at any point is simply

$$U(\theta,\phi) = \frac{P_{\text{rad}}}{4\pi}. \tag{12.11}$$

The directivity $D(\theta,\phi)$ is defined as the ratio of the radiation intensity of an actual antenna, compared to that of an isotropic antenna. That is,

$$D(\theta,\phi) = \frac{U(\theta,\phi)}{U_o} = 4\pi\frac{U(\theta,\phi)}{P_{\text{rad}}}. \tag{12.12}$$

The directivity of an isotropic antenna is unity. All physical antennas have directivities less than or more than unity. If the directivity is mentioned without the functional dependence (θ,ϕ), then it refers to the maximum directivity (i.e., $D \equiv D_{\max}$).

Example 12.4 Find $D(\theta,\phi)$ if (a) $U(\theta,\phi) = A_o \sin\theta$, and (b) $U(\theta,\phi) = A_o \sin^2\theta$. **Hint:** $\int_0^\pi \sin^2\theta\, d\theta = \pi/2$ and $\int_0^\pi \sin^3\theta\, d\theta = 4/3$.

Solution.

(a)

$$P_{\text{rad}} = \int_{\Omega} U(\theta, \phi)\, d\Omega = 2\pi A_o \int_0^{\pi} \sin^2\theta\, d\theta = \pi^2 A_o \Rightarrow D(\theta, \phi) = 1.27 \sin\theta.$$

(b)

$$P_{\text{rad}} = \int_{\Omega} U(\theta, \phi)\, d\Omega = 2\pi A_o \int_0^{\pi} \sin^3\theta\, d\theta = 8\pi A_o/3 \Rightarrow D(\theta, \phi) = 1.5 \sin^2\theta.$$

◁

12.2.5 Antenna Polarization

An antenna can be designed to transmit or receive an electromagnetic wave with a prescribed polarization. Suppose that a receiving antenna is designed to receive a wave of the form

$$\mathbf{E}_r = E_r\, \hat{\rho}_r, \tag{12.13}$$

where $\hat{\rho}_r$ defines the polarization state of antenna. Such an antenna is exposed to a transmitted wave with

$$\mathbf{E}_t = E_t\, \hat{\rho}_t, \tag{12.14}$$

where $\hat{\rho}_t$ defines the polarization state of the incident wave. Then, due to a possible polarization mismatch, we define the polarization loss factor (PLF) as

$$PLF = |\hat{\rho}_r \bullet \hat{\rho}_t|^2. \tag{12.15}$$

Such a quantity describes the power loss due to polarization mismatch.

Example 12.5 Given a transmitted electric field phasor of the form $\mathbf{E}_t = \hat{\mathbf{x}}$, and a receiving antenna with an electric field phasor of the form $\mathbf{E}_r = \hat{\mathbf{x}} + \hat{\mathbf{y}}$, obtain the polarization loss factor.

Solution After finding $\hat{\rho}_t = \hat{\mathbf{x}}$, $\hat{\rho}_r = \frac{1}{\sqrt{2}}(\hat{\mathbf{x}} + \hat{\mathbf{y}})$, we get $PLF = \frac{1}{2}$. Therefore, half of the power from the incident wave will be lost at the receiving terminal of the antenna due to the polarization mismatch. ◁

12.3 Circuit Properties

In addition to radiation aspect of the antenna, there exists a circuit aspect. An antenna has two modes of operation: a transmission mode and a receiving mode. A transmitting antenna is connected by a transmission line to a source, and power travels from the source to the antenna. A receiving antenna is connected by a transmission line to a receiver, and power travels from the antenna to the receiver. Since the source, the transmission line, and the antenna have their own internal dissipations, and since reflection takes place at the antenna terminal due to possible impedance mismatch between the antenna and the transmission line, an equivalent circuit is needed to describe all these effects. Therefore, from a circuit perspective, an antenna can be viewed as a circuit element for design purposes. An equivalent circuit is shown in Fig. 12.4 for a transmitting antenna and for a receiving antenna. In this section, we discuss circuit properties of antennas. These include input impedance, radiation resistance and efficiency, resonance frequency, reflection mismatch, and bandwidth.

12.3.1 Input Impedance

From Fig. 12.4a, we see that the accepted complex power by a transmitting antenna is

$$S_o = \frac{1}{2} V_o I_o^* = P_o + jQ_o, \tag{12.16}$$

where V_o and I_o are, respectively, the voltage and current at the antenna terminal, P_o is the real part that describes the transfer of the power to the antenna, and Q_o is the imaginary part that describes the transfer of the reactive power to the antenna. As an equivalent point of view, the input impedance Z_{in} is defined as the ratio of the terminal voltage V_o to the terminal current I_o as

$$Z_{\text{in}} = \frac{V_o}{I_o} = R_{\text{in}} + jX_{\text{in}}, \tag{12.17}$$

where R_{in} is called the input resistance, and X_{in} is called the input reactance. Therefore, Z_{in} represents the accepted complex power S_o, R_{in} represents the real power P_o, and X_{in} represents the reactive power Q_o.

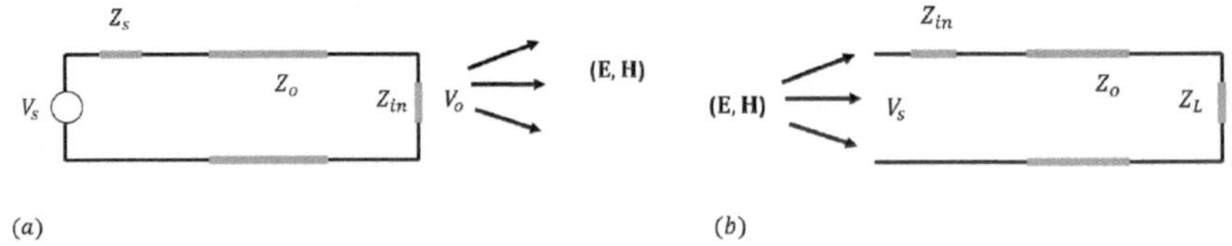

Fig. 12.4 **a** Transmitting antenna, and **b** receiving antenna

12.3.2 Radiation Resistance and Radiation Efficiency

The input real power to the antenna P_o will not be entirely radiated, but part of it will get dissipated due to the dissipation of the antenna (e.g., conduction, dielectric, and magnetic dissipations). That is,

$$P_o = P_L + P_{\text{rad}}, \tag{12.18}$$

where P_L is the dissipated power and P_{rad} is the radiated power defined in Eq. (12.7). Consequently, the input resistance can be written as

$$R_{\text{in}} = R_L + R_{\text{rad}}, \tag{12.19}$$

where R_L accounts for the dissipation, and R_{rad} accounts for the actual radiated power. This resistance is called the radiation resistance. Since

$$P_o = \frac{1}{2}|I_o|^2 R_{\text{in}} = \frac{1}{2}|I_o|^2 (R_L + R_{\text{rad}}) = P_L + P_{\text{rad}}, \tag{12.20}$$

we see that

$$R_{\text{rad}} = \frac{2P_{\text{rad}}}{|I_o|^2}. \tag{12.21}$$

The dissipation resistance R_L can be obtained depending on the current distribution on the antenna. For a half-wave dipole antenna (see Chap. 13), if the antenna is made of a cylindrical wire of length l, radius a, conductivity σ, and $kl >> 1$, the dissipation resistance can be found as

$$R_L = \frac{1}{2}\frac{l}{2\pi a}\sqrt{\frac{\omega\mu_0}{2\sigma}}. \tag{12.22}$$

The radiation efficiency e_{cd} is the ratio of the radiated power to the input real power. That is

$$e_{cd} = \frac{R_{\text{rad}}}{R_L + R_{\text{rad}}} \leq 1. \tag{12.23}$$

Also, sometimes it is called the conduction-dielectric efficiency. Using the radiation efficiency, the radiated power can be related to the power accepted by the antenna as

$$P_{\text{rad}} = e_{cd} P_o. \tag{12.24}$$

Example 12.6 Given a half-wave dipole antenna in free space with $l = \lambda/2m$, $a = 3 \times 10^{-4}\lambda$, $\sigma = 5.7 \times 10^7\,\text{S m}^{-1}$, and $f = 100\,\text{MHz}$, determine the radiation efficiency, given that the radiation resistance is $80\,\Omega$.

Solution We find $\lambda = \lambda_o = c/f = 3$. Then, $R_L = \frac{1}{2}\frac{l}{2\pi a}\sqrt{\frac{\omega\mu_0}{2\sigma}} = 0.349\,\Omega$. Therefore, $e_{cd} = 0.9956$. ◁

12.3.3 Resonance Frequency

The resonance frequency f_r is the frequency at which the input reactance $X_{\text{in}} = 0$. At such frequency, the electric and magnetic energies balance. The resonance frequency is important to determine when designing antennas for matching purposes, as to be explained next.

12.3.4 Reflection Mismatch Factor

With reference to the transmitting antenna equivalent circuit shown in Fig. 12.4a, due to a possible difference between the antenna input impedance Z_{in} and the transmission line characteristic impedance Z_o, there will be a reflection

$$\Gamma = \frac{Z_{\text{in}} - Z_o}{Z_{\text{in}} + Z_o} \tag{12.25}$$

at the antenna terminal. This reflection leads to power dissipation as follows. If a real power P_s is supplied by the source V_s, then due to the reflection, an amount of $|\Gamma|^2 P_s$ will get reflected back to the source, and an amount of $(1 - |\Gamma|^2) P_s$ will be transmitted to the antenna. Therefore, the real power of the antenna becomes

$$P_o = (1 - |\Gamma|^2) P_s. \tag{12.26}$$

The term

$$e_r = 1 - |\Gamma|^2 \tag{12.27}$$

is often called the mismatch efficiency, or the mismatch factor. The quantity $|\Gamma|^2$ is also measured in dB as

$$|\Gamma|(dB) = 10 \log_{10} |\Gamma|^2 = 20 \log_{10} |\Gamma|. \tag{12.28}$$

Notice that $-\infty < |\Gamma|(dB) \leq 0$, because $0 \leq |\Gamma| \leq 1$. An ideal situation is when no reflection takes place (i.e., $\Gamma = 0$). Then, $|\Gamma|(dB) \to -\infty$. So, a small $|\Gamma|(dB)$ is desirable. Also, since $0 \leq |\Gamma| \leq 1$, we see that the standing-wave ratio $S = \dfrac{1 + |\Gamma|}{1 - |\Gamma|}$ must be $1 \leq S < \infty$. Therefore, $S \Rightarrow 1$ is desirable. A closely-related quantity to $|\Gamma|$, called the return loss (RL), is defined as

$$RL = -10 \log_{10} |\Gamma|^2 = -20 \log_{10} |\Gamma|. \tag{12.29}$$

Notice that $RL = -|\Gamma|(dB)$, which implies that $0 \leq RL < \infty$. Also, $RL \Rightarrow \infty$ is desirable. As a summary, in a perfect situation, $|\Gamma| = 0$, $|\Gamma|(dB) \to -\infty$, $e_r = 1$, $RL \to \infty$, and $S \to 1$.

A reflection is caused by the difference between the transmission line characteristic impedance Z_o and the antenna input impedance Z_{in}. To have zero reflection, we should choose the frequency at which

$$Z_{\text{in}} = Z_o. \tag{12.30}$$

As Z_o is in many situations real, Eq. (12.30) can be satisfied when $R_{\text{in}} = Z_o$, and $X_{\text{in}} = 0$. As Z_{in} is frequency dependent, it is desirable search for the frequency in which $X_{\text{in}} = 0$, which is nothing but the resonance frequency. Even though $X_{\text{in}} = 0$ while $R_{\text{in}} \neq Z_o$ does not deliver zero reflection, matching circuits (e.g., quarter-wave transformers) can be used to match the antenna input resistance and the transmission line characteristic impedance (see Sect. 9.3.2).

12.3.5 Bandwidth

The bandwidth BW of an antenna is the range of frequencies (centered about a center frequency that is usually taken to be the resonance frequency) in which the performance of the antenna, with respect to one or more antenna properties, is within an acceptable value. As an example, Fig. 12.5 shows a plot of $|\Gamma|$ in dB versus $f \in [0, 10]$ MHz for a certain antenna. An ideal situation would be when $|\Gamma|(dB) \to -\infty$. Since this cannot be literally achieved in practice, we look for a range of frequencies in which, for instance, $|\Gamma|(dB) \leq 10$. Accordingly, there are two bands each one centring around a center (resonant) frequency: one band around $f_c = f_r = 2$ MHz, and the other band around $f_c = f_r = 6$ MHz. The bandwidth of the first resonant frequency is roughly $BW = 2.48 - 1.34 = 1.14$ MHz, while the bandwidth of the second resonant frequency is roughly $BW = 6.64 - 5.53 = 1.11$ MHz.

An antenna with one resonance frequency can be classified as a single-band antenna. If more than one resonance frequency occurs, it can be classified as a multi-band antenna. Also, an antenna can be classified as a narrow-band antenna when the fractional bandwidth (i.e., BW/f_c) is small. In contrary, a wideband or a broadband antenna is an antenna with a high fractional bandwidth.

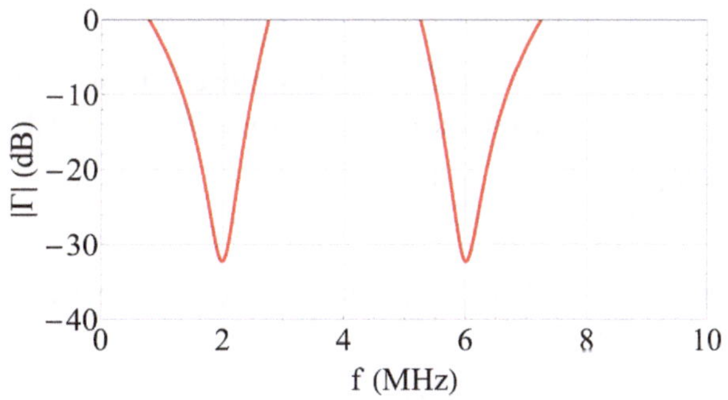

Fig. 12.5 $|\Gamma|$ in dB versus $f \in [0, 10]$ MHz

12.4 Radiation-Circuit Properties

In addition to radiation and circuit properties, there are properties combining both radiation properties and circuit properties. These include the gain and the effective area. These are discussed next.

12.4.1 Gain

The gain $G(\theta, \phi)$ of a transmitting antenna is defined as the ratio of the radiation intensity of an actual antenna, compared to that of an isotropic antenna that radiates the power accepted by the antenna (P_s). That is,

$$G(\theta, \phi) = 4\pi \frac{U(\theta, \phi)}{P_s}. \tag{12.31}$$

Since $P_{\text{rad}} = e_{cd} P_s$, the gain can be written as

$$G(\theta, \phi) = e_{cd} 4\pi \frac{U(\theta, \phi)}{P_{\text{rad}}} = e_{cd} D(\theta, \phi). \tag{12.32}$$

So, the gain is simply the directivity, taking dielectric and conduction dissipation into account. If the gain is mentioned without the functional dependence (θ, ϕ), then it refers to the maximum gain (i.e., $G \equiv G_{\max} = e_{cd} D_{\max}$).

To take reflection into account, we introduce the absolute gain (or realized gain) as

$$G_{\text{abs}}(\theta, \phi) = e_{cd} e_r D(\theta, \phi). \tag{12.33}$$

Notice that the directivity and gain are related ultimately to the radiation intensity. In other words, patterns of $U(\theta, \phi)$, $D(\theta, \phi)$, $G(\theta, \phi)$, and $G_{\text{abs}}(\theta, \phi)$ are all identical, except that each one has a different scale. Also, these are often plotted in dB scale by taking $10 \log_{10}$ of their values.

Example 12.7 Given a non dissipative antenna with $U(\theta, \phi) = \sin^3 \theta$, $Z_{\text{in}} = 73\,\Omega$, and $Z_o = 50\,\Omega$, determine (a) the maximum gain G, and (b) the maximum absolute gain G_{abs}.
Hint: $\int_0^\pi \sin^4 \theta \, d\theta = 3\pi/8$.

Solution (a) Since the antenna is non dissipative, $e_{cd} = 1$. Therefore, $G = D$. So,

$$P_{\text{rad}} = \int_0^{2\pi} \int_0^{\pi} \sin^4 \theta \, d\theta d\phi = 3\pi^2/4 \Rightarrow G(\theta, \phi) = \frac{4\pi}{P_{\text{rad}}} U(\theta, \phi) = 1.697 \sin^3 \theta \Rightarrow G = 1.697.$$

(b)

$$\Gamma = \frac{Z_{\text{in}} - Z_o}{Z_{\text{in}} + Z_o} = 0.187 \Rightarrow e_r = 1 - |\Gamma|^2 = 0.965 \Rightarrow G_{\text{abs}} = 1.638.$$

In dB, we have $G = 2.297$ dB and $G_{\text{abs}} = 2.142$ dB. ◁

12.4.2 Effective Area

Consider a plane wave with an associated Poynting vector $\mathbf{W}_i(\theta, \phi)$ is incident upon a receiving antenna with a corresponding surface $\mathcal{S}$. The power at the receiving antenna terminal P_r is given by

$$P_r = \int_S \mathbf{W}_i(\theta, \phi) \bullet d\mathbf{S}. \tag{12.34}$$

When the antenna surface is properly aligned with $\mathbf{W}_i$, the power P_r can be found from a quantity called the effective area $A_e(\theta, \phi)$ (in m^2). The effective area is defined as the ratio of the power available at the antenna terminal P_r to the power density of the plane wave incident from a certain direction θ and ϕ. That is,

$$A_e(\theta, \phi) = \frac{P_r}{|\mathbf{W}_i(\theta, \phi)|}. \tag{12.35}$$

Assuming that the receiving antenna is polarization matched to the incident plane wave, the receiving antenna and the transmission line are impedance matched, and that the receiving antenna is non dissipative, it can be shown that the effective area of an antenna can be related to its directivity through [2]

$$A_e(\theta, \phi) = \frac{\lambda^2}{4\pi} D(\theta, \phi). \tag{12.36}$$

Therefore, given an antenna with a maximum directivity D, its maximum effective area is

$$A_{em} = \frac{\lambda^2}{4\pi} D. \tag{12.37}$$

If the above assumptions are relaxed, the maximum effective area becomes

$$A_{em} = e_{cd} e_r PLF \frac{\lambda^2}{4\pi} D = PLF \frac{\lambda^2}{4\pi} G_{\text{abs}}. \tag{12.38}$$

12.4.3 Friis Transmission Equation

Consider the situation where a transmitting antenna and a receiving antenna are a distance R apart, where R is such that each antenna is in the far zone with respect to the other. The transmitting antenna is characterized by a transmitted power P_t, a directivity $D^t(\theta_t, \phi_t)$

with θ_t and ϕ_t being the transmitted wave angles, a radiation efficiency e_{cdt}, and a mismatch efficiency e_{rt}. Similarly, the receiving antenna is characterized by a received power P_r, a directivity $D^r(\theta_r, \phi_r)$ with θ_r and ϕ_r being the receiving wave angles, a radiation efficiency e_{cdr}, and a mismatch efficiency e_{rr}. The received power P_r at the receiving antenna can be related to the transmitted power P_t of the transmitting antenna through [2]

$$\frac{P_r}{P_t} = e_{cdt}e_{cdr}e_{rt}e_{rr}\left(\frac{\lambda}{4\pi R}\right)^2 D^t(\theta_t, \phi_t)D^r(\theta_r, \phi_r)PLF, \tag{12.39}$$

where PLF is the polarization mismatch between the transmitting receiving antennas, and the quantity $\left(\frac{\lambda}{4\pi R}\right)^2$, called the free-space loss factor, takes into account the losses due to the spherical spreading of the energy by the antenna. Equation (12.39) is called the Friis transmission equation. It simply tells the qualifications for maximum power transfer between the transmitter and the receiver. Alternately, Eq. (12.39) can be written more compactly as

$$\frac{P_r}{P_t} = \left(\frac{\lambda}{4\pi R}\right)^2 G^t_{\text{abs}}(\theta_t, \phi_t)G^r_{\text{abs}}(\theta_r, \phi_r)PLF, \tag{12.40}$$

Example 12.8 Two non dissipative, polarization-matched antennas are separated by a distance of 100λ. The reflection coefficients at the terminals of the transmitting and receiving antennas are 0.1 and 0.2, respectively. The maximum directivities of the transmitting and receiving antennas are 16 dB and 20 dB, respectively. Assuming that the input power of the transmitting antenna is 2W, find the maximum power delivered to the receiving antenna.
Hint: Conversion from dB scale (x) to linear scale (y) is done as

$$y = 10^{0.1x}.$$

Solution We have $R = 100\lambda$, $\Gamma_t = 0.1$ and $\Gamma_r = 0.2$. Since both antennas are non dissipative and polarization matched, $e_{cdt} = e_{cdr} = 0$ and $PLF = 1$. Also, $D^t = 10^{0.1\times16} = 39.81$, $D^r = 10^{0.1\times20} = 100$, and $P_t = 2$ W. Therefore, $P_r = 4.777$ mW. ◁

Problems

12.1 On the xz plane, find the HPBW and the FNBW for (a) $U(\theta) = \cos\theta$, (b) $U(\theta) = \cos^2\theta$, and (c) $U(\theta) = \cos 2\theta$.

12.2 Find the directivity for (a) $U(\theta) = \sin^3\theta$, (b) $U(\theta, \phi) = \sin\theta \sin^2\phi$, and (c) $U(\theta, \phi) = \cos 2\theta \sin^2\phi$.

12.3 Two non dissipative, polarization-matched antennas are separated by a distance of 200λ. The reflection coefficients at the terminals of the transmitting and receiving antennas are 0.1 and 0.2, respectively. The maximum directivity of the transmitting antenna is 25 dB. The input power to the transmitting antenna is 2 W, and the power delivered to the receiver is 2 mW. Obtain the maximum directivity of the receiving antenna in dB.

12.4 Given the far-zone Poynting vector

$$\mathbf{W}_{\text{ave}} = \hat{\mathbf{r}}\frac{\cos^4\theta}{r^2}, \qquad 0 \le \theta \le \pi, 0 \le \phi \le \pi.$$

Determine (a) P_{rad}, (b) D in dB and the HPBW, and (c) the maximum received power when the antenna is operated in the receiving mode with the following specifications: $\Gamma = 0.1$, $PLF = 0.9$, $R_{\text{rad}} = 47\,\Omega$, $R_L = 3\,\Omega$, $f = 1$ GHz, and $W_r = 10\,\text{mW m}^{-2}$.

12.5 Two antennas are separated by a distance $R = 100\lambda$. The transmitting antenna has the following specifications: $R_{\text{rad}} = 50\,\Omega$, $R_L = 5\,\Omega$, $Z_o = 50\Omega$, $D = 25\,\text{dB}$, $\mathbf{E}_t = (\hat{\mathbf{y}} + \hat{\mathbf{z}})E_t(\mathbf{r})$. The receiving antenna has the following specifications: $R_{\text{rad}} = 75\,\Omega$, $R_{\text{in}} = 85\,\Omega$, $Z_o = 75\Omega$, $D = 30\,\text{dB}$, $\mathbf{E}_r = (\hat{\mathbf{x}} + \hat{\mathbf{z}})E_r(\mathbf{r})$. Given that the transmitted power is 2 W, determine the received power.

12.6 Given an antenna with $P_{\text{in}} = 100\,W$ (i.e., the power at the antenna terminal), $Z_o = 75\,\Omega$, $R_{\text{in}} = 100\,\Omega$, $R_{\text{rad}} = 50\,\Omega$, and

$$U(\theta,\phi) = U_o \sin\theta, \qquad 0 \le \theta \le \pi, 0 \le \phi \le 2\pi,$$

where U_o is the maximum intensity. Determine (a) U_o, (b) $D(\theta,\phi)$ and the HPBW, and (c) $G(\theta,\phi)$ and $G_{\text{abs}}(\theta,\phi)$.

Appendix

The following computer program can be used to produce the three-dimensional and two-dimensional patterns in Fig. 12.1.

```
aa = 2;
Show[
 Graphics3D@{Arrow[{{0, 0, 0}, {2 aa, 0, 0}}]},
 Graphics3D@{Arrow[{{0, 0, 0}, {0, 2 aa, 0}}]},
 Graphics3D@{Arrow[{{0, 0, 0}, {0, 0, 2 aa}}]},
 Graphics3D@{Text[x, {2 aa + 0.5, 0, 0}]},
 Graphics3D@{Text[y, {0, 2 aa + 0.5, 0}]},
 Graphics3D@{Text[z, {0, 0, 2 aa + 0.5}]},
 SphericalPlot3D[
  Sin[\[Theta]]^2, {\[Theta], 0, \[Pi]}, {\[Phi], 0, 2 \[Pi]},
```

```
  AspectRatio -> 1,
  ImageSize -> 72*9,
  PlotRange -> {-10, 10},
  AxesOrigin -> {0, 0, 0},
  Mesh -> 3,
  LabelStyle -> Directive[20, Bold],
  Boxed -> False , Ticks -> None,
  PlotStyle -> {{Thickness[0.005]}},
  ColorFunction -> "Rainbow",
  PlotLegends ->
   BarLegend[{"Rainbow", {0, 3}}, LabelStyle -> {FontSize -> 20},
    LegendMargins -> 5, LegendLabel -> "U"],
  BaseStyle -> {FontFamily -> "Times New Roman", 50,
    SingleLetterItalics -> True},
  TicksStyle -> Black,
  Epilog -> {Inset[Style["(a)", FontSize -> 50], Scaled[{0.1,
     0.05}]]},
  AxesStyle -> Arrowheads[{0, 0.05}]],
Boxed -> False]

PolarPlot[Sin[\[Theta]]^2, {\[Theta], 0, 2 \[Pi]},
PlotRange -> 1.5*fmax, PolarAxesOrigin -> {0, fmax},
PolarAxes -> True,
BaseStyle -> {FontFamily -> "Arial", FontSize -> 25},
PlotRange -> {-1, 1},
PolarTicks -> {Table[{N[Pi/6 i],
    ToString[30 Abs[i]] <> "\[Degree]"}, {i, -5, 6}], Automatic
      },
PolarGridLines -> Automatic,
PlotStyle -> {Blue},
ImageSize -> 72*9]
```

References

1. E.J. Rothwell, M.J. Cloud, *Electromagnetics*, 3rd edn. (CRC Press, 2018)
2. C.A. Balanis, *Antenna Theory: Analysis and Design*, 4th ed. (Wiley, 2016)

Wire Antennas 13

In this chapter, we discuss linear wire antennas. We discuss two types of linear wire antennas: (i) straight-wire antennas (or dipole antennas), and (ii) curved-wire antennas (or loop antennas). These are discussed, respectively, in Sects. 13.1 and 13.2. For simplicity, approximated analytical expression for the current distribution will be provided for those antennas. We then consider the situation where a dipole antenna is placed over an infinite PEC in Sect. 13.3. Finally, we discuss integral equations and how they can be used to yield exact current distributions in Sect. 13.4. Useful computer programs are given in the appendix at the end of the chapter.

13.1 Dipole Antennas

A dipole antenna is a wire of length l and radius a as shown in Fig. 13.1. The antenna is made of a metallic material and is excited by a power source placed in its center. Usually the length and radius are specified in units of the operating wavelength λ. The input current from the source causes a volume electric current $\mathbf{J}_e$ to flow inside the antenna. The determination of $\mathbf{J}_e$ is done by formulating and solving an integral equation. In order to simplify matter, the following assumptions are made:

1. The metallic material is assumed to be a PEC. This implies that no volume electric current $\mathbf{J}_e$ exists inside the cylinder, but a surface electric current $\mathbf{K}_e$ exists on the cylinder surface. Also, PEC assumption automatically makes the antenna non dissipative.
2. The radius is much smaller than the length $a << l$; an approximation known as thin-dipole approximation in which the radius a is neglected. In this case, the problem reduces to linear electric current $\mathbf{I}_e$ flowing along a wire of length l.

H. M. Alkhoori, *Concise Introduction to Electromagnetic Fields*, Synthesis Lectures on Electromagnetics, https://doi.org/10.1007/978-3-031-60331-0_13

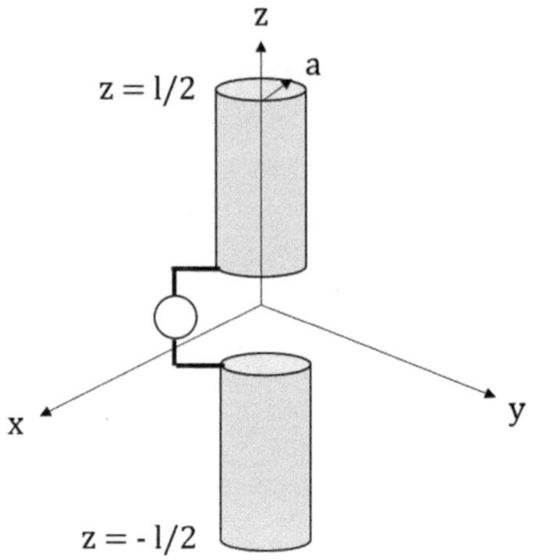

Fig. 13.1 A dipole antenna

Depending on the electrical length of the antenna l/λ, certain approximations on the current distribution can be used. These are (i) constant-current approximation (Hertzian or infinitesimal dipole) when $l << \lambda$, (ii) triangular-current approximation (or small dipole) when $l << \lambda$, and (iii) sinusoidal-current approximation (or finite-length dipole) when $l \sim \lambda$. These are discussed next.

13.1.1 Hertzian Dipole

When $l << \lambda/50$, the current can be approximated as

$$\mathbf{I}_e = I_o\,\hat{\mathbf{z}}, \tag{13.1}$$

where I_o is constant. This is called constant-current approximation, and such a dipole antenna is called a Hertzian dipole. This approximation is not realistic, because it assumes that the current at the ends $z = \pm l/2$ does not vanish. Yet, it does not cause a significant error for Hertzian dipoles.

Example 13.1 Obtain the magnetic vector potential of a Hertzian dipole.

Solution Since we have an electric line current, we use

$$\mathbf{A} = \frac{\mu}{4\pi} \int_{\mathcal{L}'} \mathbf{I}_e(\mathbf{r}')\,\frac{e^{-jkR}}{R}\,d\mathcal{L}'.$$

After noting that $\mathbf{r} = r\,\hat{\mathbf{r}}$, $\mathbf{r}' = z'\,\hat{\mathbf{z}} \approx \mathbf{0}$ $R = r$, and $d\mathcal{L}' = dz'$, we get

$$\mathbf{A} = \hat{\mathbf{z}}\,\frac{\mu I_o}{4\pi}\frac{e^{-jkr}}{r}\int_{-l/2}^{l/2} dz' = \hat{\mathbf{z}}\,\frac{\mu I_o l}{4\pi}\frac{e^{-jkr}}{r} = \frac{\mu I_o l}{4\pi}\frac{e^{-jkr}}{r}(\cos\theta\,\hat{\mathbf{r}} - \sin\theta\,\hat{\boldsymbol{\theta}}).$$

◁

Example 13.2 Obtain the electric field and magnetic field phasors of a Hertzian dipole.

Solution We first find **H** from **A**. That is,

$$\mathbf{H} = \frac{1}{\mu}\nabla\times\mathbf{A} = j\frac{kI_o l\sin\theta}{4\pi r}\left(1+\frac{1}{jkr}\right)e^{-jkr}\,\hat{\boldsymbol{\phi}},$$

where use was made for $\omega\mu = k\eta$. Then, **E** is found from Maxwell equations. That is

$$\mathbf{E} = \frac{1}{j\omega\varepsilon}\nabla\times\mathbf{H} = \eta\frac{I_o l\cos\theta}{2\pi r^2}\left(1+\frac{1}{jkr}\right)e^{-jkr}\,\hat{\mathbf{r}} + j\eta\frac{I_o kl\sin\theta}{4\pi r}\left[1+\frac{1}{jkr}-\frac{1}{(kr)^2}\right]e^{-jkr}\,\hat{\boldsymbol{\theta}},$$

where use was made for $\omega\varepsilon = k\eta^{-1}$. ◁

In time domain, the electric field $\mathbf{E}(\mathbf{r}, t)$ is related to the electric field phasor through

$$\mathscr{E}(\mathbf{r}, t) = \mathrm{Re}\{\mathbf{E}(\mathbf{r})e^{-j\omega t}\}. \tag{13.2}$$

The magnetic field can be obtained from the magnetic field phasor similarly. Figure 13.2 shows three-dimensional streamline plots of $\mathscr{E}(\mathbf{r}, 0) = \mathrm{Re}\{\mathbf{E}(\mathbf{r})\}$ and $\mathscr{H}(\mathbf{r}, 0) = \mathrm{Re}\{\mathbf{H}(\mathbf{r})\}$ for a Hertzian dipole in free space (i.e., $\varepsilon = \varepsilon_0$, $\mu = \mu_0$, and hence, $k = k_o$). Two-dimensional streamline plots of the electric field on the $y = 0$ plane and the magnetic field on the $z = 0$ plane are also depicted in Fig. 13.3. As time marches on, the electric field lines keep emanating from the dipole, whereas the magnetic field lines keep rotating about the dipole.

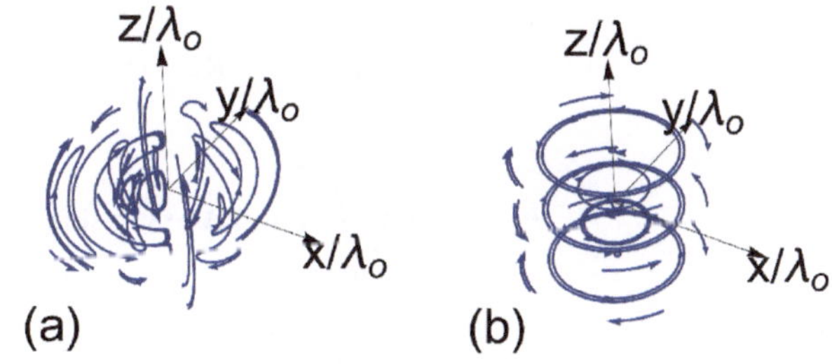

Fig. 13.2 Three-dimensional streamline plots of **a** $\mathscr{E}(\mathbf{r}, 0)$ and **b** $\mathscr{H}(\mathbf{r}, 0)$

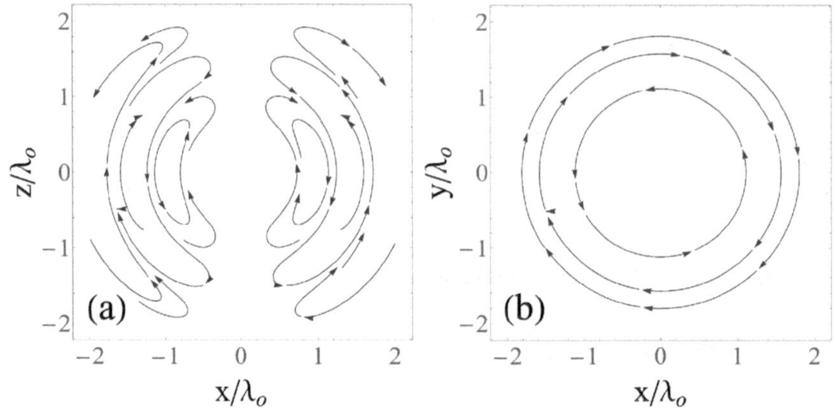

Fig. 13.3 Two-dimensional streamline plots of **a** $\mathscr{E}(\mathbf{r}, 0)$ on the $y = 0$ plane and **b** $\mathscr{H}(\mathbf{r}, 0)$ on the $z = 0$ plane

Example 13.3 Obtain (a) the Poynting vector $\mathbf{W}$, (b) the complex power $P(r)$ over an arbitrary sphere of radius r, (c) the radiated power P_{rad}, and (d) the radiation resistance R_{rad} of a Hertzian dipole.

Solution.

(a)

$$\mathbf{W} = \frac{1}{2}\mathbf{E} \times \mathbf{H}^* = \frac{1}{2}\left(\hat{\mathbf{r}} E_\theta H_\phi^* - \hat{\boldsymbol{\theta}} E_r H_\phi^*\right)$$
$$= \frac{\eta}{8}\left|\frac{I_o l}{\lambda}\right|^2 \frac{\sin^2\theta}{r^2}\left[1 - j\frac{1}{(kr)^3}\right]\hat{\mathbf{r}} + j\eta\frac{k|I_o l|^2 \cos\theta \sin\theta}{16\pi^2 r^3}\left[1 + \frac{1}{(kr)^2}\right]\hat{\boldsymbol{\theta}}.$$

(b)

$$P(r) = \int_S \mathbf{W} \cdot d\mathbf{S} = \int_0^{2\pi}\int_0^{\pi} \mathbf{W} \cdot \hat{\mathbf{r}} r^2 \sin\theta d\theta d\phi = \eta\frac{\pi}{3}\left|\frac{I_o l}{\lambda}\right|^2\left[1 - j\frac{1}{(kr)^3}\right].$$

(c)

$$P_{\text{rad}} = \lim_{r\to\infty} \text{Re}\{P(r)\} = \eta\frac{\pi}{3}\left|\frac{I_o l}{\lambda}\right|^2.$$

or

$$P_{\text{rad}} = \lim_{r\to\infty}\int_0^{2\pi}\int_0^{\pi} \text{Re}\{\mathbf{W}\} \cdot \hat{\mathbf{r}} r^2 \sin\theta d\theta d\phi = \eta\frac{\pi}{3}\left|\frac{I_o l}{\lambda}\right|^2.$$

(d)

$$R_{\text{rad}} = \frac{2P_{\text{rad}}}{|I_o|^2} = \eta\frac{2\pi}{3}\left(\frac{l}{\lambda}\right)^2 = 80\pi^2\left(\frac{l}{\lambda}\right)^2.$$

◁

The following points are noteworthy:

- The Poynting vector **W** is complex. Its real part represents the real power density carried by the field, while its reactive part represents reactive power stored in the field.
- The complex power $P(r)$ over any spherical surface only captures the $\hat{\mathbf{r}}$ component of the complex Poynting vector. The $\hat{\mathbf{r}}$ component contains real and imaginary components, while the $\hat{\boldsymbol{\theta}}$ component contains imaginary component only. Therefore, the complex power contains the total real power only, but not the total imaginary power. Consequently, the input resistance can be computed from it, but the input reactance can't be computed from it. The input reactance can be computed approximately using a method called induced EMF method [1], or exactly using integral equations [2].
- The complex power $P(r)$ close to the source (i.e., $kr \to 0$) is dominantly reactive. That is why the region close to the source is called reactive near-zone region.
- The radiation resistance R_{rad} is proportional to l/λ. However, keep in mind that the assumed current distribution is valid only when $l < \lambda/50$. So, R_{rad} is at most equal to $0.316\,\Omega$. Since most commercial coaxial lines have characteristic impedances between 50 to 75 Ω, there will be a huge amount of reflection. Therefore, a Hertzian dipole antenna is generally a poor radiator from circuit point of view.
- Since the antenna was assumed to be non dissipative, the input resistance is merely the radiation resistance.

Example 13.4 Obtain (a) the radiation intensity $U(\theta, \phi)$ and its maximum (with location where maximum occurs), (b) the directivity $D(\theta, \phi)$ and its maximum, (c) the gain $G(\theta, \phi)$, and (d) the maximum effective area A_{em} of a Hertzian dipole.

Solution.
(a)

$$U(\theta, \phi) = \lim_{r\to\infty} r^2 \mathbf{W}_{\text{ave}} \bullet \hat{\mathbf{r}} = \frac{\eta}{8}\left|\frac{I_o l}{\lambda}\right|^2 \sin^2\theta.$$

Also, $U = \frac{\eta}{8}\left|\frac{I_o l}{\lambda}\right|^2$ is the maximum intensity that occurs when $\theta = \pi/2$ (i.e., on the xy plane).
(b)

$$D(\theta, \phi) = \frac{U(\theta, \phi)}{U_o} = 4\pi\frac{U(\theta, \phi)}{P_{\text{rad}}} = \frac{3}{2}\sin^2\theta.$$

Also, $D = \frac{3}{2}$ is the maximum directivity.
(c) Since the antenna is non dissipative, $G(\theta, \phi) = D(\theta, \phi)$.
(d)

$$A_{em} = \frac{\lambda^2}{4\pi}D = \frac{3\lambda^2}{8\pi}.$$

The directivity pattern of a Hertzian dipole is similar to that in Fig. 12.1, except with a difference in scale. It is seen that no radiation takes place along the dipole axis (i.e., $\theta = 0$). The pattern is directional on the xz plane with one main lobe and one back lobe only. The associated HPBW is $\pi/2$ (or 90°). On the contrary, the pattern is non directional on the xy plane. When an antenna exhibits a directionality property on a plane, and a non directionality property on another plane, it is called omnidirectional. ◁

Example 13.5 Given the Hertzian dipole electric field and magnetic field phasors found in Example 13.2, identify (a) the reactive near-zone components $kr << 1$, (b) the radiating near-zone (intermediate) components $kr > 1$, and (c) the far-zone components $kr >> 1$.

Solution (a) When $kr << 1$, we deduce that

$$\left.\begin{aligned} E_r &\approx -j\eta \frac{I_o l \cos\theta}{2\pi k r^3} e^{-jkr} \\ E_\theta &\approx -j\eta \frac{I_o l k \sin\theta}{4\pi k^2 r^3} e^{-jkr} \\ E_\phi &= 0 \\ H_r &= H_\theta = 0 \\ H_\phi &= \frac{I_o l \sin\theta}{4\pi r^2} e^{-jkr} \end{aligned}\right\} .$$

We see that the electric field phasor has $1/r^3$ dependence, while the magnetic field phasor has $1/r^2$ dependence. Also, the electric field phasor components are in phase. Furthermore, it can be shown that **W** is purely imaginary. So fields very close to the antenna are said to be reactive because they give rise to reactive power only. Finally, aside from the term e^{-jkr} the electric field and the magnetic field phasors are exactly those of static electric dipole and magnetic dipole, respectively. Therefore, they are called quasi-static fields.

(b) When $kr > 1$, we deduce that

$$\left.\begin{aligned} E_r &\approx \eta \frac{I_o l \cos\theta}{2\pi r^2} e^{-jkr} \\ E_\theta &\approx j\eta \frac{I_o l k \sin\theta}{4\pi r} e^{-jkr} \\ E_\phi &= 0 \\ H_r &= H_\theta = 0 \\ H_\phi &= j \frac{I_o l k \sin\theta}{4\pi r} e^{-jkr} \end{aligned}\right\} .$$

We see that the electric field phasor has $1/r^2$ and $1/r$ dependences, while the magnetic field phasor has $1/r$ dependence. Also, the electric field phasor components are out of phase. Furthermore, it can be shown that **W** is complex with imaginary part dominating the real

part. So fields in the intermediate region are both radiating and reactive; yet, reactive nature is dominant. Finally, the E_r component is quasi static, while E_θ and H_ϕ are not.
(c) When $kr >> 1$, we deduce that

$$\left.\begin{aligned} E_r &\approx 0 \\ E_\theta &\approx j\eta\frac{I_o lk\sin\theta}{4\pi r}e^{-jkr} \\ E_\phi &= 0 \\ H_r &= H_\theta = 0 \\ H_\phi &= j\frac{I_o lk\sin\theta}{4\pi r}e^{-jkr} \end{aligned}\right\}.$$

In this region, both of the electric field and magnetic field phasors have $1/r$ dependence. Also, it can be shown that $\mathbf{W}$ is purely real. So fields in the far-zone region are only radiating. Notice also that the ratio between E_θ and H_ϕ is noting but the intrinsic impedance η. From this example, we see that radiation process starts gradually in the intermediate region and then becomes more significant in the far zone region. ◁

We saw that obtaining the electric field and magnetic field phasors everywhere for this simple source is tedious. For more complicated sources, the procedure becomes harder and more time consuming. However, we saw by now that radiation is more significant in the far zone region. Additionally, as far-zone calculation is way simpler than calculating the fields everywhere, we restrict our attention henceforth to far-zone calculation only. Near-zone calculation is needed in integral equation formulation, as well as when discussing coupling that occurs among various antenna elements when used to form an array.

Example 13.6 In Example 13.1, we computed $\mathbf{A}$ for a Hertzian dipole of length l, oriented along the z axis, and located at the origin $\mathbf{r}' = \mathbf{0}$. Obtain the far-zone $\mathbf{A}$, $\mathbf{E}$, and $\mathbf{H}$ when the dipole is located at $\mathbf{r}' = \mathbf{r}_o$ and carrying a current $\mathbf{I}_e = I_o\,\hat{\mathbf{p}}$, where $\hat{\mathbf{p}}$ is the dipole orientation.

Solution

$$\mathbf{A} \approx \frac{\mu}{4\pi}\frac{e^{-jkr}}{r}\int_{\mathcal{L}'}\mathbf{I}_e(\mathbf{r}')\,e^{jk\hat{\mathbf{r}}\bullet\mathbf{r}'}\,d\mathcal{L}' = \frac{\mu I_o}{4\pi}\frac{e^{-jkr}}{r}\hat{\mathbf{p}}e^{jk\hat{\mathbf{r}}\bullet\mathbf{r}_o}\int_{\mathcal{L}'}d\mathcal{L}' = \frac{\mu I_o l}{4\pi}\frac{e^{-jkr}}{r}e^{jk\hat{\mathbf{r}}\bullet\mathbf{r}_o}\,\hat{\mathbf{p}}.$$

We let $\mathbf{p} = I_o l\,\hat{\mathbf{p}}$ be the electric dipole moment. Then,

$$\mathbf{A} \approx \frac{\mu e^{-jkr}}{4\pi r}\mathbf{p}\,e^{jk\hat{\mathbf{r}}\bullet\mathbf{r}_o}.$$

The associated far-zone $\mathbf{E}$ is

$$\mathbf{E} \approx -j\omega\mathbf{A}_\perp = -\frac{jk\eta e^{-jkr}}{4\pi r}\mathbf{p}_\perp\,e^{jk\hat{\mathbf{r}}\bullet\mathbf{r}_o},$$

with $\mathbf{p}_\perp$ being $\mathbf{p}$ in spherical bases without the $\hat{\mathbf{r}}$ component. Finally, the associated far-zone $\mathbf{H}$ is

$$\mathbf{H} \approx \frac{jke^{-jkr}}{4\pi r}(\hat{\mathbf{r}} \times \mathbf{p}_\perp)\, e^{jk\hat{\mathbf{r}} \bullet \mathbf{r}_o}.$$

◁

13.1.2 Small Dipole

Recall that the constant-current approximation is not accurate, but can be used safely for Hertzian dipoles. A better approximation is the triangular current approximation, which can be used when $\lambda/50 < l \leq \lambda/10$. In particular, the current can be approximated as,

$$\mathbf{I}_e(z) = I_o\left(1 - \frac{2|z|}{l}\right)\hat{\mathbf{z}}. \tag{13.3}$$

where I_o is a constant. Notice that the current vanishes at the ends $z = \pm l/2$. Such a dipole antenna can be called a small dipole.

Example 13.7 Obtain the magnetic vector potential of a small dipole.

Solution Like for a Hertzian dipole, we use

$$\mathbf{A} = \frac{\mu}{4\pi}\int_{\mathcal{L}'} \mathbf{I}_e(\mathbf{r}')\,\frac{e^{-jkR}}{R}\,d\mathcal{L}'.$$

Then, after noting that $\mathbf{r} = r\,\hat{\mathbf{r}}$, $\mathbf{r}' = z'\,\hat{\mathbf{z}} \approx \mathbf{0}$, $R = r$, and $d\mathcal{L}' = dz'$, we get

$$\mathbf{A} = \hat{\mathbf{z}}\,\frac{\mu I_o}{4\pi}\frac{e^{-jkr}}{r}\int_{-l/2}^{l/2}\left(1 - \frac{2|z'|}{l}\right)dz' = \hat{\mathbf{z}}\,\frac{\mu I_o}{4\pi}\frac{e^{-jkr}}{r}\left[\int_{-l/2}^{0}\left(1 + \frac{2z'}{l}\right)dz' + \int_{0}^{l/2}\left(1 - \frac{2z'}{l}\right)dz'\right]$$

$$= \hat{\mathbf{z}}\,\frac{\mu I_o l}{8\pi}\frac{e^{-jkr}}{r} = \frac{\mu I_o l}{8\pi}\frac{e^{-jkr}}{r}(\cos\theta\,\hat{\mathbf{r}} - \sin\theta\,\hat{\boldsymbol{\theta}}).$$

◁

Example 13.8 Obtain the far-zone electric field and magnetic field phasors of a small dipole.

Solution These are found directly from $\mathbf{A}_\perp = \mathbf{A} - (\mathbf{A} \bullet \hat{\mathbf{r}})\hat{\mathbf{r}}$ as

$$\left.\begin{aligned} \mathbf{E} &\approx -j\omega\mathbf{A}_\perp = j\omega\frac{\mu I_o l \sin\theta}{8\pi}\frac{e^{-jkr}}{r}\hat{\boldsymbol{\theta}} = j\eta\frac{I_o lk\sin\theta}{8\pi r}e^{-jkr}\hat{\boldsymbol{\theta}} \\ \mathbf{H} &\approx \frac{1}{\eta}\hat{\mathbf{r}} \times \mathbf{E} = j\frac{I_o lk \sin\theta}{8\pi r}e^{-jkr}\hat{\boldsymbol{\phi}} \end{aligned}\right\}.$$

◁

Example 13.9 Obtain the radiation resistance of a small dipole.

Solution After finding the average Poynting vector

$$\mathbf{W}_{\text{ave}} = \frac{1}{2}\text{Re}\{\mathbf{E} \times \mathbf{H}^*\} = \frac{\eta}{32}\left|\frac{I_o l}{\lambda}\right|^2 \frac{\sin^2\theta}{r^2}\hat{\mathbf{r}},$$

we find the average power from

$$P_{\text{ave}} = \int_0^{2\pi}\int_0^{\pi} \mathbf{W}_{\text{ave}} \bullet \hat{\mathbf{r}}\, r^2 \sin\theta d\theta d\phi = \eta\frac{\pi}{12}\left|\frac{I_o l}{\lambda}\right|^2.$$

Finally, the radiation resistance is found using

$$R_{\text{rad}} = \frac{2P_{\text{rad}}}{|I_o|^2} = \eta\frac{2\pi}{12}\left(\frac{l}{\lambda}\right)^2 = 20\pi^2\left(\frac{l}{\lambda}\right)^2.$$

◁

13.1.3 Finite-Length Dipole

When $l > \lambda/10$, the constant-current and triangular-current approximations become less accurate. A better approximation is the sinusoidal approximation given by

$$\mathbf{I}_e(z) = I_o \sin k\left(\frac{l}{2} - |z|\right)\hat{\mathbf{z}}. \tag{13.4}$$

Such distribution is very close to the actual distribution determined upon solving an integral equation. Also, it ensures that the current vanishes at the ends. However, it introduces a distinction between the input resistance and the radiation resistance (see Example 13.13).

Example 13.10 Obtain the far-zone magnetic vector potential of a finite-length dipole.
Hint:

$$\int_{-l/2}^{l/2} \sin k\left(\frac{l}{2}-|z'|\right) e^{jkz'\cos\theta}\, dz' = 2\left[\frac{\cos\left(\frac{kl}{2}\cos\theta\right)-\cos\frac{kl}{2}}{k\sin^2\theta}\right].$$

Solution The far-zone magnetic vector potential is found as

$$\mathbf{A} \approx \frac{\mu}{4\pi}\frac{e^{-jkr}}{r}\int_{\mathcal{L}'} \mathbf{I}_e(\mathbf{r}')\, e^{jk\hat{\mathbf{r}}\bullet\mathbf{r}'}\, d\mathcal{L}'.$$

After noting that $\mathbf{r}' = z'\,\hat{\mathbf{z}} = z'(\cos\theta\,\hat{\mathbf{r}} - \sin\theta\,\hat{\boldsymbol{\theta}})$, $\hat{\mathbf{r}}\bullet\mathbf{r}' = z'\cos\theta$, and $d\mathcal{L}' = dz'$, we get

$$\begin{aligned}\mathbf{A} &\approx (\cos\theta\,\hat{\mathbf{r}} - \sin\theta\,\hat{\boldsymbol{\theta}})\,\frac{\mu I_o}{4\pi}\frac{e^{-jkr}}{r}\int_{-l/2}^{l/2} \sin k\left(\frac{l}{2}-|z'|\right) e^{jkz'\cos\theta}\, dz' \\ &= (\cos\theta\,\hat{\mathbf{r}} - \sin\theta\,\hat{\boldsymbol{\theta}})\,\frac{\mu I_o}{2\pi}\frac{e^{-jkr}}{krsin\theta}F(\theta;kl),\end{aligned}$$

where

$$F(\theta;kl) = \frac{\cos\left(\frac{kl}{2}\cos\theta\right)-\cos\frac{kl}{2}}{sin\theta}.$$

◁

Example 13.11 Obtain the far-zone electric field and magnetic field phasors of a finite-length dipole.

Solution These are found directly from $\mathbf{A}_\perp = \mathbf{A} - (\mathbf{A}\bullet\hat{\mathbf{r}})\hat{\mathbf{r}}$ as

$$\left.\begin{aligned}\mathbf{E} &\approx -j\omega\mathbf{A}_\perp = \frac{j\eta I_o e^{-jkr}}{2\pi r}F(\theta;kl)\,\hat{\boldsymbol{\theta}} \\ \mathbf{H} &\approx \frac{1}{\eta}\hat{\mathbf{r}}\times\mathbf{E} = \frac{jI_o e^{-jkr}}{2\pi r}F(\theta;kl)\,\hat{\boldsymbol{\phi}}\end{aligned}\right\}.$$

◁

Example 13.12 Obtain the radiation intensity $U(\theta,\phi)$ of a finite-length dipole.

Solution After obtaining

$$\mathbf{W}_{\text{ave}} = \frac{1}{2}\text{Re}\{\mathbf{E}\times\mathbf{H}^*\} = \eta\frac{|I_o|^2}{8\pi^2r^2}F^2(\theta;kl)\,\hat{\mathbf{r}},$$

we get

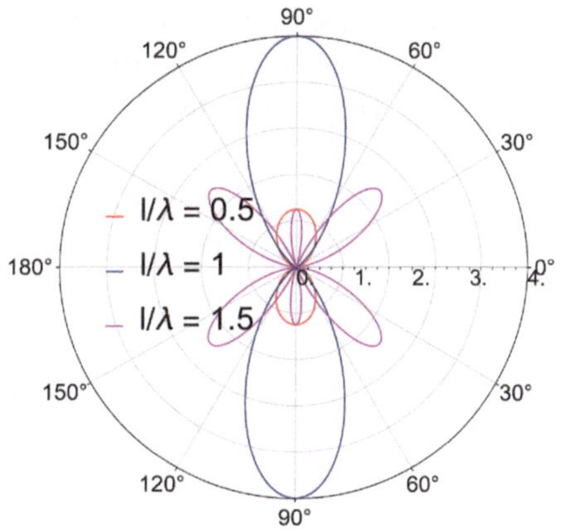

Fig. 13.4 Two-dimensional pattern of $U(\theta, 0°)$ versus $\theta \in [0, 180°]$ when $l/\lambda = \{0.5, 1, 1.5\}$

$$U(\theta, \phi) = \lim_{r\to\infty} r^2 \mathbf{W}_{\text{ave}} \bullet \hat{\mathbf{r}} = \eta \frac{|I_o|^2}{8\pi^2} F^2(\theta; kl).$$

A two-dimensional pattern of the radiation intensity is shown in Fig. 13.4 when $l/\lambda = \{0.5, 1, 1.5\}$. When $l/\lambda = 0.5$, the pattern consists of two lobes. As $l/\lambda < 1$ increases, the pattern becomes more directive and the associated HPBW of the lobes decreases. Beyond $l/\lambda > 1$, side lobes begin to emerge. The formation of side lobes as the electrical length increases is common in all antenna structures.

◁

Example 13.13 Obtain (a) the radiation power P_{rad}, (b) the radiation resistance R_{rad}, (c) the input resistance R_{in}, and (d) the directivity $D(\theta, \phi)$ of a finite-length dipole.

Solution.
(a)

$$P_{\text{rad}} = \int_0^{2\pi}\int_0^{\pi} U(\theta, \phi) \sin\theta d\theta d\phi = \eta \frac{|I_o|^2}{4\pi} \int_0^{\pi} F^2(\theta; kl) \sin\theta d\theta = \eta \frac{|I_o|^2}{4\pi} C(kl),$$

where

$$C(kl) = C + \ln(kl) - Ci(kl) + 0.5 \sin(kl)[Si(2kl) - 2Si(kl)] \\ + 0.5 \cos(kl)[C + \ln(kl/2) + Ci(2kl) - 2Ci(kl)].$$

Here, $C = 0.5772$ is called Euler number, $Si(x)$ is the sine integral and $Ci(x)$ is the cosine integral.
(b)

$$R_{\text{rad}} = \frac{2P_{\text{rad}}}{|I_o|^2} = \frac{\eta}{2\pi} C(kl).$$

(c) Because of the sinusoidal current assumption, we see that the input current $I_{\text{in}} = |\mathbf{I}_e(z = 0)| = I_o \sin\frac{kl}{2}$ is not necessarily I_o. Since the antenna is non dissipative, we must have $P_{\text{in}} = P_{\text{rad}}$. This implies that

$$\frac{1}{2}|I_{\text{in}}|^2 R_{\text{in}} = \frac{1}{2}|I_o|^2 R_{\text{rad}}.$$

Since $I_{\text{in}} = I_o \sin\frac{kl}{2}$, we get

$$R_{\text{in}} = \frac{R_{\text{rad}}}{\sin^2\dfrac{kl}{2}} = \frac{\eta\, C(kl)}{2\pi \sin^2\dfrac{kl}{2}}.$$

(d)

$$D(\theta,\phi) = \frac{4\pi}{P_{\text{rad}}} U(\theta,\phi) = \frac{2}{C(kl)} F^2(\theta, kl).$$

◁

Example 13.14 Given a half-wave dipole ($l/\lambda = 1/2$), obtain (a) far-zone $\mathbf{E}$ and $\mathbf{H}$, (b) $U(\theta,\phi)$, (c) $D(\theta,\phi)$ and D, and (d) R_{in}. **Hint:** $C(\pi) = 1.22$.

Solution The electrical length is $kl = 2\pi l/\lambda = \pi$. Then,

$$F(\theta;\pi) = \frac{\cos\left(\frac{\pi}{2}\cos\theta\right)}{\sin\theta}.$$

(a) The far-zone field phasors of a half-wave dipole can be readily found from those of a finite-length dipole as

$$\left.\begin{aligned} \mathbf{E} &= \frac{j\eta I_o e^{-jkr}}{2\pi r}\frac{\cos\left(\frac{\pi}{2}\cos\theta\right)}{\sin\theta}\hat{\boldsymbol{\theta}} \\ \mathbf{H} &= \frac{j I_o e^{-jkr}}{2\pi r}\frac{\cos\left(\frac{\pi}{2}\cos\theta\right)}{\sin\theta}\hat{\boldsymbol{\phi}} \end{aligned}\right\}.$$

(b)

$$U(\theta,\phi) = \eta\frac{|I_o|^2}{8\pi^2}\frac{\cos^2\left(\frac{\pi}{2}\cos\theta\right)}{\sin^2\theta}.$$

(c)

$$D(\theta,\phi) = \frac{2}{C(\pi)}\frac{\cos^2\left(\frac{\pi}{2}\cos\theta\right)}{\sin^2\theta} = 1.64\frac{\cos^2\left(\frac{\pi}{2}\cos\theta\right)}{\sin^2\theta}.$$

Also, $D = 1.64$.

(d)

$$R_{\text{in}} = \frac{\eta\, C(\pi)}{2\pi\sin^2\dfrac{\pi}{2}} = 73\,\Omega.$$

◁

13.2 Loop Antennas

In this section, we discuss curved-wire antennas. Consider a toroid of mean radius a and cross-sectional radius b (see Fig. 13.5). Same as before, the antenna surface is made of a PEC. To simplify analysis, we assume $a >> b$, and hence, the toroid becomes a loop of a radius a. The current distribution can be approximated as (i) a constant current, (ii) a sinusoidal current, (iii) a travelling-wave current, or (iv) a general non-uniform current using Fourier series. We limit our discussion here to constant-current distribution in the far-zone region.

Example 13.15 The loop $\{\rho = a, z = 0\}$ carries a current $\mathbf{I}_e(\mathbf{r}) = I_o\,\hat{\boldsymbol{\phi}}$. Obtain the far-zone magnetic vector potential.

Solution

$$\mathbf{A} \approx \frac{\mu}{4\pi}\frac{e^{-jkr}}{r}\int_{\mathcal{L}'} \mathbf{I}_e(\mathbf{r}')\, e^{jk\hat{\mathbf{r}}\,\bullet\,\mathbf{r}'}\, d\mathcal{L}'.$$

After noting that $\mathbf{r}' = \rho'\,\hat{\boldsymbol{\rho}}' = \rho'(\cos\phi'\,\hat{\mathbf{x}} + \sin\phi'\,\hat{\mathbf{y}})$, we get

$$\hat{\mathbf{r}}\bullet\mathbf{r}' = \rho'(\cos\phi'\,\hat{\mathbf{r}}\bullet\hat{\mathbf{x}} + \sin\phi'\,\hat{\mathbf{r}}\bullet\hat{\mathbf{y}})$$

Using

$$\begin{pmatrix}\hat{\mathbf{x}}\\ \hat{\mathbf{y}}\\ \hat{\mathbf{z}}\end{pmatrix} = \begin{pmatrix}\sin\theta\cos\phi & \cos\theta\cos\phi & -\sin\phi\\ \sin\theta\sin\phi & \cos\theta\sin\phi & \cos\phi\\ \cos\theta & -\sin\theta & 0\end{pmatrix}\begin{pmatrix}\hat{\mathbf{r}}\\ \hat{\boldsymbol{\theta}}\\ \hat{\boldsymbol{\phi}}\end{pmatrix},$$

we get

$$\hat{\mathbf{r}}\bullet\mathbf{r}' = \rho'(\sin\theta\cos\phi\cos\phi' + \sin\theta\sin\phi\sin\phi') = \rho'\sin\theta\cos(\phi - \phi').$$

Also, since $\mathbf{I}_e(\mathbf{r}') = I_o\,\hat{\boldsymbol{\phi}}'$, $d\mathcal{L}' = \rho'\,d\phi'$, and the integral is evaluated on $\rho' = a$, we get

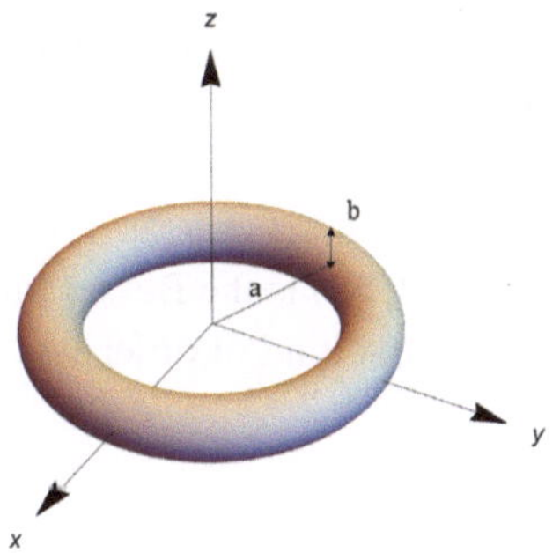

Fig. 13.5 A toroidal antenna

$$\mathbf{A} \approx \frac{\mu I_o}{4\pi} \frac{e^{-jkr}}{r} \int_0^{2\pi} \hat{\boldsymbol{\phi}}' \, e^{jka \sin\theta \cos(\phi-\phi')} \, ad\phi'.$$

Since we are interested only in the far-zone region, we need only to obtain A_θ and A_ϕ. These are given by

$$A_\theta \approx \frac{\mu I_o a}{4\pi} \frac{e^{-jkr}}{r} \int_0^{2\pi} \hat{\boldsymbol{\theta}} \bullet \hat{\boldsymbol{\phi}}' \, e^{jka \sin\theta \cos(\phi-\phi')} \, d\phi',$$

and

$$A_\phi \approx \frac{\mu I_o a}{4\pi} \frac{e^{-jkr}}{r} \int_0^{2\pi} \hat{\boldsymbol{\phi}} \bullet \hat{\boldsymbol{\phi}}' \, e^{jka \sin\theta \cos(\phi-\phi')} \, d\phi'.$$

After noting that

$$\hat{\boldsymbol{\theta}} \bullet \hat{\boldsymbol{\phi}}' = \hat{\boldsymbol{\theta}} \bullet (-\sin\phi' \, \hat{\mathbf{x}} + \cos\phi' \, \hat{\mathbf{y}}) = -\sin\phi' \cos\theta \cos\phi + \cos\phi' \cos\theta \sin\phi = \cos\theta \sin(\phi - \phi'),$$

and

$$\hat{\boldsymbol{\phi}} \bullet \hat{\boldsymbol{\phi}}' = \hat{\boldsymbol{\phi}} \bullet (-\sin\phi' \, \hat{\mathbf{x}} + \cos\phi' \, \hat{\mathbf{y}}) = \sin\phi' \sin\phi + \cos\phi' \cos\phi = I_o \cos(\phi - \phi'),$$

we get

$$\left.\begin{aligned} A_\theta &\approx \frac{\mu I_o a \cos\theta}{4\pi} \frac{e^{-jkr}}{r} \int_0^{2\pi} \sin(\phi - \phi') \, e^{jka \sin\theta \cos(\phi-\phi')} \, d\phi' \\ A_\phi &\approx \frac{\mu I_o a}{4\pi} \frac{e^{-jkr}}{r} \int_0^{2\pi} \cos(\phi - \phi') \, e^{jka \sin\theta \cos(\phi-\phi')} \, d\phi' \end{aligned}\right\}.$$

For A_θ, the term $\sin(\phi - \phi')$ is an odd function of ϕ', while $e^{jka \sin\theta \cos(\phi-\phi')}$ is an even function of ϕ'. Using change of variables, the integral can be made from $\phi' = -\pi$ to $\phi' = \pi$, and thus , $A_\theta = 0$. The integral in A_ϕ can be carried out by a change of variables $u = \phi - \phi'$ to get [1]

$$A_\phi \approx \frac{\mu I_o a}{4\pi} \frac{e^{-jkr}}{r} j2\pi J_1(ka \sin\theta),$$

where $J_1(\bullet)$ is the Bessel function of the first kind of order 1. In general, a Bessel function of the first kind of order ν is given by [3]

$$J_\nu(x) = \sum_{n=0}^{\infty} \frac{(-1)^n}{n!\Gamma(\nu + n + 1)} \left(\frac{x}{2}\right)^{2n+\nu},$$

where $\Gamma(\bullet)$ is the Gamma function [4]. Plots of few orders of Bessel functions are shown in Fig. 13.6.

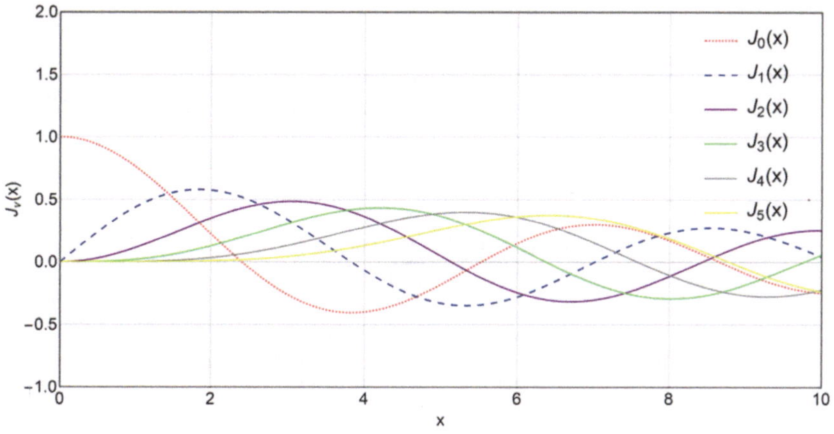

Fig. 13.6 $J_\nu(x)$ versus $x \in [0, 10]$ when $\nu = \{0, 1, 2, 3, 4, 5\}$

Therefore,

$$\mathbf{A}_\perp \approx \frac{j\mu I_o a}{2r} e^{-jkr} J_1(ka \sin\theta)\,\hat{\boldsymbol{\phi}}.$$

◁

Example 13.16 Obtain the faz-zone electric field and magnetic field phasors of a loop antenna.

Solution These are found readily as

$$\left.\begin{aligned} \mathbf{E} &\approx -j\omega\mathbf{A}_\perp = \frac{\eta k I_o a}{2r} e^{-jkr} J_1(ka \sin\theta)\,\hat{\boldsymbol{\phi}} \\ \mathbf{H} &\approx \frac{1}{\eta}\hat{\mathbf{r}} \times \mathbf{E} = -\frac{k I_o a}{2r} e^{-jkr} J_1(ka \sin\theta)\,\hat{\boldsymbol{\theta}} \end{aligned}\right\}.$$

◁

Example 13.17 Obtain (a) $\mathbf{W}_{\text{ave}}$, (b) $U(\theta, \phi)$, (c) P_{rad}, (d) $D(\theta, \phi)$, and (e) R_{rad} of a loop antenna.

Solution.

(a)

$$\mathbf{W}_{\text{ave}} = \frac{1}{2}\text{Re}\{\mathbf{E} \times \mathbf{H}^*\} = \frac{\eta k^2 I_o^2 a^2}{8r^2} J_1^2(ka \sin\theta)\,\hat{\mathbf{r}}.$$

(b)

$$U(\theta, \phi) = \frac{\eta k^2 I_o^2 a^2}{8} J_1^2(ka \sin\theta).$$

(c)

$$P_{\text{rad}} = \int_0^{2\pi}\int_0^{\pi} U(\theta,\phi)\,\sin\theta d\theta d\phi = \frac{\eta k^2 I_o^2 \pi a^2}{4}\int_0^{\pi} J_1^2(ka\sin\theta)\,\sin\theta d\theta = \frac{\eta k^2 I_o^2 \pi a^2}{4} C(ka),$$

where

$$C(ka) = \frac{2}{ka}\sum_{n=0}^{\infty} J_{2n+3}(2ka).$$

(d)

$$D(\theta,\phi) = \frac{4\pi}{P_{\text{rad}}} U(\theta,\phi) = \frac{2}{C(ka)} J_1^2(ka\sin\theta).$$

(e)

$$R_{\text{rad}} = \frac{2P_{\text{rad}}}{|I_o|^2} = \frac{\eta\pi(ka)^2}{2} C(ka).$$

◁

13.2.1 An Equivalence Theorem

In electromagnetics, equivalence theorem can be used to simplify a problem by (i) constructing an equivalent problem with a mapping between the original and equivalent problems, (ii) solving the equivalent problem, and (iii) returning back to the original problem using the mapping already established. Equivalence theorem has many forms. Here, we establish an equivalence theorem showing that an electrically-small electric loop is equivalent to a Hertzian magnetic dipole.

Example 13.18 Given that $J_1(x) \approx \frac{x}{2}$ when $x << 1$, obtain the far-zone electric field and magnetic field phasors for an electrically small loop (i.e., $a << \lambda \Rightarrow a/\lambda << 1$) carrying a current I_e.

Solution In general,

$$\left.\begin{aligned} \mathbf{E} &\approx= \frac{\eta k I_e a}{2r} e^{-jkr} J_1(ka\sin\theta)\,\hat{\boldsymbol{\phi}} \\ \mathbf{H} &\approx -\frac{k I_e a}{2r} e^{-jkr} J_1(ka\sin\theta)\,\hat{\boldsymbol{\theta}} \end{aligned}\right\}.$$

An electrically-small loops implies that $ka\sin\theta = 2\pi\frac{a}{\lambda}\sin\theta << 1$. Therefore,

$$\left.\begin{aligned} \mathbf{E} &\approx \frac{\eta k^2 I_e a^2}{4r} e^{-jkr}\sin\theta\,\hat{\boldsymbol{\phi}} = \frac{\eta k^2 I_e \pi a^2}{4\pi r} e^{-jkr}\sin\theta\,\hat{\boldsymbol{\phi}} \\ \mathbf{H} &\approx -\frac{k^2 I_e a^2}{4r} e^{-jkr}\sin\theta\,\hat{\boldsymbol{\theta}} = -\frac{k^2 I_e \pi a^2}{4\pi r} e^{-jkr}\sin\theta\,\hat{\boldsymbol{\theta}} \end{aligned}\right\}.$$

◁

Example 13.19 Obtain the far-zone electric vector potential, electric field phasors, and magnetic field phasor for a Hertzian magnetic dipole with magnetic current $\mathbf{I}_m = I_m \hat{\mathbf{z}}$ and of length l.

Solution These are found, respectively, as

$$\left.\begin{aligned} \mathbf{F} &\approx \frac{\varepsilon}{4\pi}\frac{e^{-jkr}}{r}\int_{\mathcal{L}'} \mathbf{I}_m(\mathbf{r}')\, e^{jk\hat{\mathbf{r}}\bullet\mathbf{r}'}\, d\mathcal{L}' = \frac{\varepsilon I_m l}{4\pi}\frac{e^{-jkr}}{r}(\cos\theta\,\hat{\mathbf{r}} - \sin\theta\,\hat{\boldsymbol{\theta}}) \\ \mathbf{H} &\approx -j\omega\mathbf{F}_\perp = \frac{j\omega\varepsilon I_m l}{4\pi}\frac{e^{-jkr}}{r}\sin\theta\,\hat{\boldsymbol{\theta}} = \frac{jkI_m l}{4\pi\eta r}e^{-jkr}\sin\theta\,\hat{\boldsymbol{\theta}} \\ \mathbf{E} &\approx -\eta\,\hat{\mathbf{r}}\times\mathbf{H} = -\frac{jkI_m l}{4\pi r}e^{-jkr}\sin\theta\,\hat{\boldsymbol{\phi}} \end{aligned}\right\}.$$

◁

On comparing

$$\mathbf{E} \approx \frac{\eta k^2 I_e \pi a^2}{4\pi r}e^{-jkr}\sin\theta\,\hat{\boldsymbol{\phi}} \tag{13.5}$$

for an electrically-small loop with

$$\mathbf{E} \approx -\frac{jkI_m l}{4\pi r}e^{-jkr}\sin\theta\,\hat{\boldsymbol{\phi}} \tag{13.6}$$

for a Hertzian magnetic dipole, we see that both fields are identical provided that

$$\eta k^2 I_e \pi a^2 = -jkI_m l \Rightarrow j\eta k I_e \pi a^2 = I_m l \Rightarrow j\omega\mu I_e \pi a^2 = I_m l. \tag{13.7}$$

Therefore, an electrically-small loop with current I_e and radius a is equivalent to a Hertzian magnetic dipole with current I_m and length l when $j\omega\mu I_e \pi a^2 = I_m l$.

13.3 Hertzian-Dipole Reflection by an Infinite PEC

In this section, we consider the situation where an antenna is located above the ground. Typically, this is the situation with communication base stations. Recall that the procedure of determining the field phasors of Chap. 12 assumes that the antenna is in located in an empty, unbounded space. When ground (an obstacle) is introduced, this formulation is no more valid.

Physically, a source radiates a wave everywhere. When the wave reaches the ground, part of it gets reflected and part gets transmitted. So, the wave above the ground is the sum of the incident (radiated) wave and the reflected wave. By the principle of superposition, we solve the problem by decomposing it into two parts: (i) a part where the source exists while the ground is removed in which we compute the incident wave from the source, and

(ii) a part where the ground exists while the source is removed in which we compute the reflected wave from the ground treating it as a source of the reflected wave. The solution of the overall problem is summed to get the total wave. In what follows, we assume that (i) the ground is infinite in extent, (ii) the ground is a PEC (i.e., the field is zero inside it), (iii) the source is a Hertzian dipole, and (iv) the ground is located at a distance h from the source, where $h > 2D^2/\lambda$ (i.e., the ground is in the far-zone region with respect to the source). The assumptions we made allow us using the image theory, which simplifies the solution procedure. This problem can be called Hertzian-dipole reflection by an infinite PEC.

Consider an electric dipole with dipole moment $\mathbf{p}$ located at $\mathbf{r}_o = h\,\hat{\mathbf{z}}$. Such a dipole is located a distance h above the ground. Due to the symmetry of the problem, the three-dimensional picture can be made two-dimensional either on the xz plane, or the yz plane. If the dipole moment is normal to the ground plane, it is called a vertical electric dipole (VED). Otherwise, it is called a horizontal electric dipole (HED). If the electric dipole $\mathbf{p}$ is replaced by a magnetic dipole $\mathbf{m}$, then it can be vertical magnetic dipole (VMD), or horizontal magnetic dipole (HMD), depending on the orientation of $\mathbf{m}$ with respect to the normal of the ground plane.

An electric dipole radiates a wave in the presence of ground. What is the total wave above the ground? According to boundary conditions, the tangential electric field must vanish at the boundary $z = 0$ (e.g., $\mathbf{E}^{\parallel} = \mathbf{0}$ when $z = 0$) since the field below it is zero (PEC assumption). It happens that this can be accomplished, provided that the ground plane is removed, and an image source of dipole moment $\mathbf{p}^{\text{img}}$ is placed beneath the actual ground by a distance h from the origin. The orientation of the image source is dictated by that of the actual source. If the actual source is VED, then for the tangential electric field to vanish at the boundary, the image source must have the same orientation as that of the VED. If the actual source is HED, then for the tangential electric field to vanish at the boundary, the image source must have the opposite orientation as that of the HED. The converse is true for VMD and HMD [5]. Dipoles and their image sources are shown in Fig. 13.7. Electric dipoles are represented by a single arrow, whereas magnetic dipoles by double arrows. Also, actual sources are solid lines, whereas image sources are dashed lines.

As a summary, when an electric dipole $\mathbf{p}$ located at $\mathbf{r}_o$ is placed above the PEC $z = 0$, the problem is solved by

- replacing the PEC plane by an image source with dipole moment $\mathbf{p}^{\text{img}}$ located at $\mathbf{r}_o^{\text{img}} = -\mathbf{r}_o$. For VED, $\mathbf{p}^{\text{img}} = \mathbf{p}$, whereas for HED, $\mathbf{p}^{\text{img}} = -\mathbf{p}$,
- finding the incident field phasor $\mathbf{E}_i$ from $\mathbf{p}$, as if the PEC plane does not exist,
- finding the reflected field phasor $\mathbf{E}_r$ from $\mathbf{p}^{\text{img}}$, as if the PEC plane does not exist, and
- summing the two field phasors to get $\mathbf{E}$.

This procedure yields $\mathbf{E}^{\parallel} = \mathbf{0}$ at $z = 0$, and it works only when the ground is a PEC and of an infinite extent.

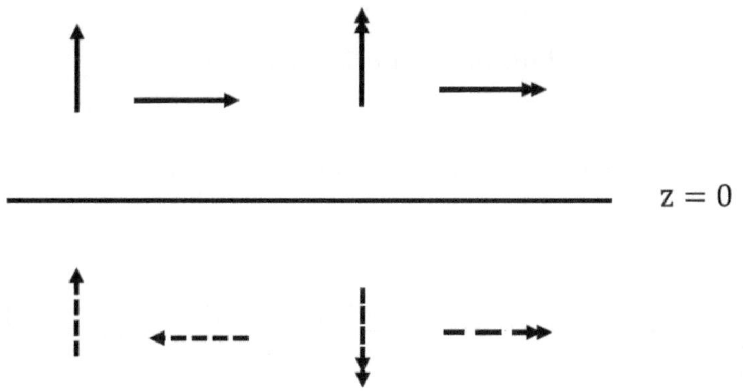

Fig. 13.7 Dipoles and their image sources

Example 13.20 Obtain far-zone electric field and magnetic field phasors for a VED with $\mathbf{p} = I_o l\, \hat{\mathbf{z}}$ located at $\mathbf{r}_o = h\, \hat{\mathbf{z}}$ above the PEC $z = 0$.

Solution The electric field phasor can be written as $\mathbf{E} = \mathbf{E}_i + \mathbf{E}_r$, where

$$\mathbf{E}_i = -\frac{jk\eta e^{-jkr}}{4\pi r}\mathbf{p}_\perp\, e^{jk\hat{\mathbf{r}}\bullet\mathbf{r}_o} = \frac{jkI_o l\eta e^{-jkr}}{4\pi r}\sin\theta\, e^{jkh\cos\theta}\,\hat{\boldsymbol{\theta}},$$

and

$$\mathbf{E}_r = -\frac{jk\eta e^{-jkr}}{4\pi r}\mathbf{p}_\perp^{\text{img}}\, e^{jk\hat{\mathbf{r}}\bullet\mathbf{r}_o^{\text{img}}} = -\frac{jk\eta e^{-jkr}}{4\pi r}\mathbf{p}_\perp\, e^{-jk\hat{\mathbf{r}}\bullet\mathbf{r}_o} = \frac{jkI_o l\eta e^{-jkr}}{4\pi r}\sin\theta\, e^{-jkh\cos\theta}\,\hat{\boldsymbol{\theta}}.$$

Therefore,

$$\mathbf{E} = \frac{jkI_o l\eta e^{-jkr}}{4\pi r}\sin\theta[2\cos(kh\cos\theta)]\,\hat{\boldsymbol{\theta}}.$$

Notice that, this is valid only in $z \geq 0$. In $z < 0$, $\mathbf{E} = \mathbf{0}$ since PEC assumption was made. For $\mathbf{H}$,

$$\mathbf{H} = \frac{1}{\eta}\hat{\mathbf{r}} \times \mathbf{E} = \frac{jI_o lke^{-jkr}}{4\pi r}\sin\theta[2\cos(kh\cos\theta)]\,\hat{\boldsymbol{\phi}}.$$

◁

Example 13.21 Obtain (a) $\mathbf{W}_{\text{ave}}$, (b) $U(\theta, \phi)$, (c) P_{rad}, (d) $D(\theta, \phi)$, and (e) R_{rad} for the VED above the PEC $z = 0$.

Solution (a) Using far-zone field phasors, we get

$$\mathbf{W}_{\text{ave}} = \frac{1}{2}\text{Re}\{\mathbf{E} \times \mathbf{H}^*\} = \eta\frac{|I_o l|^2 k^2}{32\pi^2 r^2}F^2(\theta; kh)\,\hat{\mathbf{r}},$$

where

$$F(\theta; kh) = \sin\theta[2\cos(kh\cos\theta)].$$

(b)

$$U(\theta,\phi) = \eta\frac{|I_o l|^2 k^2}{32\pi^2} F^2(\theta; kh) = \frac{\eta}{8}\left|\frac{I_o l}{\lambda}\right|^2 F^2(\theta; kh).$$

(c)

$$P_{\text{rad}} = \int_0^{2\pi}\int_0^{\pi/2} U(\theta,\phi)\,\sin\theta d\theta d\phi = \frac{\eta\pi}{4}\left|\frac{I_o l}{\lambda}\right|^2 \int_0^{\pi/2} F^2(\theta; kh)\,\sin\theta d\theta = \pi\eta\left|\frac{I_o l}{\lambda}\right|^2 C(kh),$$

where

$$C(kh) = \frac{1}{3} - \frac{\cos(2kh)}{(2kh)^2} + \frac{\sin(2kh)}{(2kh)^3}.$$

(d)

$$D(\theta,\phi) = \frac{4\pi}{P_{\text{rad}}} U(\theta,\phi) = \frac{1}{2C(kh)} F^2(\theta; kh).$$

(e)

$$R_{\text{rad}} = \frac{2P_{\text{rad}}}{|I_o|^2} = 2\pi\eta\left(\frac{l}{\lambda}\right)^2 C(kh).$$

◁

Example 13.22 Obtain $\mathbf{E}$ for an HED with $\mathbf{p} = I_o l\,\hat{\mathbf{y}}$ located at $\mathbf{r}_o = h\,\hat{\mathbf{z}}$ above the ground plane $z = 0$.

Solution The electric field phasor can be written as $\mathbf{E} = \mathbf{E}_i + \mathbf{E}_r$, where

$$\mathbf{E}_i = -\frac{jk\eta e^{-jkr}}{4\pi r}\mathbf{p}_\perp\, e^{jk\hat{\mathbf{r}}\bullet\mathbf{r}_o} = -\frac{jkI_o l\eta e^{-jkr}}{4\pi r}(\cos\theta\sin\phi\,\hat{\boldsymbol{\theta}} + \cos\phi\,\hat{\boldsymbol{\phi}})\, e^{jkh\cos\theta},$$

and

$$\begin{aligned}\mathbf{E}_r &= -\frac{jk\eta e^{-jkr}}{4\pi r}\mathbf{p}_\perp^{\text{img}}\, e^{jk\hat{\mathbf{r}}\bullet\mathbf{r}_o^{\text{img}}} = \frac{j\eta e^{-jkr}}{4\pi r}\mathbf{p}_\perp\, e^{-jk\hat{\mathbf{r}}\bullet\mathbf{r}_o} \\ &= \frac{jkI_o l\eta e^{-jkr}}{4\pi r}(\cos\theta\sin\phi\,\hat{\boldsymbol{\theta}} + \cos\phi\,\hat{\boldsymbol{\phi}})\, e^{-jkh\cos\theta}.\end{aligned}$$

Therefore,

$$\mathbf{E} = -\frac{jkI_o l\eta e^{-jkr}}{4\pi r}\sin\theta[2j\sin(kh\cos\theta)]\,(\cos\theta\sin\phi\,\hat{\boldsymbol{\theta}} + \cos\phi\,\hat{\boldsymbol{\phi}}).$$

Notice that, this is valid only in $z \geq 0$. In $z < 0$, $\mathbf{E} = \mathbf{0}$ since PEC assumption was made. ◁

13.4 Integral Equations

In electromagnetic, integral equations are frequently encountered whenever the unknown appears inside an integral operator. They occur in electrostatics, magnetostatics, as well as in time-dependent electromagnetics. Analytical solutions are rare for such equations; hence, they are subject to numerical treatments. In essence, the method of moments (MoM) has been used extensively to solve integral equations. First, we discuss the solution procedure in an abstract manner. Then, we illustrate the solution procedure in electrostatics; the simplest setting in electromagnetics subject. Finally, we discuss integral equations in radiation problems.

13.4.1 The Method of Moments

Consider the equation

$$\mathcal{L}f(x) = g(x), \qquad a < x < b, \tag{13.8}$$

where $\mathcal{L}$ is an integral operator, $g(x)$ is known, and $f(x)$ in unknown to be determined. To find $f(x)$, let us expand it as

$$f(x) = \sum_{n=1}^{N} a_n f_n(x), \tag{13.9}$$

where a_n are the expansion coefficients to be determined, $f_n(x)$ are known basis functions, and N is taken sufficiently large. Substituting Eqs. (13.9) into (13.8) gives

$$\sum_{n=1}^{N} a_n \mathcal{L} f_n(x) = g(x). \tag{13.10}$$

Now, introducing weighting functions $w_m(x)$, $m = 0, 1, \ldots, N$, multiplying both sides of Eq. (13.10) by them, and integrating both sides over $x \in [a, b]$, we get

$$\sum_{n=1}^{N} a_n \int_a^b w_m(x) \mathcal{L} f_n(x)\, dx = \int_a^b w_m(x) g(x)\, dx, \; m = 1, 2, \ldots, N. \tag{13.11}$$

On defining

$$A_{mn} = \int_a^b w_m(x) \mathcal{L} f_n(x)\, dx, \tag{13.12}$$

and

$$b_m = \int_a^b w_m(x) g(x)\, dx, \tag{13.13}$$

Equation (13.11) can be written as

$$\sum_{n=1}^{N} A_{mn}a_n = b_m,\ m = 1, 2, \ldots, N. \tag{13.14}$$

It is to be noted that Eq. (13.14) is not a single equation, but a set of N equations. Those equations can be written in matrix notation as

$$\begin{bmatrix} A_{11} & A_{12} & \ldots & A_{1N} \\ A_{21} & A_{22} & \ldots & A_{2N} \\ \vdots & \vdots & \ldots & \vdots \\ A_{N1} & A_{N2} & \ldots & A_{NN} \end{bmatrix} \begin{bmatrix} a_1 \\ a_2 \\ \vdots \\ a_N \end{bmatrix} = \begin{bmatrix} b_1 \\ b_2 \\ \vdots \\ b_N \end{bmatrix}, \tag{13.15}$$

or in compact form as

$$\underline{\underline{A}} \bullet \mathbf{a} = \mathbf{b}. \tag{13.16}$$

Therefore, **a** can be found from

$$\mathbf{a} = \underline{\underline{A}}^{-1} \bullet \mathbf{b}, \tag{13.17}$$

where $\underline{\underline{A}}^{-1}$ is the inverse of $\underline{\underline{A}}$. Hence, once the basis functions $f_n(x)$ and the weighting functions $w_n(x)$ are decided, the expansion coefficients a_n can be found using Eq. (13.17). This is called the method of moments (MoM). It is to be noted that the MoM generalizes eigenfunction expansion method. Before discussing the MoM in radiation problems, let us use it in solving a simple problem from electrostatics.

13.4.2 An Example from Electrostatics

13.4.2.1 Formulation

Consider the cylinder $\{\rho = a, z \in [0, l]\}$ carrying a surface charge ρ_s (unknown) on its curved surface $\rho = a$. The electric scalar potential $V(\mathbf{r})$ at any point in space is given by

$$V(\mathbf{r}) = \frac{1}{\varepsilon} \int_{S'} \rho_s(\mathbf{r}') G(\mathbf{r}, \mathbf{r}')\, dS', \tag{13.18}$$

where

$$G(\mathbf{r}, \mathbf{r}') = \frac{1}{4\pi R} = \frac{1}{4\pi |\mathbf{r} - \mathbf{r}'|} \tag{13.19}$$

is the Green function. In cylindrical coordinates, this can be written as

$$V(\rho, \phi, z) = \frac{1}{\varepsilon} \int_0^l \int_0^{2\pi} \rho_s(\phi', z') G(\rho, \phi, z; a, \phi', z')\, a d\phi' dz'. \tag{13.20}$$

Now, let us assume that the potential is known on the cylindrical surface $\rho = a$ [i.e., $V(\rho, \phi, z) = V_o(a, \phi, z)$, where $V_o(a, \phi, z)$ is known]. Then,

$$\varepsilon V_o(a,\phi,z) = \int_0^l \int_0^{2\pi} \rho_s(\phi',z')G(a,\phi,z;a,\phi',z')\, ad\phi' dz' \tag{13.21}$$

is an integral equation for $\rho_s(\phi', z')$. In order to simplify matter, let us assume that $l >> a$. Then, we approximate $\rho = a \approx 0$, but we will not implement $\rho' = a \approx 0$ at this time. This is to avoid having a singularity in the integral, which can be treated, but its treatment is beyond the scope of this book. Then,

$$\varepsilon V_o(0,\phi,z) = \varepsilon V_o(\phi,z) = \int_0^l \int_0^{2\pi} \rho_s(\phi',z')G(0,\phi,z;a,\phi',z')\, ad\phi' dz'. \tag{13.22}$$

The Green function is then simplified to

$$G(0,\phi,z;a,\phi',z') \approx \frac{1}{4\pi\sqrt{a^2+(z-z')^2}} \equiv G(z,z'). \tag{13.23}$$

The integral equation now becomes

$$\varepsilon V_o(\phi,z) = \int_0^l \int_0^{2\pi} \rho_s(\phi',z')G(z,z')\, ad\phi' dz'. \tag{13.24}$$

Let us further assume that $V_o(\phi, z) \equiv V_o(z)$. Then, by symmetry, $\rho_s(\phi', z') \equiv \rho_s(z')$. Therefore,

$$\varepsilon V_o(z) = \int_0^l \int_0^{2\pi} \rho_s(z')G(z,z')\, ad\phi' dz' = \int_0^l 2\pi a \rho_s(z')G(z,z')\, d\phi' dz'. \tag{13.25}$$

After noting that $2\pi a \rho_s = \rho_l$, where ρ_l is the line charge, we finally get

$$\varepsilon V_o(z) = \int_0^l \rho_l(z')G(z,z')\, dz'. \tag{13.26}$$

13.4.2.2 MoM Solution

We expand $\rho_l(z')$ into basis functions as

$$\rho_l(z') = \sum_{n=1}^{N} a_n f_n(z'). \tag{13.27}$$

Then,

$$\varepsilon V_o(z) = \sum_{n=1}^{N} a_n \int_0^l f_n(z')G(z, z')\, dz'. \tag{13.28}$$

After introducing the weighting functions $w_m(z)$, we get

$$\int_0^l \varepsilon w_m(z) V_o(z)\, dz = \int_0^l \left(w_m(z) \sum_{n=1}^{N} a_n \int_0^l f_n(z')G(z, z')\, dz' \right) dz, \tag{13.29}$$

which becomes

$$\varepsilon \int_0^l w_m(z) V_o(z)\, dz = \sum_{n=1}^{N} a_n \int_0^l \left(w_m(z) \int_0^l f_n(z')G(z, z')\, dz' \right) dz. \tag{13.30}$$

On defining

$$A_{mn} = \int_0^l \left(w_m(z) \int_0^l f_n(z')G(z, z')\, dz' \right) dz, \tag{13.31}$$

and

$$b_m = \varepsilon \int_0^l V_o(z) w_m(z)\, dz, \tag{13.32}$$

the integral equation becomes

$$\sum_{n=1}^{N} A_{mn} a_n = b_m,\ m = 1, 2, \ldots, N. \tag{13.33}$$

Let us make the basis functions $f_n(z')$ the pulse functions, given by

$$f_n(z') = \begin{cases} 1,\ (n-1)\Delta \le z' \le n\Delta, \\ 0,\ \text{otherwise,} \end{cases} \tag{13.34}$$

where $\Delta = l/N$ is the pulse width. Then,

$$A_{mn} = \int_0^l \left(\int_{(n-1)\Delta}^{n\Delta} G(z, z')\, dz' \right) w_m(z)\, dz. \tag{13.35}$$

Let us make the weighting functions $w_m(z)$ the delta functions, give by

$$w_m(z) = \delta(z - z_m). \tag{13.36}$$

Then,

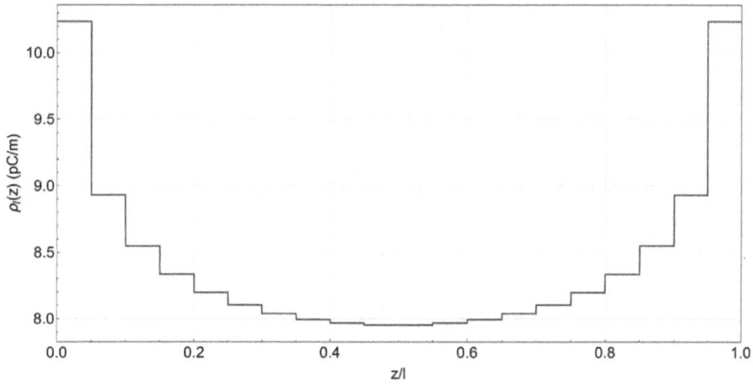

Fig. 13.8 Charge distribution

$$A_{mn} = \int_0^l \left(\int_{(n-1)\Delta}^{n\Delta} G(z, z')\, dz' \right) \delta(z - z_m)\, dz = \int_{(n-1)\Delta}^{n\Delta} G(z_m, z')\, dz', \tag{13.37}$$

and

$$b_m = \varepsilon \int_0^l V_o(z) \delta(z - z_m)\, dz = \varepsilon V_o(z_m). \tag{13.38}$$

It is to be noted that more accurate results can be obtained as N becomes larger. However, extreme choice may result in the divergence of the series. A computer program is written to compute the line charge ρ_l when $V_o(z) = 1$ after discretizing the wire length l into N segments with a distance Δ between each segment. The potential at the center of each segment is made unity, and the line charge coefficients a_n are obtained. Results when $l = 1$ m, $a = 1$ mm, $N = 20$, and $\Delta = l/N = 0.2$ mm is shown in Fig. 13.8.

13.4.3 An Example from Radiation Problems

13.4.3.1 Formulation

Consider the cylinder $\{\rho = a, z \in [-l/2, l/2]\}$ made of a PEC, and is exposed to an incident electric field phasor $\mathbf{E}^i$ (known). This incident field phasor will interact with the cylinder, which produces an induced electric surface current phasor $\mathbf{K}_e$ (unknown), which in turn produces a scattered electric field phasor $\mathbf{E}^s$ (unknown). Assuming $l >> a$ and $\lambda >> a$, the cylinder ends can be neglected. Then, on the surface $\rho = a$, the tangential total electric field phasor $\mathbf{E} = \mathbf{E}^i + \mathbf{E}^s$ vanishes. That is,

$$\hat{\rho} \times (\mathbf{E}^i + \mathbf{E}^s)\Big|_{\rho=a} = \mathbf{0}. \tag{13.39}$$

The tangential components are E_ϕ and E_z. Assuming that $E_z >> E_\phi$, which holds true for rotationally symmetric $\mathbf{E}^i$, we get

$$E_z^s = -E_z^i, \qquad \rho = a. \tag{13.40}$$

This equation allows finding $\mathbf{K}_e$ as to be shown. Once $\mathbf{K}_e$ is found, we can find the scattered magnetic vector potential from

$$\mathbf{A}^s(\mathbf{r}) = \frac{\mu}{4\pi} \int_{S'} \mathbf{K}_e(\mathbf{r}') \frac{e^{-jkR}}{R} \, dS', \tag{13.41}$$

which can be written as

$$\mathbf{A}^s(\mathbf{r}) = \mu \int_{S'} \mathbf{K}_e(\mathbf{r}') \, G(\mathbf{r}, \mathbf{r}') \, dS', \tag{13.42}$$

where

$$G(\mathbf{r}, \mathbf{r}') = \frac{e^{-jkR}}{4\pi R} \tag{13.43}$$

is the Green function. Before proceeding to $\mathbf{E}^s$, we notice that we have already implemented the conditions $l >> a$ and $\lambda >> a$. This implies that

$$2\pi a \, \mathbf{J_s}(\phi', z') \rightarrow I(z') \, \hat{\mathbf{z}}, \tag{13.44}$$

and

$$G(\mathbf{r}, \mathbf{r}') = \frac{e^{-jkR}}{4\pi R} \approx \frac{e^{-jk\sqrt{a^2+(z-z')^2}}}{4\pi\sqrt{a^2+(z-z')^2}} = G(z, z'). \tag{13.45}$$

Therefore,

$$A_z^s(z) = \mu \int_{-l/2}^{l/2} I(z') \, G(z, z') \, dz'. \tag{13.46}$$

Then, the scattered electric field phasor is found using

$$\mathbf{E}^s(\mathbf{r}) = -j\omega \mathbf{A}^s - \frac{j}{\omega\mu\varepsilon} \nabla(\nabla \cdot \mathbf{A}^s). \tag{13.47}$$

Since $\mathbf{A}^s(\mathbf{r}) = A_z^s(z) \, \hat{\mathbf{z}}$ in our case, this becomes

$$E_z^s(z) = -j\omega A_z^s - \frac{j}{\omega\mu\varepsilon} \frac{d^2}{dz^2} A_z^s = -j\omega \left(1 + \frac{1}{k^2} \frac{d^2}{dz^2}\right) A_z^s. \tag{13.48}$$

But $E_z^s = -E_z^i$ on $\rho = a$. Therefore,

$$\left(1 + \frac{1}{k^2} \frac{d^2}{dz^2}\right) A_z^s(z) = \frac{E_z^i(z)}{j\omega}, \qquad \rho = a, \tag{13.49}$$

which can be written as

$$\left(\frac{d^2}{dz^2}+k^2\right)A_z^s(z)=-\frac{jk^2}{\omega}E_z^i(z), \qquad \rho=a. \tag{13.50}$$

This is a second-order nonhomogeneous ODE with a solution [1]

$$\int_{-l/2}^{l/2} I(z')\,G(z,z')\,dz'=a_o\cos kz+b_o\sin kz-\frac{j8\pi^2}{\eta}\int_{-l/2}^{z} E_z^i(u)\sin k(z-u)\,du, \tag{13.51}$$

where a_o and b_o are unknown constants. This is called Hallén integral equation. Once $E_z^i(z')$ is specified, and two more conditions on the current are specified due to the presence of a_o and b_o, the current $I(z)$ can be found. Under the assumption that $I(-l/2)=I(l/2)=0$, and a voltage is V_o is applied at $z=0$, the Hallén integral equation becomes

$$\int_{-l/2}^{l/2} I(z')\,G(z,z')\,dz'=\frac{-j}{\eta}\left(a_o\cos kz+\frac{V_o}{2}\sin k|z|\right). \tag{13.52}$$

13.4.3.2 MoM Solution

We expand $I(z')$ into basis functions as

$$I(z')=\sum_{n=1}^{N} a_n f_n(z'). \tag{13.53}$$

Then,

$$\frac{-j}{\eta}\left(a_o\cos kz+\frac{V_o}{2}\sin k|z|\right)=\sum_{n=1}^{N} a_n\int_{-l/2}^{l/2} G(z,z')\,dz'. \tag{13.54}$$

After introducing the weighting functions $w_m(z)$, $m=1,2,\ldots N+1$, we get

$$\frac{-j}{\eta}\int_{-l/2}^{l/2} w_m(z)\left(a_o\cos kz+\frac{V_o}{2}\sin k|z|\right)dz=\int_{-l/2}^{l/2}\left(w_m(z)\sum_{n=1}^{N} a_n\int_{-l/2}^{l/2} f_n(z')G(z,z')\,dz'\right)dz, \tag{13.55}$$

which becomes

$$\frac{-jV_o}{2\eta}\int_{-l/2}^{l/2} w_m(z)\sin k|z|\,dz = \sum_{n=1}^{N} a_n \int_{-l/2}^{l/2}\left(w_m(z)\int_{-l/2}^{l/2} f_n(z')G(z,z')\,dz'\right)dz$$
$$+\,a_o\frac{j}{\eta}\int_{-l/2}^{l/2} w_m(z)\cos kz\,dz. \tag{13.56}$$

On defining

$$A_{mn} = \int_{-l/2}^{l/2}\left(w_m(z)\int_{-l/2}^{l/2} f_n(z')G(z,z')\,dz'\right)dz, \tag{13.57}$$

$$\tilde{A}_m = \frac{j}{\eta}\int_{-l/2}^{l/2} w_m(z)\cos kz\,dz, \tag{13.58}$$

and

$$b_m = \frac{-jV_o}{2\eta}\int_{-l/2}^{l/2} w_m(z)\sin k|z|\,dz, \tag{13.59}$$

the integral equation becomes

$$\tilde{A}_m\, a_o + \sum_{n=1}^{N} A_{mn}a_n = b_m,\ m = 1, 2, \ldots, N+1. \tag{13.60}$$

On making use of pulse functions as basis functions and delta functions as weighing functions, we get

$$A_{mn} = \int_{(n-1)\Delta}^{n\Delta} G(z_m, z')\,dz', \tag{13.61}$$

$$\tilde{A}_m = \frac{j}{\eta}\cos kz_m, \tag{13.62}$$

and

$$b_m = \frac{-jV_o}{2\eta}\sin k|z_m|. \tag{13.63}$$

It is to be noticed that the matrix equation has the form

$$\begin{bmatrix}\tilde{A}_1\\ \tilde{A}_2\\ \vdots\\ \tilde{A}_{N+1}\end{bmatrix} a_o + \begin{bmatrix} A_{11} & A_{12} & \ldots & A_{1N}\\ A_{21} & A_{22} & \ldots & A_{2N}\\ \vdots & \vdots & \ldots & \vdots\\ A_{N+1,1} & A_{N+1,2} & \ldots & A_{N+1,N}\end{bmatrix}\begin{bmatrix}a_1\\ a_2\\ \vdots\\ a_N\end{bmatrix} = \begin{bmatrix}b_1\\ b_2\\ \vdots\\ b_{N+1}\end{bmatrix}. \tag{13.64}$$

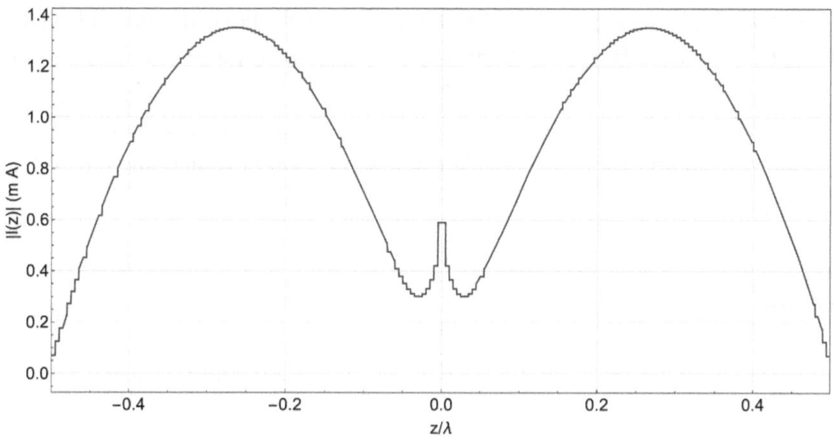

Fig. 13.9 Current distribution

In order to make it in the form $\mathbf{A} \bullet \mathbf{a} = \mathbf{b}$, we rewrite it as

$$\begin{bmatrix} A_{11} & A_{12} & \dots & A_{1N} & \tilde{A}_1 \\ A_{21} & A_{22} & \dots & A_{2N} & \tilde{A}_2 \\ \vdots & \vdots & \dots & \vdots & \vdots \\ A_{N,1} & A_{N,2} & \dots & A_{N,N} & \tilde{A}_N \\ A_{N+1,1} & A_{N+1,2} & \dots & A_{N+1,N} & \tilde{A}_{N+1} \end{bmatrix} \begin{bmatrix} a_1 \\ a_2 \\ \vdots \\ a_N \\ a_o \end{bmatrix} = \begin{bmatrix} b_1 \\ b_2 \\ \vdots \\ b_N \\ b_{N+1} \end{bmatrix}. \qquad (13.65)$$

Notice that the square matrix has the dimension $(N + 1) \times (N + 1)$, whereas the column matrices have the dimension $(N + 1) \times 1$. Figure 13.9 shows the magnitude of the current distribution for a half-wave dipole when $a = 0.001l$. The number of terms used is $N = 100$.

Problems

13.1 Given a dipole with travelling-wave current

$$\mathbf{I}_e(z) = I_o e^{-jkz}\,\hat{\mathbf{z}}, \qquad 0 < z < l,$$

where I_o is a constant, obtain (a) far-zone **E** and **H**, and (b) the directivity $D(\theta)$.

13.2 Given a small magnetic dipole with magnetic current

$$\mathbf{I}_m(z) = I_o\left(1 - \frac{|z|}{l}\right)\hat{\mathbf{z}}, \qquad -l < z < l,$$

obtain (a) far-zone **E** and **H**, and (b) the directivity $D(\theta)$.

13.3 Given an HED with $\mathbf{p} = I_o l\,\hat{\mathbf{y}}$ located at $\mathbf{r}_o = h\,\hat{\mathbf{x}}$ above the ground plane $x = 0$, obtain (a) far-zone **E** and **H**, (b) the directivity $D(\theta, \phi)$, and (c) the radiation resistance R_{rad}.

13.4 Derive an expression of the magnetic vector potential, valid everywhere, for a loop of radius a carrying an arbitrary current $\mathbf{I}_e(\phi) = I(\phi)\,\hat{\boldsymbol{\phi}}$. Don't evaluate the integrals.

Appendix

The following computer program implements the MoM procedure of Sect. 13.4.2.

```
nmax = 20;
l = 1;
\[CapitalDelta] = l/nmax;
a = 10^-3 l;
\[Epsilon]o = (10^-9)/(36 \[Pi]);
zmat = ConstantArray[0, {nmax, 1}];
Module[{i}, For[i = 1, i <= nmax, i++,
  zmat[[i, 1]] = ( (i - 1) \[CapitalDelta] + i \[CapitalDelta])/2;]]
fn[n_, z_] :=
  Which[z < (n - 1) \[CapitalDelta],
   0, (n - 1) \[CapitalDelta] < z < n \[CapitalDelta], 1,
   n \[CapitalDelta] < z, 0];
G[z_, zp_] := 1/(4 \[Pi] Sqrt[(z - zp)^2 + a^2]);
Vo[z_] := 1;

Amat = ConstantArray[0, {nmax, nmax}];

Module[{i, j}, For[i = 1, i <= nmax, i++, For[j = 1, j <= nmax, j++,
   Amat[[i, j]] =
     NIntegrate[fn[j, zp] G[zmat[[i, 1]], zp], {zp, 0, l}];]]]
bmat = ConstantArray[0, {nmax, 1}];
Module[{i}, For[i = 1, i <= nmax, i++,
  bmat[[i, 1]] = \[Epsilon]o Vo[zmat[[i, 1]]] ;]]


amat = Inverse[Amat] . bmat;
\[Rho]l[z_] := Sum[amat[[n, 1]] fn[n, z], {n, 1, nmax}];

Rasterize[Plot[10^12*\[Rho]l[z], {z, 0, l},
   Frame -> True,
  RotateLabel -> True,
  ExclusionsStyle -> {Red},
  FrameStyle -> Directive[20, Black],
  PlotRange -> {{0, l}, Automatic},
  FrameTicks -> {{Automatic, None}, {Automatic, None}},
```

```
  FrameLabel -> {Style["z/l", 20, Black],
    Style["\!\(\*SubscriptBox[\(\[Rho]\), \(l\)]\)(z) (pC/m)", 20,
     Black]},
  PlotStyle -> {Blue},
  BaseStyle -> {FontFamily -> "Arial", 20,
    SingleLetterItalics -> True},
  AspectRatio -> 1/2,
  ImageSize -> 90*11,
  GridLines -> {Automatic, Automatic}],
 ImageResolution -> 300]
```

The following computer program implements the MoM procedure of Sect. 13.4.3.

```
\[Lambda]o = 600 10^(-9);
ko = 2 \[Pi]/\[Lambda]o;
nmax = 100;
l = 0.5 \[Lambda]o;
\[CapitalDelta] = l/nmax;
a = 10^-5 \[Lambda]o;
\[Eta]o = 120 \[Pi];
zmat = ConstantArray[0, {nmax + 1, 1}];
Module[{i}, For[i = 1, i <= nmax + 1, i++,
  zmat[[i, 1]] = ( (i - 1) \[CapitalDelta] + i \[CapitalDelta])/2;]]
fn[n_, z_] :=
  Which[z < (n - 1) \[CapitalDelta],
   0, (n - 1) \[CapitalDelta] < z < n \[CapitalDelta], 1,
   n \[CapitalDelta] < z, 0];
G[z_, zp_] := Exp[-I ko Sqrt[(z - zp)^2 + a^2]]/(
  4 \[Pi] Sqrt[(z - zp)^2 + a^2]);
Vo[z_] := 1;

Amat = ConstantArray[0, {nmax + 1, nmax + 1}];

Module[{i, j}, For[i = 1, i <= nmax + 1, i++, For[j = 1, j <= nmax, j++,
   Amat[[i, j]] =
    NIntegrate[
     G[zmat[[i, 1]], zp], {zp, (j - 1) \[CapitalDelta] ,
      j \[CapitalDelta]}];
   Amat[[i, nmax + 1]] = I/\[Eta]o  Cos[ko zmat[[i, 1]]];
   ]]]


bmat = ConstantArray[0, {nmax + 1, 1}];
Module[{i}, For[i = 1, i <= (nmax + 1), i++,
  bmat[[i, 1]] =- I/(2 \[Eta]o)
       Vo[zmat[[i, 1]]] Sin[ko Abs[zmat[[i, 1]]]] ;]]

amat = Inverse[Amat] . bmat;
Iz[z_] := Sum[amat[[n, 1]] fn[n, Abs[z]], {n, 1, nmax}];
```

```
fig = Rasterize[Plot[10^3 Abs[Iz[\[Lambda]o z]], {z, −1/2, 1/2},
    Frame -> True,
   RotateLabel -> True,
   ExclusionsStyle -> {Red},
   FrameStyle -> Directive[20, Black],
   PlotRange -> {{−1/2, 1/2}, Automatic},
   FrameTicks -> {{Automatic, None}, {Automatic, None}},
   FrameLabel -> {Style["z/\[Lambda]", 20, Black],
     Style["|I(z)| (mA)", 20, Black]},
   PlotStyle -> {Blue},
   BaseStyle -> {FontFamily -> "Arial", 20,
     SingleLetterItalics -> True},
   AspectRatio -> 1/2,
   ImageSize -> 90*11,
   GridLines -> {Automatic, Automatic}],
  ImageResolution -> 300]
```

References

1. E.J. Rothwell, M.J. Cloud, *Electromagnetics*, 3rd edn. (CRC Press, 2018)
2. R.F. Harrignton, *Field Computation by Moment Methods* (Wiley-IEEE Press, 1993)
3. O'Neil, *Engineering Mathematics* (Thomson, 2003)
4. G.B. Arfken, H.J. Weber, F.E. Harris, *Mathematical Methods for Physicists* (Elsevier, 2013)
5. J.A. Stratton, *Electromagnetic Theory* (McGraw–Hill, 1941)

Antenna Arrays

14

Enhancing the antenna directivity can be done, for instance, by increasing its electrical length. However, this leads to increasing the dissipation, and thus, decreasing the radiation efficiency. That is, there will be a trade-off between directivity and radiation efficiency. As an alternative solution, a group of antennas, called an array, can be formed to enhance the overall radiation characteristics. Each individual antenna is called an element. Usually, an array consists of alike elements. Those elements are spaced by a certain displacement. Also, each element is excited by a current with an amplitude and a phase. An array configuration can be one-dimensional, two-dimensional, or three-dimensional (see Fig. 14.1). Therefore, desired radiation characteristics can be achieved by controlling

- the element type (e.g., dipole, loop, spiral, horn, dish, etc.),
- the geometrical configuration (e.g., 1-D, 2-D, or 3-D),
- the displacement between the elements, and
- excitation currents amplitudes and phases.

We first discuss preliminary concepts related to array fundamentals in Sect. 14.1. Then, we discuss the simplest array possible; one-dimensional, equally-spaced, uniform array in Sect. 14.2.

14.1 Preliminaries

To facilitate the presentation, let us consider electric currents only. Once the idea is grasped, extension to magnetic currents can be made readily. We first derive an expression of the radiation intensity related directly to the electric radiation vector for a single element centred

H. M. Alkhoori, *Concise Introduction to Electromagnetic Fields*, Synthesis Lectures on Electromagnetics, https://doi.org/10.1007/978-3-031-60331-0_14

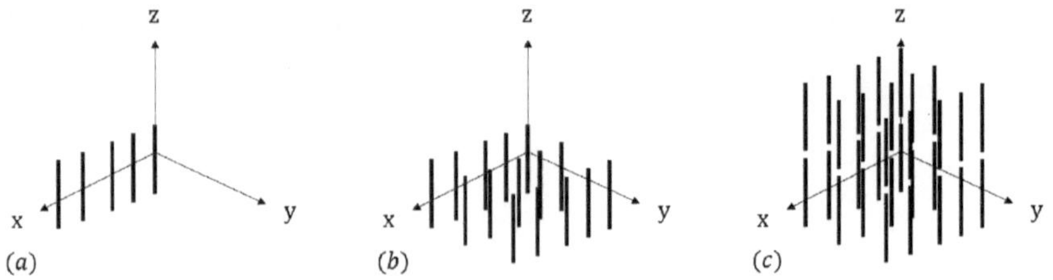

Fig. 14.1 **a** One-dimensional array, **b** two-dimensional array, and **c** three-dimensional array. The elements are represented by vertical solid lines

about the origin carrying a current $\mathbf{J}_e$. Then, after understanding the effect of shifting the current $\mathbf{J}_e$ a distance $\mathbf{d}$ from the origin, as well as its consequence on the radiation intensity, an array is introduced as a group of shifted current elements, and the associated radiation intensity is derived, which leads to the so-called array factor.

14.1.1 Single-Element Radiation Intensity

Recall from Sect. 11.7 that the far-zone magnetic vector potential of an electric current $\mathbf{J}_e$ is

$$\mathbf{A}(\mathbf{r}) \approx \frac{\mu}{4\pi}\frac{e^{-jkr}}{r}\mathbf{a}(\theta,\phi), \tag{14.1}$$

where

$$\mathbf{a}(\theta,\phi) = \int_{\mathcal{V}'} \mathbf{J}_e(\mathbf{r}')\, e^{j\mathbf{k}\bullet\mathbf{r}'}\, d\mathcal{V}' \tag{14.2}$$

is the electric radiation vector. Since the radial component does not contribute in field quantities, let us define the perpendicular electric radiation vector as

$$\mathbf{a}_\perp(\theta,\phi) = \mathbf{a}(\theta,\phi) - [\mathbf{a}(\theta,\phi)\bullet\hat{\mathbf{r}}]\hat{\mathbf{r}}. \tag{14.3}$$

Then, the electric field and magnetic field phasors become

$$\left.\begin{aligned} \mathbf{E} &\approx -j\omega\mathbf{A}_\perp = \frac{-j\eta k e^{-jkr}}{4\pi r}\mathbf{a}_\perp(\theta,\phi) \\ \mathbf{H} &\approx \frac{1}{\eta}\hat{\mathbf{r}}\times\mathbf{E} = \frac{-jke^{-jkr}}{4\pi r}\hat{\mathbf{r}}\times\mathbf{a}_\perp(\theta,\phi) \end{aligned}\right\}. \tag{14.4}$$

Using these, the Poynting vector becomes

$$\mathbf{W}_{\text{ave}} = \frac{1}{2}\text{Re}\{\mathbf{E}\times\mathbf{H}^*\} = \frac{1}{2\eta}|\mathbf{E}|^2\,\hat{\mathbf{r}} = \frac{\eta k^2}{32\pi^2 r^2}|\mathbf{a}_\perp(\theta,\phi)|^2\,\hat{\mathbf{r}}. \tag{14.5}$$

Finally, the radiation intensity becomes

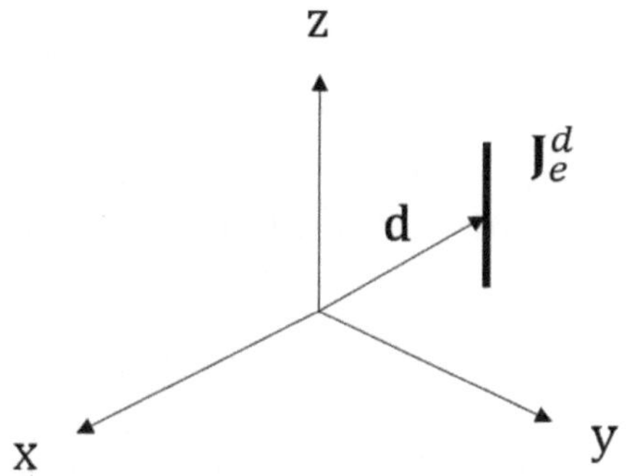

Fig. 14.2 Displaced electric current

$$U(\theta, \phi) = \frac{\eta k^2}{32\pi^2} |\mathbf{a}_\perp(\theta, \phi)|^2. \tag{14.6}$$

Therefore, the radiation intensity is nothing but the square of the magnitude of the perpendicular electric radiation vector.

14.1.2 Displaced, Single-Element Radiation Intensity

Given a current $\mathbf{J}_e(\mathbf{r})$ located at the origin, the associated radiation vector $\mathbf{a}$ is given by Eq. (14.2). Now, suppose that the current is displaced a distance $\mathbf{d}$ from the origin [i.e., $\mathbf{J}_e^d(\mathbf{r}) = \mathbf{J}_e(\mathbf{r} - \mathbf{d})$] (see Fig. 14.2). Then, the displaced radiation vector $\mathbf{a}_d$ becomes

$$\mathbf{a}_d(\theta, \phi) = \int_{\mathcal{V}'} \mathbf{J}_e^d(\mathbf{r}')\, e^{j\mathbf{k} \bullet \mathbf{r}'}\, d\mathcal{V}' = \int_{\mathcal{V}'} \mathbf{J}_e(\mathbf{r}' - \mathbf{d})\, e^{j\mathbf{k} \bullet \mathbf{r}'}\, d\mathcal{V}' = e^{j\mathbf{k} \bullet \mathbf{d}} \int_{\mathcal{V}''} \mathbf{J}_e(\mathbf{r}'')\, e^{j\mathbf{k} \bullet \mathbf{r}''}\, d\mathcal{V}'', \tag{14.7}$$

where $\mathbf{r}' = \mathbf{r}'' + \mathbf{d}$. On implementing $\mathbf{r}'' \to \mathbf{r}'$ in the integrand, we get

$$\mathbf{a}_d(\theta, \phi) = e^{j\mathbf{k} \bullet \mathbf{d}} \int_{\mathcal{V}'} \mathbf{J}_e(\mathbf{r}')\, e^{j\mathbf{k} \bullet \mathbf{r}'}\, d\mathcal{V}' = e^{j\mathbf{k} \bullet \mathbf{d}} \mathbf{a}(\theta, \phi). \tag{14.8}$$

This result shows that a displacement in the current position yields a phase shift of an amount $e^{j\mathbf{k} \bullet \mathbf{d}}$ in the radiation vector. So, we have the correspondence

$$\mathbf{J}_e(\mathbf{r}) \to \mathbf{J}_e(\mathbf{r} - \mathbf{d}) \Rightarrow \mathbf{a}(\theta, \phi) \to e^{j\mathbf{k} \bullet \mathbf{d}} \mathbf{a}(\theta, \phi). \tag{14.9}$$

14.1.3 Multi-element Radiation Intensity

Next, consider the situation where an array of N current elements, each located at $\mathbf{d}_n$. Then, the array's current can be written as

$$\mathbf{J}_{\text{array}}(\mathbf{r}) = \sum_{n=0}^{N-1} a_n \mathbf{J}_e(\mathbf{r} - \mathbf{d}_n), \tag{14.10}$$

where a_n is the nth complex current coefficient (which contains an amplitude and a phase). The corresponding radiation vector using the correspondence in Eq. (14.9) is then

$$\mathbf{a}_{\text{array}}(\theta, \phi) = \left(\sum_{n=0}^{N-1} a_n e^{j\mathbf{k} \bullet \mathbf{d}_n} \right) \mathbf{a}(\theta, \phi), \tag{14.11}$$

where $\mathbf{a}(\theta, \phi)$ is the radiation vector of a single element. The quantity

$$\sum_{n=0}^{N-1} a_n e^{j\mathbf{k} \bullet \mathbf{d}_n} \tag{14.12}$$

can be called the array factor $[AF(\theta, \phi)]$. Notice that the array factor depends on (i) the number of elements (N), (ii) the relative distances between the elements ($\mathbf{d}_n$), and (iii) the currents' coefficients (a_n). As the radiation intensity of a single element is

$$U(\theta, \phi) = \frac{\eta k^2}{32\pi^2} |\mathbf{a}_\perp(\theta, \phi)|^2, \tag{14.13}$$

the associated radiation intensity of an array is then

$$U_{\text{array}}(\theta, \phi) = |AF(\theta, \phi)|^2 U(\theta, \phi). \tag{14.14}$$

So, an array can manipulate the radiation intensity of a single element through the array factor term.

Example 14.1 An array is formed by two isotropic elements (i.e., $U(\theta, \phi) = U_o$, where we make U_o unity for simplicity) located at $\mathbf{d}_0 = \mathbf{0}$ and $\mathbf{d}_1 = d\,\hat{\mathbf{z}}$. Obtain the array radiation intensity $U_{\text{array}}(\theta, \phi)$.

Solution We first find the array factor as

$$AF(\theta) = \sum_{n=0}^{1} a_n e^{j\mathbf{k} \bullet \mathbf{d}_n} = a_0 + a_1 e^{jkd\,\hat{\mathbf{r}} \bullet \hat{\mathbf{z}}} = a_0 + a_1 e^{jkd\cos\theta}.$$

Then, the radiation intensity is found as

$$U_{\text{array}}(\theta) = |AF(\theta)|^2 = \left| a_0 + a_1 e^{jkd\cos\theta} \right|^2.$$

Figure 14.3a shows the intensity pattern when $a_o = a_1 = 1$ and $d/\lambda = \{0.1, 0.5, 1.5\}$. It should be evident that increasing the electrical spacing does produce more main lobes.

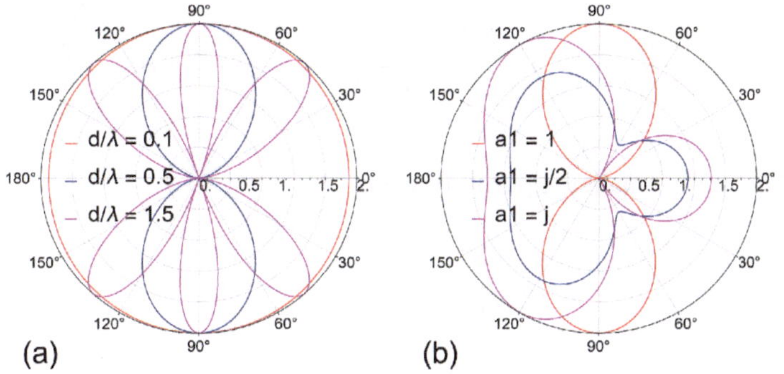

Fig. 14.3 $U_{\text{array}}(\theta)$ versus $\theta \in [0, 180°]$ **a** when $a_o = a_1 = 1$ and $d/\lambda = \{0.1, 0.5, 1.5\}$, and **b** when $a_o = 1$, $d/\lambda = 0.5$, and $a_1 = \{1, j/2, j\}$

These are called grating lobes, which are undesirable in most situations. In order to examine the effect of the relative phase on the intensity pattern, we fix $d/\lambda = 0.5$, $a_o = 1$, and make $a_1 = \{1, j/2, j\}$ in Fig. 14.3b. It is seen that the pattern is clearly affected by the phase shift. This example shows that the radiation intensity of a single element can be manipulated by introducing more elements. Desired and undesired factors could show up upon adjusting the array parameters. Hence, a judicious choice of those parameters should be made for an optimal performance. ◁

14.2 One-Dimensional, Equally-Spaced, Uniform Array

Consider an N-element, one-dimensional array. The elements are equally-spaced (i.e., $\mathbf{d}_n = nd\,\hat{\mathbf{d}}$), and are excited by uniform currents (i.e., $a_n = ae^{jn\beta}$, where a is the current amplitude and β is the progressive phase shift between the elements). With these, the array factor becomes

$$AF(\theta, \phi) = a \sum_{n=0}^{N-1} e^{jn\Psi}, \tag{14.15}$$

where

$$\Psi = kd\,\hat{\mathbf{r}} \cdot \hat{\mathbf{d}} + \beta. \tag{14.16}$$

After the substitution $z = e^{j\Psi}$ is made, and after noting that

$$\sum_{n=0}^{N-1} z^n = \frac{1 - z^N}{1 - z}, \qquad |z| < 1, \tag{14.17}$$

the array factor becomes

$$AF(\theta, \phi) = a\frac{1 - e^{jN\Psi}}{1 - e^{j\Psi}} = a\frac{e^{jN\Psi/2}}{e^{j\Psi/2}}\frac{e^{-jN\Psi/2} - e^{jN\Psi/2}}{e^{-j\Psi/2} - e^{j\Psi/2}} = a\frac{\sin\frac{N\Psi}{2}}{\sin\frac{\Psi}{2}}e^{j(N-1)\Psi/2}. \tag{14.18}$$

We let $a = 1/N$ for normalizing purpose. Then,

$$AF(\theta, \phi) = \frac{\sin\frac{N\Psi}{2}}{N\sin\frac{\Psi}{2}}e^{j(N-1)\Psi/2}. \tag{14.19}$$

Hence, the radiation intensity becomes

$$U_{\text{array}}(\theta, \phi) = \left|\frac{\sin\frac{N\Psi}{2}}{N\sin\frac{\Psi}{2}}\right|^2 U(\theta, \phi). \tag{14.20}$$

It should be noted that only the magnitude of the array factor affects the radiation intensity. As a special case, when Ψ is small, $\sin\Psi/2 \approx \Psi/2$. Then,

$$|AF(\theta, \phi)| \approx \frac{\sin\frac{N\Psi}{2}}{\frac{N\Psi}{2}} = \text{sinc}\frac{N\Psi}{2}. \tag{14.21}$$

Notice that because we made $a = 1/N$, the maximum of $|AF(\theta, \phi)|$ is unity.

14.2.1 FNBW and HPBW

Without loss of generality, let us assume that the array is oriented along the z axis. Then,

$$\Psi = kd\cos\theta + \beta. \tag{14.22}$$

An array can be designed to yield specific FNBW and HPBW. Assuming symmetric pattern, the FNBW is determined as

$$FNBW = 2|\theta_m - \theta_n|, \tag{14.23}$$

where θ_m is the location of $U_{\text{array}} = 1$, and θ_n is the location of $U_{\text{array}} = 0$. Similarly, the HPBW can be defined as

$$HPBW = 2|\theta_m - \theta_h|, \tag{14.24}$$

where θ_h is the location of $U_{\text{array}} = 1/2$. Notice that $U_{\text{array}}(\theta) = |AF(\theta)|^2$ assuming isotropic elements. Therefore, θ_n is determined from $|AF(\theta)| = 0$, θ_m from $|AF(\theta)| = 1$, and θ_h from $|AF(\theta)| = 1/\sqrt{2}$. We next derive expressions for θ_n, θ_m, and θ_h in terms of array's parameters (N, d, and β).

To find θ_n, we enforce

$$|AF(\theta)| = \frac{\sin\dfrac{N\Psi}{2}}{\dfrac{N\Psi}{2}} = 0, \tag{14.25}$$

which is satisfied when

$$\sin\frac{N\Psi}{2} = 0 \Rightarrow \frac{N\Psi}{2} = \pm n\pi, n = 1, 2, \ldots, n \neq 0, N, 2N, \ldots. \tag{14.26}$$

Notice that $n \neq 0, N, 2N, \ldots$ because then the original array factor in Eq. (14.19) will be become non zero. Since $\Psi = kd\cos\theta + \beta$, we get

$$\theta_n = \cos^{-1}\left[\frac{\lambda}{2\pi d}\left(-\beta \pm \frac{2n}{N}\pi\right)\right], n = 1, 2, \ldots, n \neq 0, N, 2N, \ldots. \tag{14.27}$$

To find θ_m, we enforce

$$|AF(\theta)| = \frac{\sin\dfrac{N\Psi}{2}}{\dfrac{N\Psi}{2}} = 1, \tag{14.28}$$

which is satisfied when

$$N\Psi/2 = 0 \Rightarrow \theta_m = \cos^{-1}\left(-\frac{\lambda\beta}{2\pi d}\right). \tag{14.29}$$

Finally, to find θ_h, we enforce

$$|AF(\theta)| = \frac{\sin\dfrac{N\Psi}{2}}{\dfrac{N\Psi}{2}} = \frac{1}{\sqrt{2}}, \tag{14.30}$$

which results in a non-linear equation for $N\Psi/2$. When solved, we get

$$\theta_h = \cos^{-1}\left[\frac{\lambda}{2\pi d}\left(-\beta \pm \frac{2.782}{N}\right)\right]. \tag{14.31}$$

Therefore, once $FNBW$ or $HPBW$ is specified, one can choose d, N, and β such that the required $FNBW$ or $HPBW$ is attained.

14.2.2 Scanning Arrays

The main lobe of an array can be directed (steered) upon adjusting d, N, and β. Such an array is called a scanning array. Usually, N is chosen large for more directive patterns, while d is kept small to avoid grating lobes. Then, it is the progressive phase shift β that is used to steer the main lobe. Suppose that we want the main lobe to be directed toward θ_o, where $\theta_o \in [0°, 180°]$. Then, what should be β? This can be done by enforcing $|AF(\theta)| = 1$ at θ_o, which then yields the desired β. Since $|AF(\theta)| = 1$ when $\Psi = 0$, this implies that

$$kd\cos\theta_o + \beta = 0. \tag{14.32}$$

Then, β is determined as

$$\beta = -kd\cos\theta_o. \tag{14.33}$$

In order to avoid grating lobes, it can be shown that the maximum spacing should satisfy [1]

$$d_{\text{max}} < \frac{\lambda}{1 + |\cos\theta_o|}. \tag{14.34}$$

Below we mention two special cases:

- Ordinary end-fire array: An ordinary end-fire array is attained when the main lobe is directed coparallel or antiparallel to the array axis (i.e., $\theta_o = 0°$ for coparallel and $\theta_o = 180°$ for antiparallel). Therefore, $\beta = -kd$ for $\theta_o = 0°$ and $\beta = kd$ for $\theta_o = 180°$.
- Broadside array: A broadside array is attained when the main lobe is directed normal to the array axis (i.e., $\theta_o = 90°$). Therefore, $\beta = 0$. That is, we make all elements in phase.

Using Eq. (14.34), we see that $d_{\text{max}} < \lambda$ for broadside radiation, while $d_{\text{max}} < \lambda/2$ for end-fire radiation. Examples for ordinary end-fire and broadside patterns are shown in Fig. 14.4

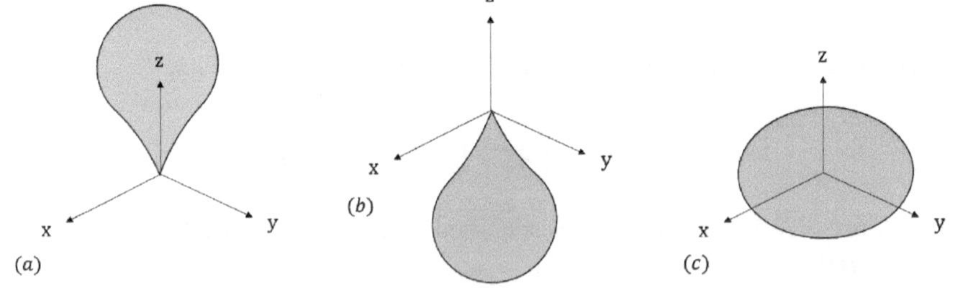

Fig. 14.4 **a** Ordinary end-fire ($\theta_o = 0°$), **b** ordinary end-fire ($\theta_o = 180°$), and **c** broadside ($\theta_o = 90°$)

As an alternative choice yielding end-fire radiation with a higher directivity, the Hansen and Woodyard conditions

$$\beta \approx \mp \left(kd + \frac{\pi}{N}\right), \tag{14.35}$$

where $-$ is used for $\theta_o = 0°$, and $+$ is used for $\theta_o = 180°$, can be implemented [2].

Example 14.2 Obtain the maximum directivity D_o for an N-element broadside array ($\theta_o = 90°$) with spacing d.

Solution For the chosen array, $\beta = 0$. So, $\Psi = kd\cos\theta$. Hence,

$$|AF(\theta)| = \frac{\sin\left(\frac{N}{2}kd\cos\theta\right)}{\frac{N}{2}kd\cos\theta},$$

from which we find

$$U = |AF(\theta)|^2 = \left[\frac{\sin\left(\frac{N}{2}kd\cos\theta\right)}{\frac{N}{2}kd\cos\theta}\right]^2.$$

Therefore,

$$P_{\text{rad}} = \int_\Omega U\,d\Omega = 2\pi\int_0^\pi \left[\frac{\sin\left(\frac{N}{2}kd\cos\theta\right)}{\frac{N}{2}kd\cos\theta}\right]^2 \sin\theta\,d\theta = \frac{4\pi^2}{Nkd}.$$

Finally,

$$D_o = 2N\frac{d}{\lambda}.$$

◁

Example 14.3 Obtain the maximum directivity D_o for an N-element ordinary end-fire array ($\theta_o = 0°$) with spacing d.

Solution For the chosen array, $\beta = -1$. So, $\Psi = kd(\cos\theta - 1)$. Hence,

$$|AF(\theta)| = \frac{\sin\left[\frac{N}{2}(\cos\theta - 1)\right]}{\frac{N}{2}(\cos\theta - 1)},$$

and

$$U = |AF(\theta)|^2 = \left\{\frac{\sin\left[\frac{N}{2}kd(\cos\theta - 1)\right]}{\frac{N}{2}kd(\cos\theta - 1)}\right\}^2.$$

Therefore,

$$P_{\text{rad}} = \int_{\Omega} U \, d\Omega = 2\pi \int_{0}^{\pi} \left\{ \frac{\sin\left[\frac{N}{2}kd(\cos\theta - 1)\right]}{\frac{N}{2}kd(\cos\theta - 1)} \right\}^2 \sin\theta \, d\theta = \frac{2\pi^2}{Nkd}.$$

Finally,

$$D_o = 4N\frac{d}{\lambda} \approx 4\frac{L}{\lambda},$$

where $L = (N-1)d$ is the length of the array. For an N-element Hansen–Woodyard end-fire array ($\theta_o = 0°$) with spacing d, it can be shown that

$$D_o = 1.805 \times 4\frac{L}{\lambda}.$$

Hence, a higher maximum directivity is obtained when using Hansen–Woodyard end-fire array as compared to when using the ordinary end-fire array. ◁

Example 14.4 Design a broadside array with $D_o = 20\,\text{dB}$. In particular, determine a_n, d_{max}, and N_{min}.

Solution We make $a_n = ae^{jn\beta}$, where $a = 1/N$. Since $\theta_o = 90°$, the phase β is found as $\beta = 0$, the maximum spacing as $d_{\text{max}} < \lambda$, and the minimum number of elements as $N_{\text{min}} = 10^2/2 = 50$. ◁

Problems

14.1 An array is formed by two isotropic elements located at $\mathbf{d}_0 = d\,\hat{\mathbf{x}}$, and $\mathbf{d}_1 = -d\,\hat{\mathbf{x}}$. Both elements carry the same current (i.e., $|a_0| = |a_1| = a$), but are out of phase by 180°. Obtain the array radiation intensity $U_{\text{array}}(\theta, \phi)$. You must get a real-valued expression.

14.2 Given a z-oriented phased array with isotropic elements, suppose that we want $U_{\text{array}} = 0.25$ hold at θ_o. Derive an expression for θ_o as a function of the array parameters d, N, and β. **Hint:** $\sin x = x/2 \Rightarrow x = 1.895$.

14.3 A reflector antenna is formed by placing a source (usually a horn antenna) in front of a large curved surface (dish). The equivalent problem of the reflector is determined according to the orientation of the surface. A 90°-corner reflector can be though of as an array with four elements placed at distance d with respect of the origin, as presented in Fig. 14.5. The currents are all unity in magnitude. A point out of the page corresponds to a current with zero phase, and a point into the page corresponds to a current with a phase of 180°. Show that the array factor is given by

$$AF = 2[\cos(kd\sin\theta\cos\phi) - \cos(kd\sin\theta\sin\phi)].$$

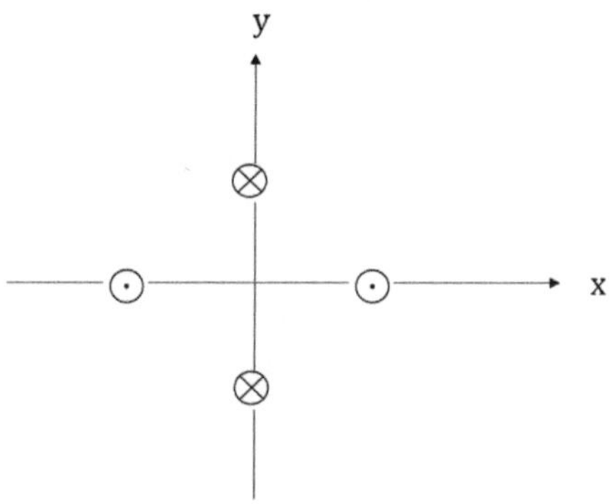

Fig. 14.5 For Problem 14.3

References

1. S.J. Orfanidis, *Electromagnetic Waves and Antennas* (2016). https://www.ece.rutgers.edu/orfanidi/ewa/
2. W.W. Hansen, J.R. Woodyard, A new principle in directional antenna design. Proc. Inst. Radio Eng. **26**(3), 333–345 (1938). https://doi.org/10.1109/JRPROC.1938.228128